JPL PUBLICATION 82-44, VOLUME II

Planetary Geometry Handbook
Venus Positional Data, 1988-2020

Andrey B. Sergeyevsky
Gerald C. Snyder
Barbara L. Paulson
Ross A. Cunniff

October 1, 1983

National Aeronautics and
Space Administration

Jet Propulsion Laboratory
California Institute of Technology
Pasadena, California

Abstract

This document contains graphical data necessary for the analysis of planetary exploration missions to Venus. Positional and geometric information spanning the time period from 1988 through 2020 is presented; the text explains the data and its usage. This volume is one in a planned series covering planetary mission targets.

Preface

This publication is one in a series of volumes devoted to planetary positional and geometric data. The present Volume II provides information describing the planet Venus. Other volumes to be published in 1983 are III, IV, and V, describing Mars, Jupiter, and Saturn geometric data, respectively. The presentations of planetary positional and geometric data for Mercury (Volume I), Uranus (Volume VI), Neptune (Volume VII), and the Pluto/Charon System (Volume VIII) will be published later.

Contents

I. Introduction 1

II. Description of the Data 1
    A. General Comments 1
    B. Data Presentation 1
    C. Definition of Terms 1
    D. Description of Individual Plots 4

III. Application of the Data Presented 6
    A. Geocentric Declination of Target Body 6
    B. Planetocentric Declination of Earth and Sun ... 7
    C. Right Ascension of Target Planet, Sun, and Earth ... 7
    D. Sun- or Earth-to-Planet Distance 7
    E. Heliocentric Longitude 7
    F. Angles Between Sun, Earth, and Planet (SEP and ESP) ... 7
    G. Cone and Clock Angles 8
    H. Station Rise and Set 8

IV. Astrodynamic Constants 8

Acknowledgments 10

References 11

Figures

    1. Definition of vernal equinox 2
    2. Definition of Sun- and Earth-to-planet distance, and Earth-Sun- and Sun-Earth-planet angles at any epoch ... 3
    3. Definition of cone and clock angles 3
    4. Schematic of a distant object's rise and set geometry, with a mask; polar view ... 4
    5. Circumpolar and inaccessible regions of the sky for a station observer; meridional view ... 5
    6. Definition of opposition and conjunction events involving Sun, Earth, and planet ... 8

7. Definition of the mean elements of a planetary orbit 9
8. IAU definition of planetary pole and prime meridian at time d_{50} . 10

Tables

1. Deep Space Network tracking stations 5
2. Non-DSN tracking stations 5

Positional Data
Venus
1988–2020 13

1988 15
1989 21
1990 27
1991 33
1992 39
1993 45
1994 51
1995 57
1996 63
1997 69
1998 75
1999 81
2000 87
2001 93
2002 99
2003 105
2004 111
2005 117
2006 123
2007 129
2008 135
2009 141
2010 147
2011 153
2012 159

2013	165
2014	171
2015	177
2016	183
2017	189
2018	195
2019	201
2020	207

Positional Data
Earth
1985–2020 213

I. Introduction

The purpose of this series of planetary geometry handbooks is to provide mission and science planners, as well as trajectory designers, with graphical information sufficient for preliminary mission design and evaluation. It is intended to be used in conjunction with the appropriate volumes and parts of the Mission Design Handbooks, Ref. 1, which describe trajectory aspects of the particular missions.

In most respects, this planetary geometry series represents a continuation of the second parts of each of the three volumes of Ref. 2. It extends their coverage by commencing in 1985 and continuing to the year 2020. This time span was chosen to provide sufficient mission duration flexibility for all Earth departures through 2005, as presented in Ref. 1.

The series will consist of eight volumes, each describing a planet as follows:

Volume	Target planet
I	Mercury
II	Venus
III	Mars
IV	Jupiter
V	Saturn
VI	Uranus
VII	Neptune
VIII	Pluto/Charon System

The present Volume II is devoted to data on the positions of the planet Venus. It presents information characterizing the geocentric and heliocentric positional geometry of a spacecraft while it is in the vicinity of the subject planetary body. The spacecraft could encompass a lander, an entry probe, or a flyby vehicle at encounter. Such wide usage of the data is possible because planetary positions change but slowly, while the distances to Earth or Sun are always large compared to those between the spacecraft and the target planet during the encounter.

II. Description of the Data

A. General Comments

All plots for this handbook series were computer generated, to reduce production cost. A program originally developed for Mariner 10 (MVM) and later used to produce the positional plots for Ref. 2 has been used, with some modifications, on this series as well.

The major change in the new version of the program was the incorporation of the numerically integrated JPL Planetary DE-118 Ephemeris (Ref. 3) as the source for positional planetary information. An analytical and thus less-precise ephemeris, based on the mean elements of Ref. 4, was used in the preceding set of volumes, Ref. 2.

B. Data Presentation

The data in this volume are grouped into 36 subsections, each covering a specific calendar year. The data commence each year on January 1.0, 0^h Greenwich mean time (GMT) and continue through the year until "Day 380," thus providing a two-week overlap into the next year. There are eight plots presented for each year, labeled as follows:

(1) *Declin*ation (DECLIN).

(2) *Right ascension* (RT.ASC).

(3) *Distance* (DISTANCE).

(4) *Ecliptic longitude* (EC.LON).

(5) *Sun-Earth-planet* angles (SUN-EARTH-PLANET).

(6) *Cone* and *clock* angles (CA and KA, respectively).

(7) *Station rise/set* times (STATION RISE/SET) for the three DSN stations (GMT, HRS).

(8) Same as (7) for three other stations (Weilheim, Nobeyama, and Arecibo).

Twenty-six variables, relating positions of Earth, Sun, target planet (Venus), the star Canopus, and six tracking stations, are depicted on these plots for each subsection.

C. Definition of Terms

Planetary positional geometric data can be represented by directions to specified bodies and by angles between some of these directions, both shown as functions of calendar time. To distinguish among these bodies, the following subscripts are introduced:

(1) E = Earth.

(2) S = Sun.

(3) P = planet (Venus in this volume).

Two fundamental planes are used in astrodynamics: that of the Earth's orbit about the Sun and that of the Earth's equator. The same two geometric concepts can be extended to all celestial bodies.

The plane of the mean Earth's orbit (EMO) is also called the ecliptic; the term "mean" refers to an averaging process, which eliminates minor short-period oscillations of the orbit pole; these oscillations are caused by repetitive perturbing accelerations caused by solar system objects other than the Sun.

The rotational, angular-momentum vector of each planet defines the polar-axis orientation of that celestial body, but not necessarily its north pole. The plane normal to this axis of rotation is defined as the planetary equator of that body. The latest (1983) International Astronomical Union (IAU) definition of planetary poles is such that the *north pole* of a body is that pole that points north of the Invariable (or Laplacian) Plane, defined as the plane normal to the total angular momentum vector of the solar system. This plane is very close to the ecliptic, and deviates from the latter by less than 0.5 deg (Ref. 5).

The point on the celestial sphere where the ecliptic crosses the Earth's mean equator (EME) from south to north is called the *ascending node* of the Ecliptic, while the line passing from the Earth's center through that point is called *Earth's vernal equinox* (♈). The mean Sun is positioned at that point of the sky (as seen from Earth's center) at the time of each spring equinox.

Both the equator and the orbit plane of each planet slowly precess on their respective reference planes, driven by "secular" perturbing effects due to other solar system bodies.

The mean ecliptic regresses, i.e., its nodes slide backwards, on the Invariable Plane at the very slow rate of about 47 seconds of arc per century, while the Earth mean equator regresses on the ecliptic at about 50.26 seconds of arc per annum.

Clearly these motions affect the position of vernal equinox, which is therefore specified as being "of" an epoch, e.g., "of date" (i.e., the present instant of time), or "of 1950.0." The Besselian (tropical) new year has for a long while been considered a convenient defining epoch, usually advanced each half-century, e.g., 1950.0, recently renamed B1950.0, which occurred on January 0.923357, 1950 (i.e., slightly before civil new year 1950, celebrated at 0^h GMT, on January 1.0). The acronym EME50 therefore refers to the Earth mean equator and equinox of B1950.0, while EMO50 specifies the ecliptic plane of that epoch.

Rather than refer to dates frequently quoted in differing calendars, it is customary to specify dates in Julian Days (JD), consecutively counted from an arbitrary date in antiquity (January 1.0, 4713 B.C.). Thus the epoch JD 243 3282.423357 is equivalent to B1950.0.

For each celestial body, a *vernal equinox* may be similarly defined as the ascending node of that body's mean orbit plane (about its primary) upon the body's equator plane. This preferred direction is usually taken to be the X-axis of the inertial coordinate system, centered on that body. The Z-axis coincides with the north-pole axis of the body, and the Y-axis completes a *right-handed* coordinate system.

The equations and related constants defining the above planes are presented in Section IV and may be found in Refs. 4 and 5. Data in Ref. 5 reflect the latest IAU definitions.

A useful way to describe celestial directions is by means of a spherical coordinate system, based either on the equator or else on the mean orbit plane of a body, e.g., that of the Earth or of the target planet. Two angles, declination and right ascension, define the position of any point on the celestial sphere, referred to a planetary equator (Fig. 1).

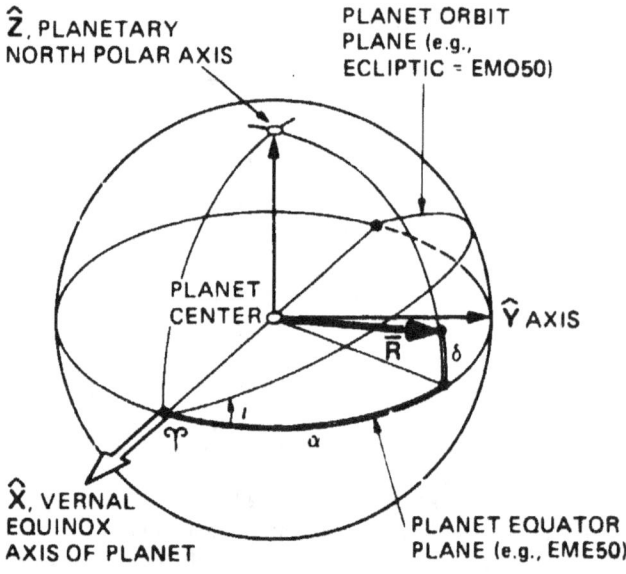

$\bar{R}(x, y, z) = (\cos \delta \cdot \cos \alpha, \cos \delta \cdot \sin \alpha, \sin \delta)$

Fig. 1. Definition of vernal equinox; a position vector R, its declination δ, and right ascension α

Declination (δ) is the central angle formed by the object's position vector with the equator plane, i.e., it is an elevation angle with respect to that plane. It is a form of latitude measure, and is positive towards the north pole and negative to the south of the equator.

Right ascension (α) of an object is the angle in the equator plane between vernal equinox and the projection of the object's position vector into the equator plane. It is positive in the eastward (i.e., right-handed, counterclockwise, when seen from north) direction, regardless of the planet's (celestial body's) own sense of rotation.

Both of these angles are expressed on the unit sphere. Actual *distances* between Sun, Earth, and planet can be presented separately and are measured in astronomical units (AU) — a defined mean distance between Sun and Earth (1 AU = 149,597,871 km; see Astrodynamic Constants, Section IV) — or also in kilometers. The distance plot scale in this volume of the handbook is in units of millions of kilometers ($10^6 \times$ km).

Position angles on the celestial sphere expressed with respect to a planetary orbit plane (e.g., the ecliptic) are referred to as latitudes and longitudes.

Ecliptic longitude (L) of an object is the angular distance between Earth's vernal equinox and the projection of the object's position vector into the ecliptic (EMO50), measured in an eastward (i.e., counterclockwise) positive direction.

Sun-Earth-planet (SEP) angle and the corresponding *Earth-Sun-planet* (ESP) angle are defined by the two respective vector pairs: the Earth-Sun and Earth-planet vectors for the SEP-angle, and the Sun-Earth and Sun-planet vectors for the ESP-angle (Fig 2). The angles are important aids in avoiding periods of communication blackout.

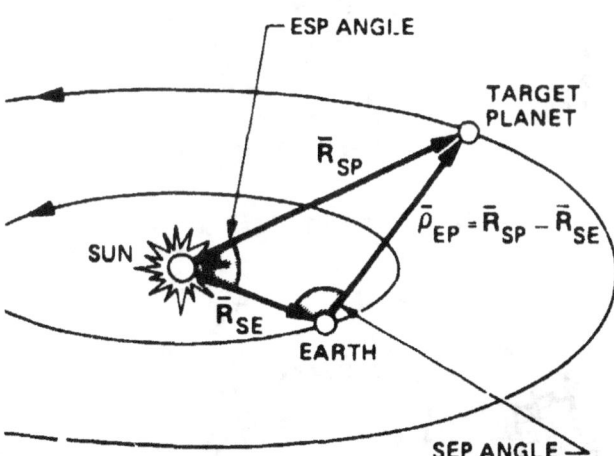

Fig. 2. Definition of Sun- and Earth-to-planet distance, and Earth-Sun- and Sun-Earth-planet angles at any epoch

Cone and clock angles (CA and KA, respectively) are spacecraft-centered coordinates of an object. The polar Z-axis of the system corresponds to the longitudinal axis of the spacecraft. It is usually pointed either towards the Sun (solar panels must generally be normal to the Sun direction), or to the Earth (the parabolic antenna bore axis should point along the communication link, i.e., towards the Earth). Within this book, the planet-Sun direction is taken to be the Z-axis orientation of the CA, KA system (Fig. 3).

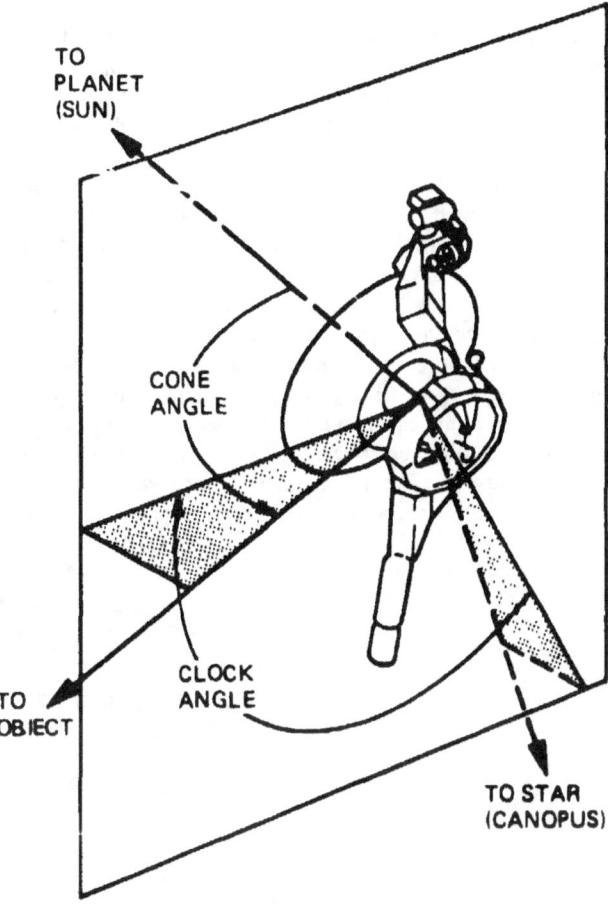

Fig. 3. Definition of cone and clock angles

The prime meridian of this system contains the viewing slot of the spacecraft star tracker, which is usually, but not necessarily, oriented toward the bright southern star Canopus (α-Carinae). Its position on January 1.0, 1950, was at right ascension α = 95.7103875 and declination δ = -52.667625 deg, expressed with respect to the EME50. Stellar proper motions slowly change star locations; the method for updating stellar positions as well as data for other stars are provided in the FK4 Star Catalogue (Ref. 6).

The cone angle of an object, described by its spacecraft-centered position vector, is its angular distance from the Z-axis; i.e., it is the Sun-planet-object angle.

The clock angle of an object is measured in the X-Y plane, which is akin to a "spacecraft equator" and normal to the spacecraft Z-axis. Clock angles are measured from the star-tracker (Canopus) meridian; they are positive clockwise when looking along the Z-axis towards the Sun.

Although the clock angle of Canopus is zero by definition (when the spacecraft is Canopus oriented), the Canopus cone angle generally differs from 90 deg, because the position of Canopus is offset several degrees from the ecliptic pole.

Station rise/set times are determined by the daily rotation of the Earth, which takes all ground tracking stations around the polar axis on the small circles that correspond to their terrestrial latitudes. As a result, every object on the celestial sphere rises and sets upon the local horizon of the station (Fig. 4), provided the object's declination is within ±90 deg of the station latitude (e.g., an object at -55 deg declination cannot be seen by a 40-deg latitude station). Only objects whose declinations equal the station latitude rise to the zenith, in this daily spectacle (Fig. 5). The angular height of an object above the local horizon is termed *elevation* (Γ). Within this book, an object is considered to have risen if it reaches above an elevation mask of 6 deg, which accounts for obstructions by local terrain features, atmospheric signal refraction, and radio noise.

Rise and set time information is presented for the Deep Space Network (DSN), the stations for which are listed in Table 1, as well as for three other stations (Table 2).

The Arecibo radio telescope, located within a natural bowl, is not movable. The slewing of a dish, therefore, is replaced by horizontal motion of a cable-mounted cab, which translates the antenna feed to track a celestial object. The resulting daily coverage of this station is clearly very limited.

D. Description of Individual Plots

The individual plots within each annual subsection present the following 26 planetary geometric characteristics as functions of calendar time (recall that for the purposes of this volume "planet" refers to Venus):

Plot 1. DECLIN. This plot shows three curves:

(1) The geocentric declination of the target planet, referred to the Earth mean equator and equinox of B1950.0 (EME50); it is labeled "P". All angular measures on the plots are in degrees.

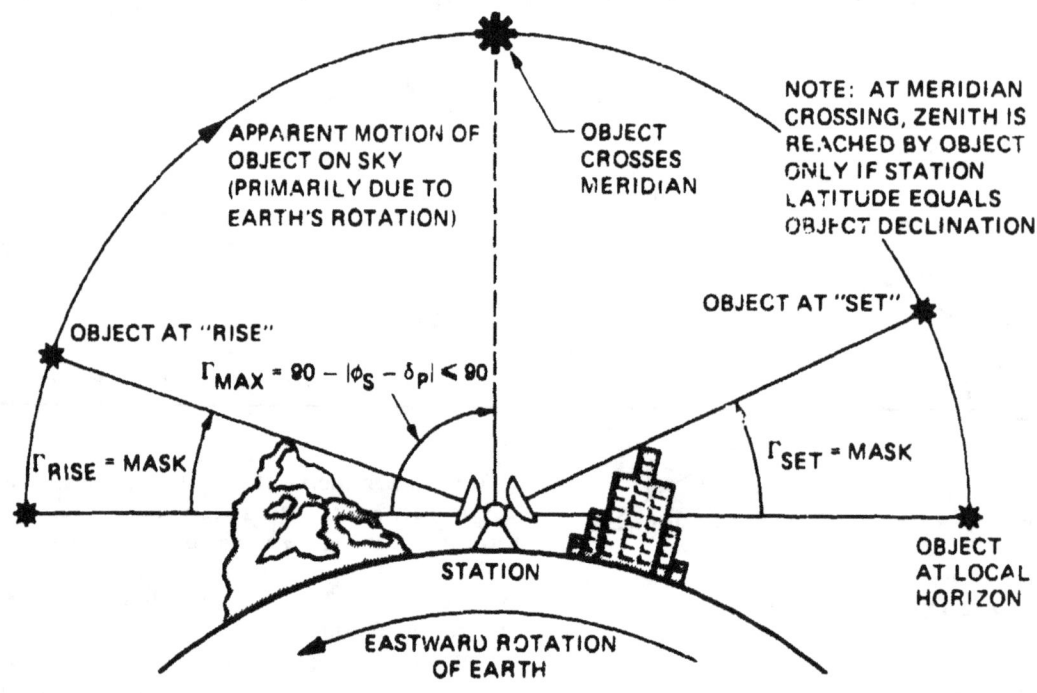

Fig. 4 Schematic of a distant object's rise and set geometry, with a mask; polar view

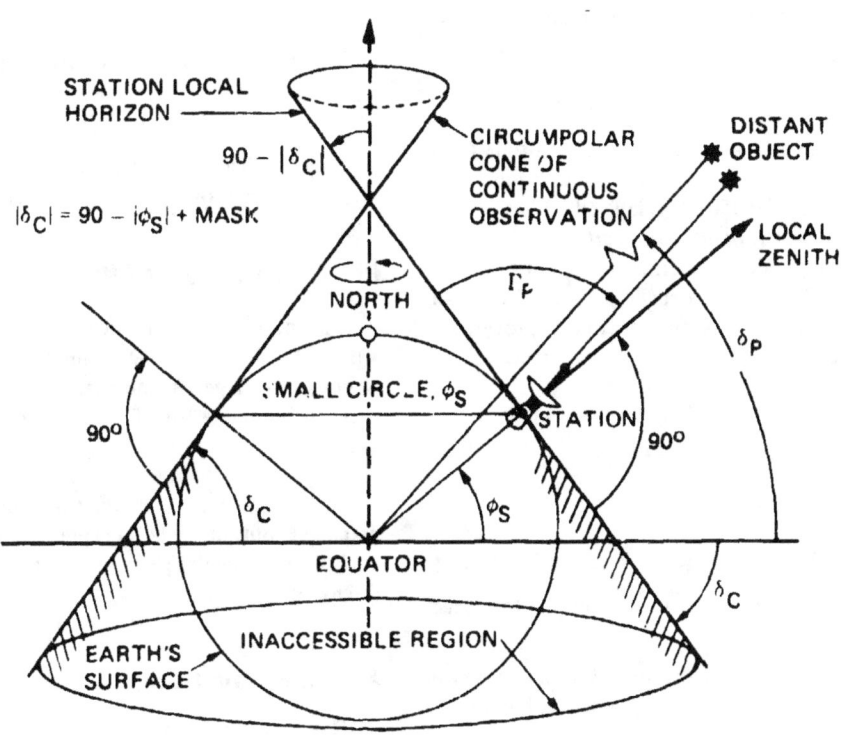

Fig. 5. Circumpolar and inaccessible regions of the sky for a station observer; meridional view

Table 1. Deep Space Network tracking stations

Name	Geocentric latitude ϕ_s	Geocentric longitude (west) λ_s
DSS 14, Goldstone, California	35.24	116.89
DSS 43, Canberra, Australia	-35.22	211.02
DSS 63, Madrid, Spain	40.24	4.25

Table 2. Non-DSN tracking stations

Name	Geocentric latitude ϕ_s	Geocentric longitude (west) λ_s
Weilheim, Germany	47.69	348.92
Nobeyama, Japan	35.94	221.53
Arecibo, Puerto Rico	18.34	66.75

(2) The planetocentric (i.e., as seen from the center of the target planet) declination of the Earth, referred to the planet's mean equator and equinox of date (PME/DA). It is labeled "E".

(3) The planetocentric declination of the Sun, also referred to the PME/DA. It is labeled "S".

Plot II. RT.ASC Three curves are shown on this plot

(4) The geocentric right ascension of the planet with respect to EME50. It is labeled "P".

(5) The planetocentric right ascension of the Earth with respect to PME/DA. It is labeled "E".

(6) The planetocentric right ascension of the Sun with respect to PME/DA. It is labeled "S".

Plot III. DISTANCE: Two curves are presented on this plot.

(7) The Earth-planet distance, in this volume expressed in units of $10^6 \times$ km (millions of km). It is labeled "EP".

(8) The Sun-planet distance, also in millions of km. It is labeled "SP".

Plot IV. EC.LON. A single curve is shown on this plot:

(9) The planet's heliocentric, ecliptic longitude (with respect to the mean ecliptic and equinox of B1950.0, i.e., EMO50). It is labeled "P".

The annual variation in Earth's heliocentric ecliptic longitude is repetitive to a fair degree of accuracy, hence only a single plot, valid throughout the 1985 to 2020 time period, is provided at the end of the plotted positional data section.

Plot V. SUN-EARTH-PLANET: Two curves are presented.

(10) The "SEP" angle (Sun-Earth-planet)

(11) The "ESP" angle (Earth-Sun-planet)

Plot VI. CA, KA: Three curves are depicted on the plot:

(12) The cone angle of Earth for a Sun-oriented spacecraft, labeled "ECA".

(13) The clock angle of Earth, for a Sun-Canopus two-axis-oriented spacecraft. It is labeled "EKA".

(14) The cone angle of Canopus, in any Sun-oriented spacecraft attitude. It is labeled "CCA".

Plot VII. STATION RISE/SET: This plot shows six curves, each representing the Greenwich mean time (GMT) of the following daily Deep Space Network events (all based on a 6-deg mask above the local horizon).

(15) Planet rise at DSS 14 (Goldstone), labeled "RISE 14".

(16) Planet set at DSS 14, labeled "SET 14".

(17) Planet rise at DSS 43 (Canberra), labeled "RISE 43".

(18) Planet set at DSS 43, labeled "SET 43".

(19) Planet rise at DSS 63 (Madrid), labeled "RISE 63".

(20) Planet set at DSS 63, labeled "SET 63".

Plot VIII. STATION RISE/SET: The plot presents up to six curves, representing the GMT of the following tracking station events (6-deg mask, unless specified otherwise):

(21) Planet rise at Weilheim, Germany. It is labeled "R WEIL".

(22) Planet set at Weilheim, labeled "S WEIL".

(23) Planet rise at Nobeyama, Japan. It is labeled "R JAPN".

(24) Planet set at Nobeyama, labeled "S JAPN".

(25) Planet rise at the Arecibo radio telescope, Puerto Rico, above the 70-deg special horizon mask, if realizable. It is labeled "R AREC".

(26) Planet set at Arecibo, under the same conditions as those in (25). It is labeled "S AREC".

III. Application of the Data Presented

The number of practical mission analysis problems to which the information presented in this volume is applicable is obviously large. By no means is it intended to enumerate herein a complete list of such applications. Rather, only some examples, typical of such usage, shall be discussed.

The comments will be arranged in the same order as the graphical output data are presented. Problems in mission analysis to which a particular output type is relevant will be discussed accordingly.

A. Geocentric Declination of Target Body

The spacecraft orbit determination process uses DSN radiometric range and range-rate (Doppler) data. Whenever the observed body as seen from Earth is at near-equatorial (≈ 0 deg) declination, a situation develops that seriously degrades orbit determination accuracy. As a consequence, the timing of critical maneuvers or planetary encounter events should either utilize more sophisticated data types (e.g., ΔDOR, a method based on long-base interferometry) or avoid periods when the target planet is near its node with the Earth's equator (i.e., the times of near-zero declination of the target body).

Communication with the spacecraft greatly depends on a ground station's ability to receive the data signal. An elevation angle high over the horizon, coupled with as long a listening period as possible, are two important assets in this respect. The highest elevation angle of an object at each station occurs at the meridian passage of the object; its value is:

$$\Gamma_{p\,max} = 90 - |\phi_s - \delta_p|$$

Its maximum of $\Gamma_p = 90$ deg can be realized only at a station whose latitude, ϕ_s, equals the declination δ_p of the target. Exact equality, however, may cause some data loss: to continue tracking, some antennas may have to swing through a 180-deg arc in azimuth as soon as the object reaches zenith (i.e., at elevation $\Gamma_p = 90$ deg). As already mentioned, the maximum elevation angle drops with an increase in the difference between station latitude, ϕ_s, and object declination δ_p. At a difference of

$$|\phi_s - \delta_p| + \text{mask} \geq 90 \text{ deg}$$

coverage ceases altogether (Fig. 5). The mask, it will be recalled, is an angle of about 6 deg, designed to assure good signal strength at low elevations.

A notable aspect of this is that, as a station moves in latitude toward either pole (north or south), objects located between the nearer pole and the circumpolar latitude circle δ_c of the station, defined by

$$|\delta_c| = 90 - |\phi_s|$$

will never set and communication may be continuous, if so desired. Knowledge of target-planet declination allows one to choose the particular station for which most critical events would appear at high elevation angles or, conversely, to locate latitudes at which ships and portable or permanent stations should be positioned to support a given difficult mission tracking assignment.

B. Planetocentric Declination of Earth and Sun

These data are based on the target planet's own equator and provide a means of determining the solar lighting conditions at a landing site or during approach. They also allow one to obtain the Earth-oriented downlink elevation angle for a lander on the surface at a given latitude.

For ringed planets (e.g., Saturn), the data can indicate which side (north or south) of the rings is illuminated, which side is visible from Earth, and where on the planetary disk the ring will project its shadow. Observability of disk features and phenomena (e.g., the spokes) may also depend on this information.

C. Right Ascension of Target Planet, Sun, and Earth

This information, when combined with the declination data already discussed, provides positional reference directions toward these bodies, in spherical coordinates as presented, or in Cartesian unit vector form, if transformed as follows:

$$X = \cos \delta_0 \cos \alpha_0$$
$$Y = \cos \delta_0 \sin \alpha_0$$
$$Z = \sin \delta_0$$

where α_0 = right ascension and δ_0 = declination of vector to object.

A number of mission-related problems in spherical trigonometry may be solved using the above information.

D. Sun- or Earth-to-Planet Distance

When the spacecraft is near the planet, Sun-to-planet distance data help in computing such solar effects upon the spacecraft as solar heating and thermal balance, solar-panel electrical output, and spacecraft trajectory and attitude perturbation magnitudes due to the solar-radiation pressure.

Earth-to-planet distance information allows estimation of maximum available data transmission rates in the radio up- or downlink, as well as the one- and two-way light times required in communicating with the spacecraft.

When combined with information on direction unit vectors (X, Y, Z) from paragraph C above, position vectors for Sun, Earth, and planet may be constructed in the respective coordinate systems

E. Heliocentric Longitude

The heliocentric longitudes of the departure and arrival planetary positions allow sketching the transfer trajectory trace and its orientation with respect to the ecliptic nodes or the perihelion of the target-planet orbit. The relevant planetary information, e.g., the planetary mean orbital elements and related constants, may be found in Section IV.

F. Angles Between Sun, Earth, and Planet (SEP and ESP)

These angles are of paramount importance to spacecraft communication planning. For example, the radio downlink signal becomes distorted and contaminated with noise, if the radio beam passes through the solar corona. This effect is observed near "superior conjunction" between Sun and the target planet. It occurs when the Sun-Earth-planet angle (SEP) reaches near-zero values (commonly, $\leq \pm 5$ deg) while at the same time the Earth-Sun-planet angle (ESP) reaches its maximum value near 180 deg. This unfavorable condition as well as the two other extreme configurations of "opposition" and "inferior conjunction" (*neither* of which significantly affects up- and downlink communication capability) are shown in Fig. 6.

It is essential to proper mission design that important science, orbit determination, or engineering events be *avoided*

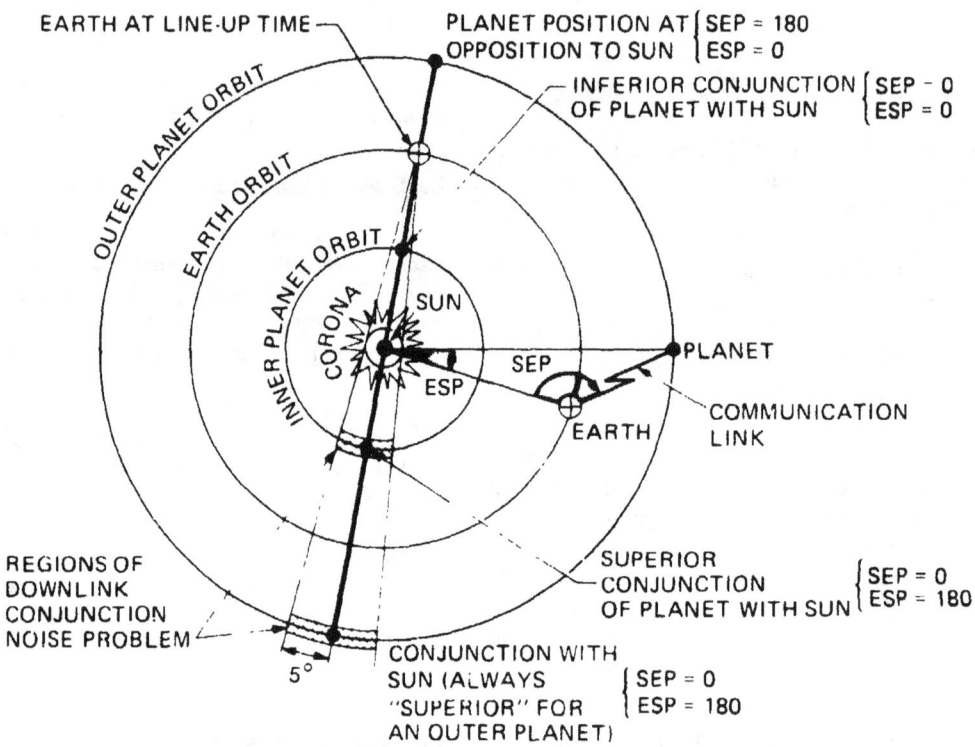

Fig. 6. Definition of opposition and conjunction events involving Sun, Earth, and planet

during the 10 to 20 days that the spacecraft and/or target planet are near superior conjunction. On the other hand, solar physicists do delight in studying the corona's signature superimposed on the distorted signal. This period of time then becomes an opportunity for corona radio-occultation experiments and related solar studies, including relativity investigations.

G. Cone and Clock Angles

As outlined earlier, the cone- and clock-angle system is primarily useful because it is a spacecraft-fixed set of coordinates. As such, and together with data and equations provided in the Mission Design Handbook series (Ref. 1), the cone and clock angles presented on the plots can be used to resolve a number of geometry problems, such as:

(1) The angular travel range required by the movable antenna to track the Earth, if Sun-oriented.

(2) The extent of the Sun-sensor bias required, if the Earth-orientation of the entire spacecraft, together with its antenna, is desired.

(3) The impact of lighting and thermal constraints upon spacecraft orientation options.

(4) The Star tracker slot size in cone angle (CA) required to observe Canopus at different times during an orbiter mission.

H. Station Rise and Set

Station rise and set times, given in GMT, greatly affect mission design planning. After consideration is given to the one-way light time required by the radio signal to traverse the Earth-to-planet distance (and back if needed), significant spacecraft events can be positioned in universal (GMT) time using the rise/set information provided, such as to assure that a given station is "up," i.e., that the spacecraft is above mask in elevation angle at a suitable tracking station. The width (in hours) of the up-time band of a station is in itself an indication of the maximum elevation angle available during the station's pass "under" the spacecraft (i.e., the planet) — the greater the width, the higher the maximum elevation angle.

IV. Astrodynamic Constants

This section presents information necessary for relating the various coordinate systems used in the handbook. For other astrodynamic constants, see Refs. 1, 4, 5, and 7.

Mean orbital elements for the planets are *averaged* analytical models of planetary motion, adopted for reference usage by the astronomical community (e.g., H. M. Nautical Almanac Office or the U.S. Naval Observatory). Mean elements change as functions of time, T_{50}.

T_{50} is defined as the time interval, in Julian centuries, elapsed at date of interest, since epoch January 1.0, 1950, 0^h ephemeris time (ET) (i.e., Julian epoch J1950.0 = JD2433282.5). A Julian century contains 36525.0 mean solar (i.e., calendar) days. Ephemeris time is an atomically controlled uniform time, approximated as: ET = GMT + DUT, where the time increment DUT = +54.2 s (as of July 1983). DUT is currently increasing at about one second per year.

Six elements completely define the mean orbit (Fig. 7), as follows (all angular measures in degrees):

(a) CELESTIAL SPHERE VIEW

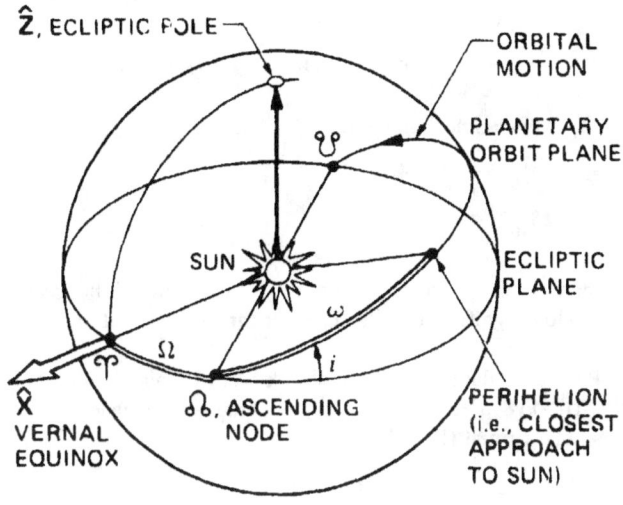

(b) ORBIT PLANE VIEW

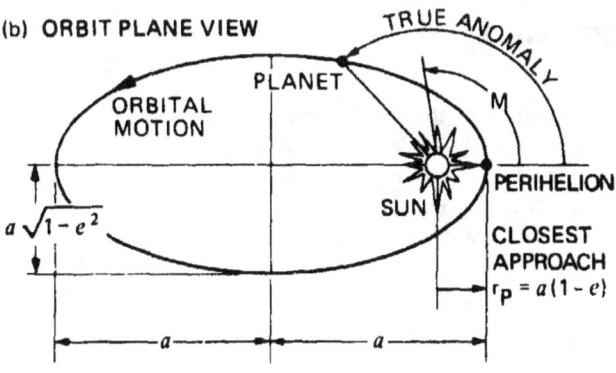

Fig. 7. Definition of the mean elements of a planetary orbit

a = semimajor axis (in astronomical units, 1 AU = 149,597,871 km).

e = eccentricity of orbit (dimensionless).

M = mean anomaly, a mean motion angular position from perihelion (i.e., closest approach to Sun) at time T_{50} (for numerical convenience also expressed as d_{50}, days from January 1.0, 1950 0^h ET). True anomaly (actual position angle from perihelion) can be obtained from M by way of iteration using Kepler's transcendental equation.

i_{50} = inclination angle of mean orbit plane at time T_{50}, with respect to the mean ecliptic of epoch B1950.0 (EMO50).

Ω_{50} = longitude of ascending node of mean orbit upon the ecliptic of B1950.0, measured from Earth's vernal equinox.

ω_{50} = argument of perihelion, an angle measured from the ascending node of the mean orbit on the ecliptic of B1950.0 to the perihelion of the mean orbit of the planet.

As presented by F. Sturms (Ref. 4) and based on information in the Explanatory Supplement to the Ephemeris (Ref. 7), the mean elements are:

(1) Earth:

a = 1.00000023 AU = 149,597,905 km

e = 0.0167301085 − 0.000041926 T_{50} − 0.000000126 T_{50}^2

M = 358.000682 + 0.9856002628 d_{50} − 0.0001550000 T_{50}^2 − 0.0000033333 T_{50}^3

i_{50} = 0.013076 T_{50} − 0.000009 T_{50}^2

Ω_{50} = 174.40956 − 0.24166 T_{50} + 0.00006 T_{50}^2

ω_{50} = 287.67097 + 0.56494 T_{50} + 0.00009 T_{50}^2

(2) Venus:

a = 0.7233316 AU = 108,208,867 km

e = 0.00679684275 − 0.000047649 T_{50} + 0.000000091 T_{50}^2

M = 311.505478 + 1.6021301892 d_{50} + 0.0012860555 T_{50}^2

i_{50} = 3.39413 − 0.00086 T_{50} − 0.00003 T_{50}^2

Ω_{50} = 76.22967 − 0.27785 T_{50} − 0.00014 T_{50}^2

ω_{50} = 54.63793 + 0.28818 T_{50} − 0.00115 T_{50}^2

The equatorial plane of a celestial body at an epoch of interest, T_{50}, was defined by the International Astronomical Union Working Group on Cartographic Coordinates and Rotational Elements (Ref 5) by the right ascension α_{50} and declination δ_{50} of the body's "north" polar axis direction with respect to the Earth mean equator of epoch B1950.0 (EME50), as shown in Fig. 8. The position of the rotating prime meridian of the body is defined by an angle W, measured positive counterclockwise in that body's equatorial plane from the ascending node of that equator upon the EME50, to the prime meridian position at epoch d_{50}. According to the Ref. 5, Table I.

(1) Sun pole:

$\alpha_{50} = 285.90$ (deg)

$\delta_{50} = 63.90$

$W = 240.90 + 14.1844000\, d_{50}$

(2) Earth pole:

$\alpha_{50} = 0.00 - 0.640\, T_{50}$

$\delta_{50} = 90.00 - 0.557\, T_{50}$

$W = 99.87 + 360.9856123\, d_{50}$

(3) Venus pole:

$\alpha_{50} = 272.80$

$\delta_{50} = 67.20$

$W = 213.63 - 1.4814205\, d_{50}$ (retrograde)

(4) Invariable Plane pole:

$\alpha_{50} = 272.40$

$\delta_{50} = +66.99$

Note that in the computation of W, care should be taken not to lose significant digits during arithmetic calculations.

Positional data in this handbook were obtained using the JPL DE-118 precision planetary ephemeris on magnetic tape in tabular format (Ref. 3).

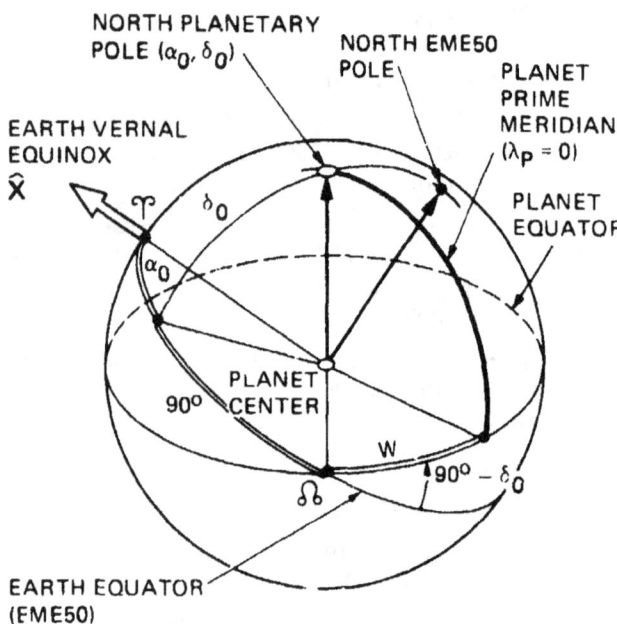

Fig. 8. IAU definition of planetary pole and prime meridian at time d_{50}, elapsed since January 1.0, 1950 0^h ET

Acknowledgments

The contributions, reviews, and suggestions by members of the Handbook Advisory Committee, especially those of K. T. Nock, R. E. Diehl, W. I. McLaughlin, W. E. Bollman, P. A. Penzo, R. A. Wallace, D. F. Bender, D. V. Byrnes, L. A. D'Amario, T. H. Sweetser, and R. S. Schlaifer are acknowledged and greatly appreciated. The authors would like to thank Mary Fran Buehler and David E. Fulton for their editorial contribution.

References

1. Sergeyevsky, A. B., et al., *Interplanetary Mission Design Handbook*, JPL Publication 82-43, Volume 1, Parts 1-4. Jet Propulsion Laboratory, Pasadena, Calif., March 1, 1983.

2. Sergeyevsky, A. B., *Mission Design Data for Venus, Mars, and Jupiter through 1990*, Technical Memorandum 33-736. Jet Propulsion Laboratory, Pasadena, Calif., Sept. 1, 1975.

3. Newhall, XX, Standish, E. M., and Williams, J. G., "DE-102: A Numerically Integrated Ephemeris of the Moon and Planets, Spanning 44 Centuries," *Astronomy and Astrophysics*, 1983 (in press).

4. Sturms, F. M., *Polynomial Expressions for Planetary Equators and Orbit Elements With Respect to the Mean 1950.0 Coordinate System*, Technical Report 32-1508, pp. 6-9. Jet Propulsion Laboratory, Pasadena, Calif., Jan. 15, 1971.

5. Davies, M. E., et al., "Report of the IAU Working Group on Cartographic Coordinates and Rotational Elements of the Planets and Satellites: 1982," *Celestial Mechanics*, Vol 29, No. 4, pp. 309-321, April 1983.

6. Fricke, W., et al., *Fourth Fundamental Catalogue (FK 4)*, No. 10 in the series, *Veröffentlichungen des Astronomischen Rechen-Instituts Heidelberg*, Verlag L. Braun. Karlsruhe, 1963.

7. "Explanatory Supplement to the Ephemeris", Her Majesty's Stationery Office, London, 1961.

Positional Data

**Venus
1988—2020**

Venus

1988

DECLIN RT.ASC 1988

VENUS 1988

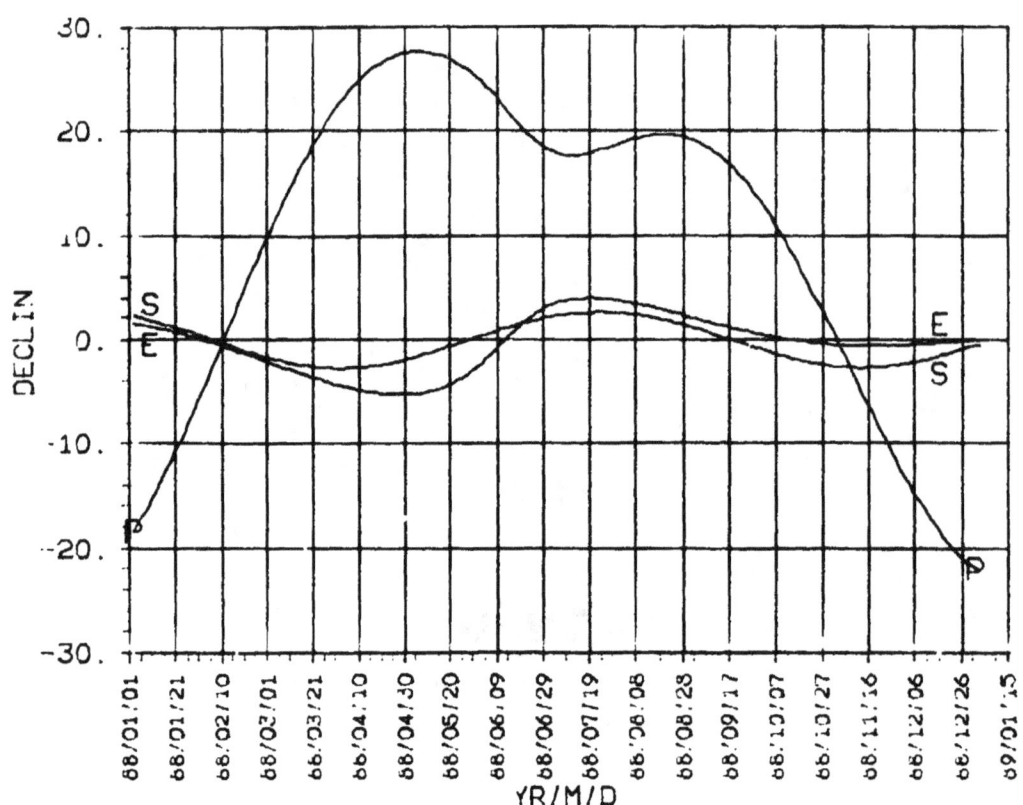

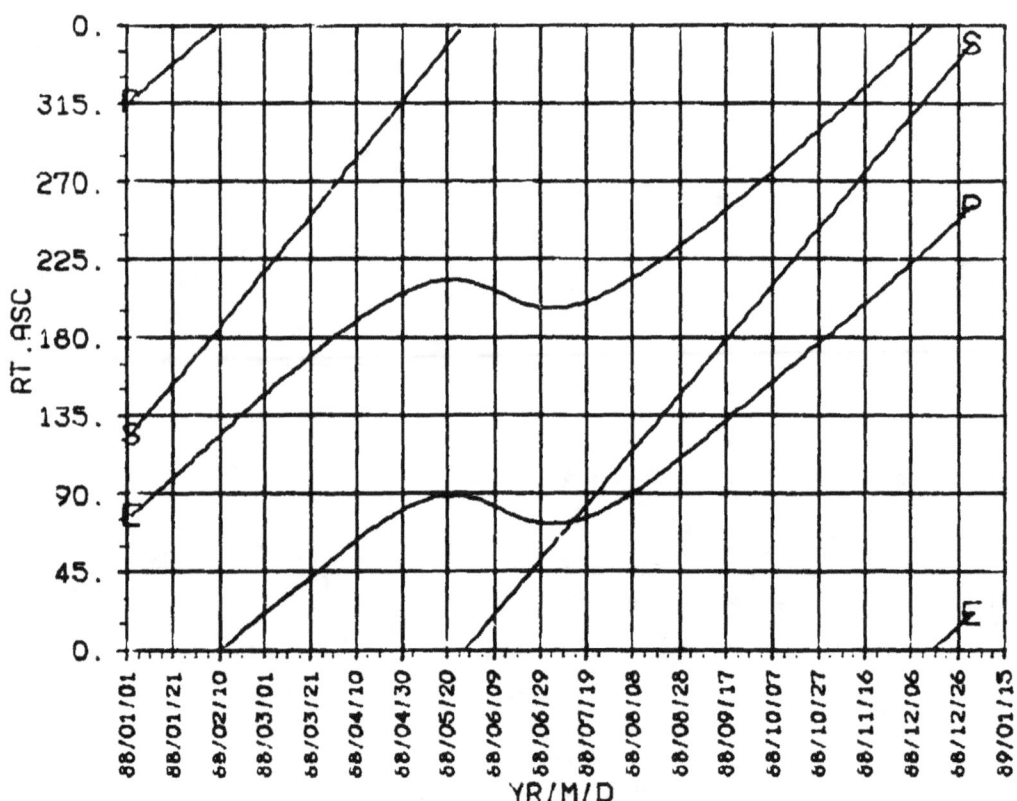

DISTANCE EC.LON 1988

VENUS 1988

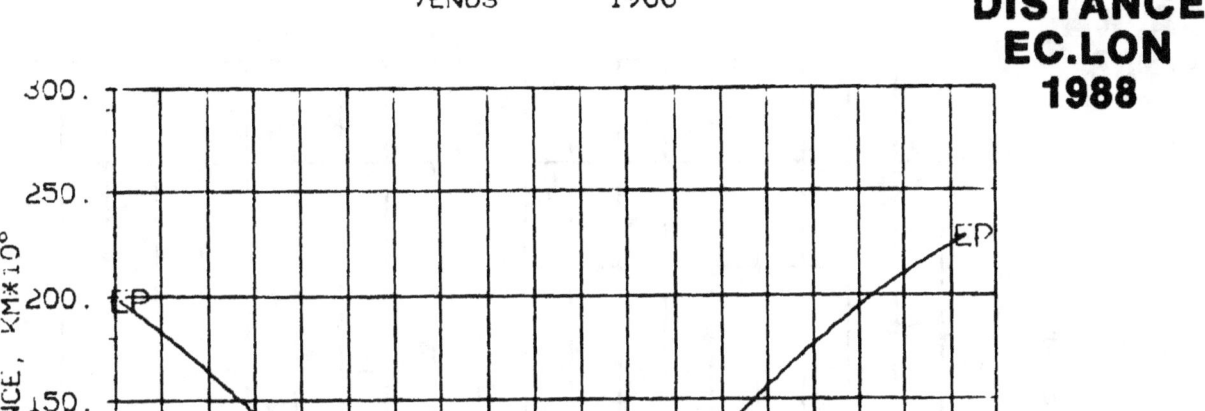

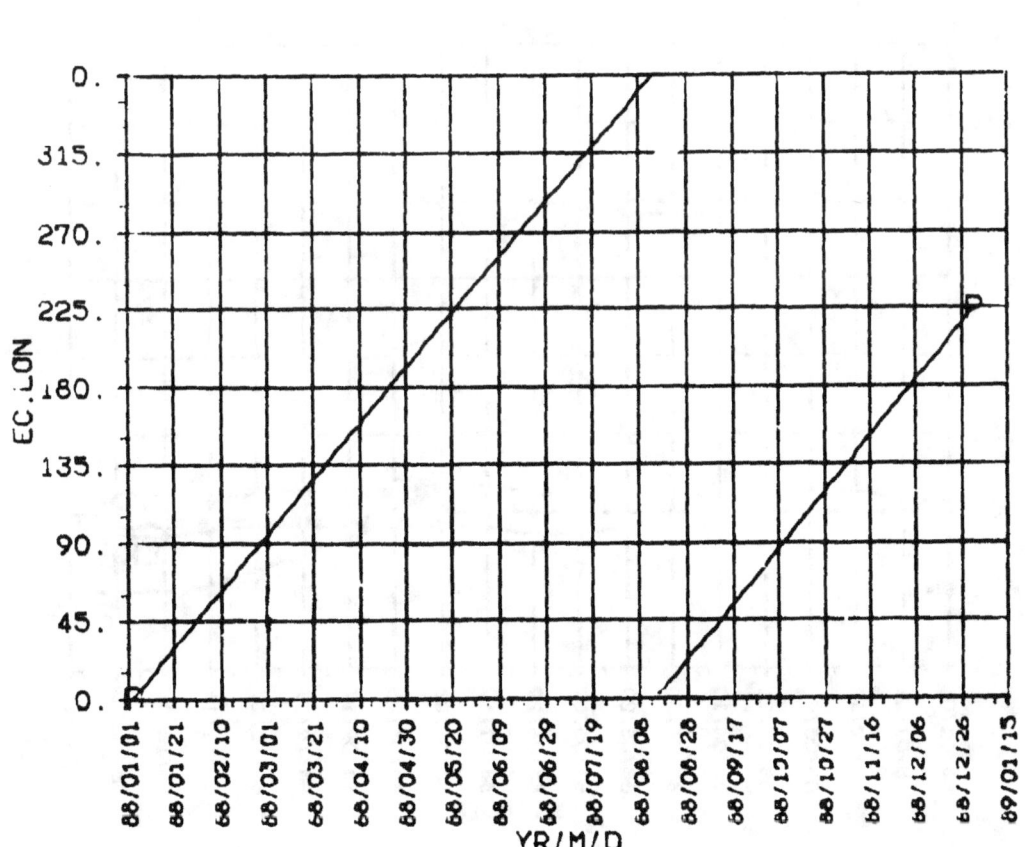

SEP, ESP CA, KA 1988

VENUS 1986

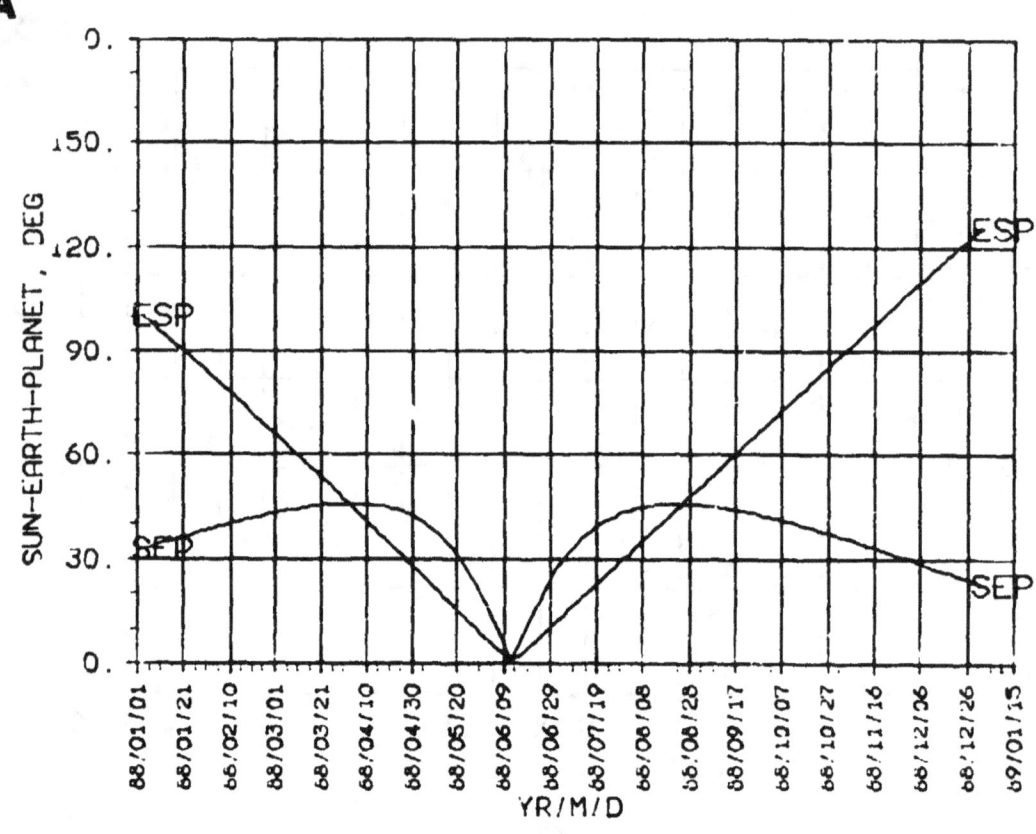

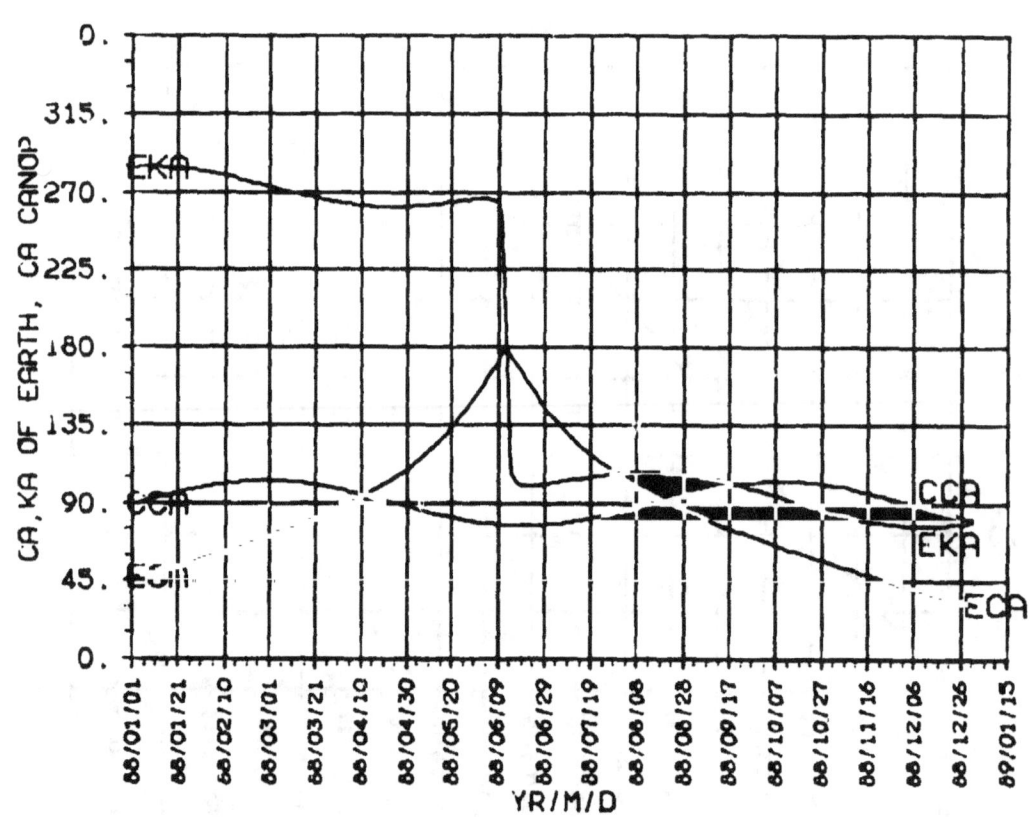

STA R/S DSN 1988

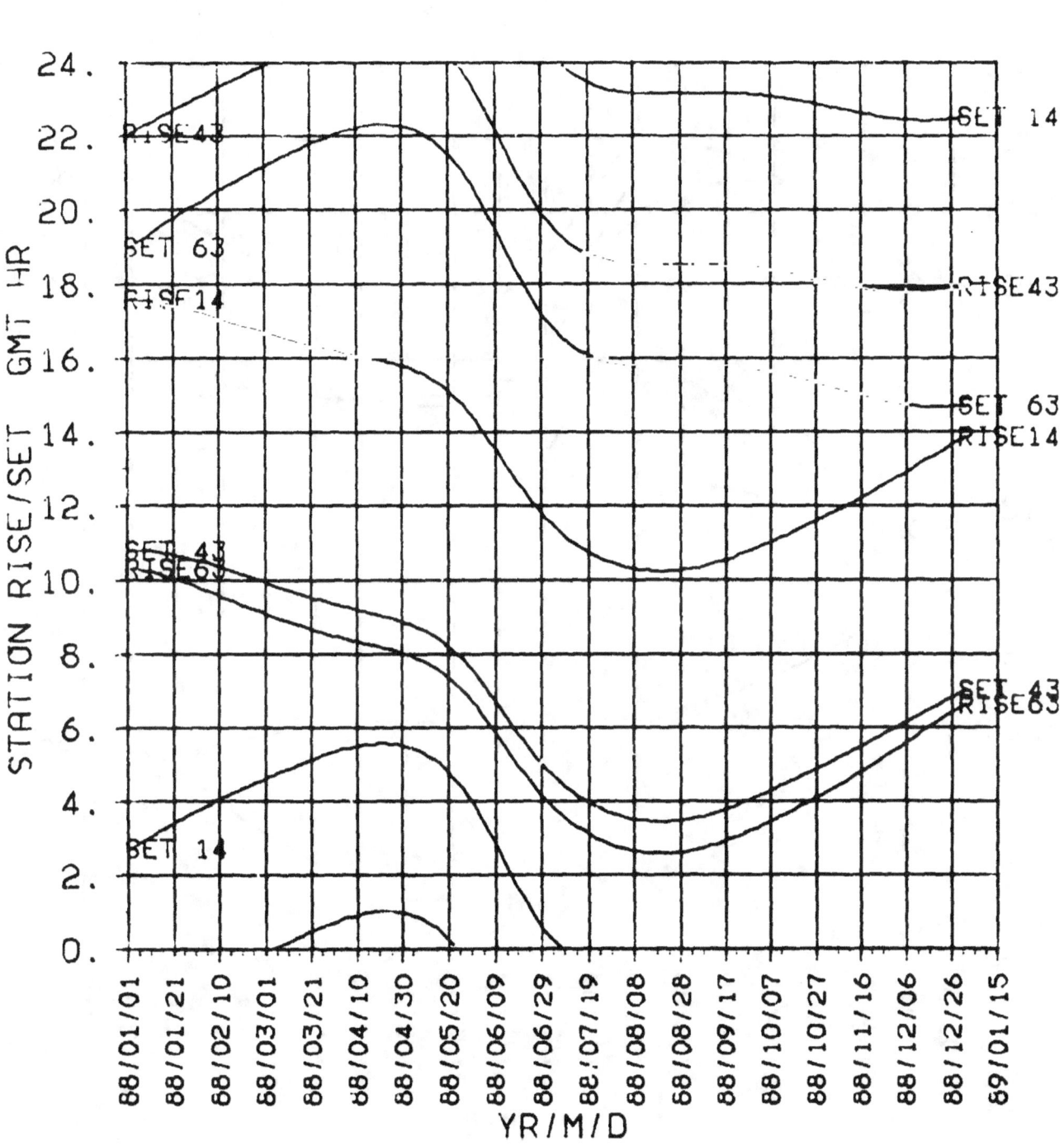

STA R/S
NON-DSN
1988

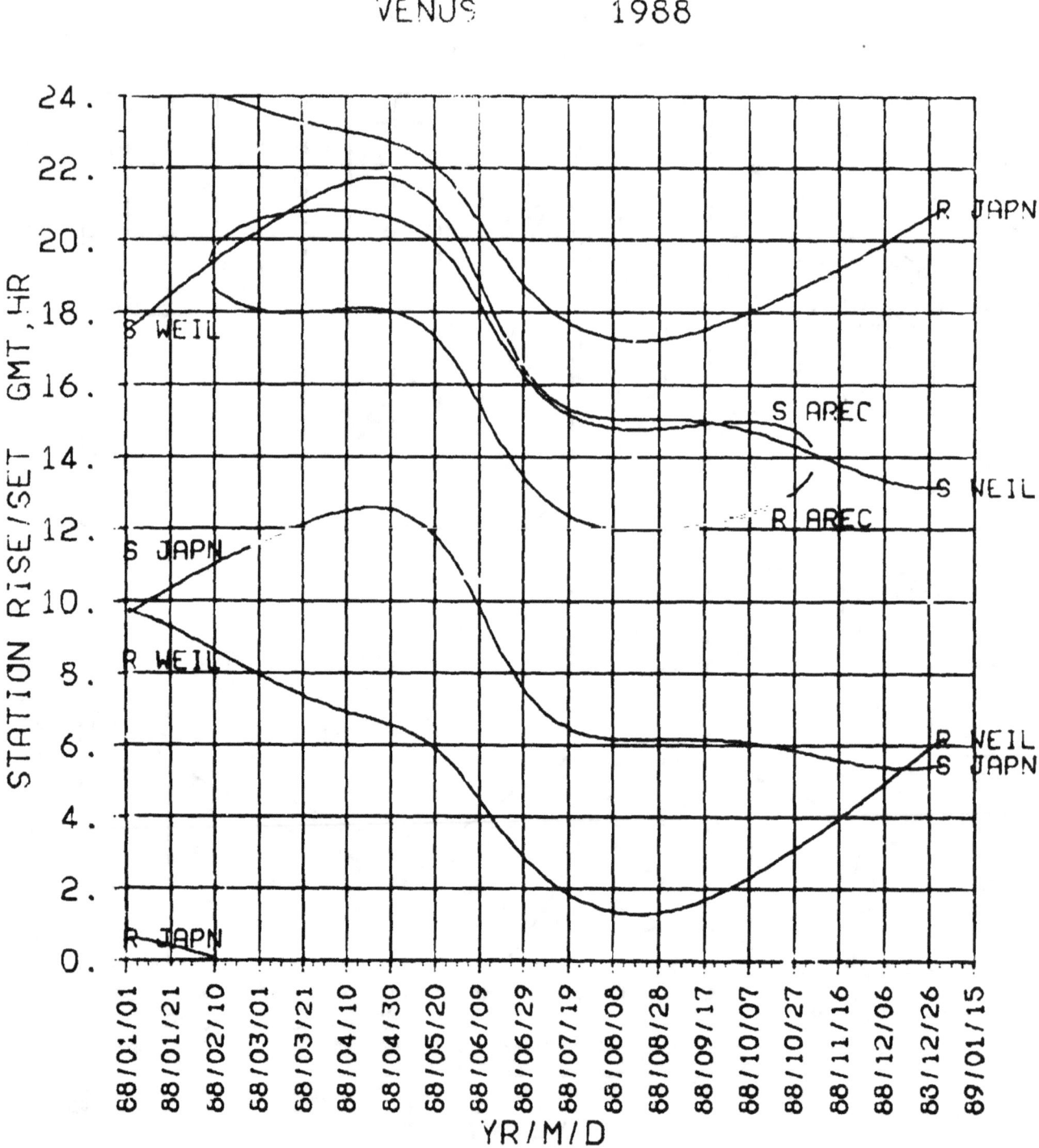

Venus

1989

DECLIN RT.ASC 1989

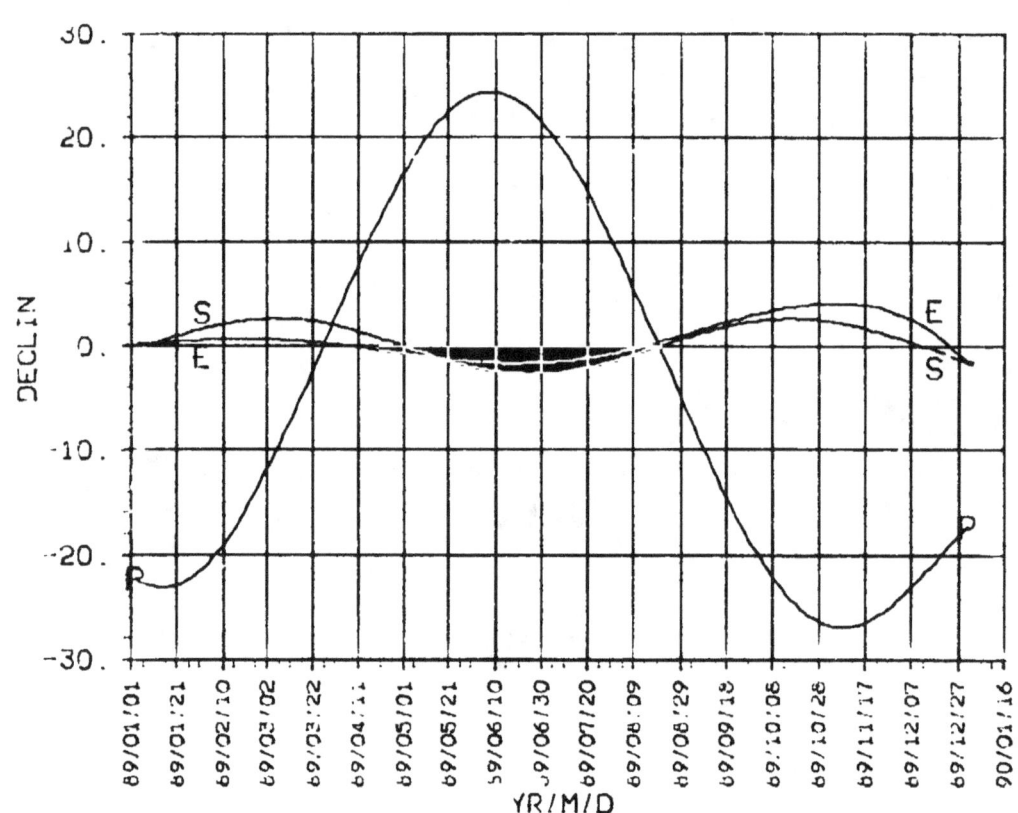

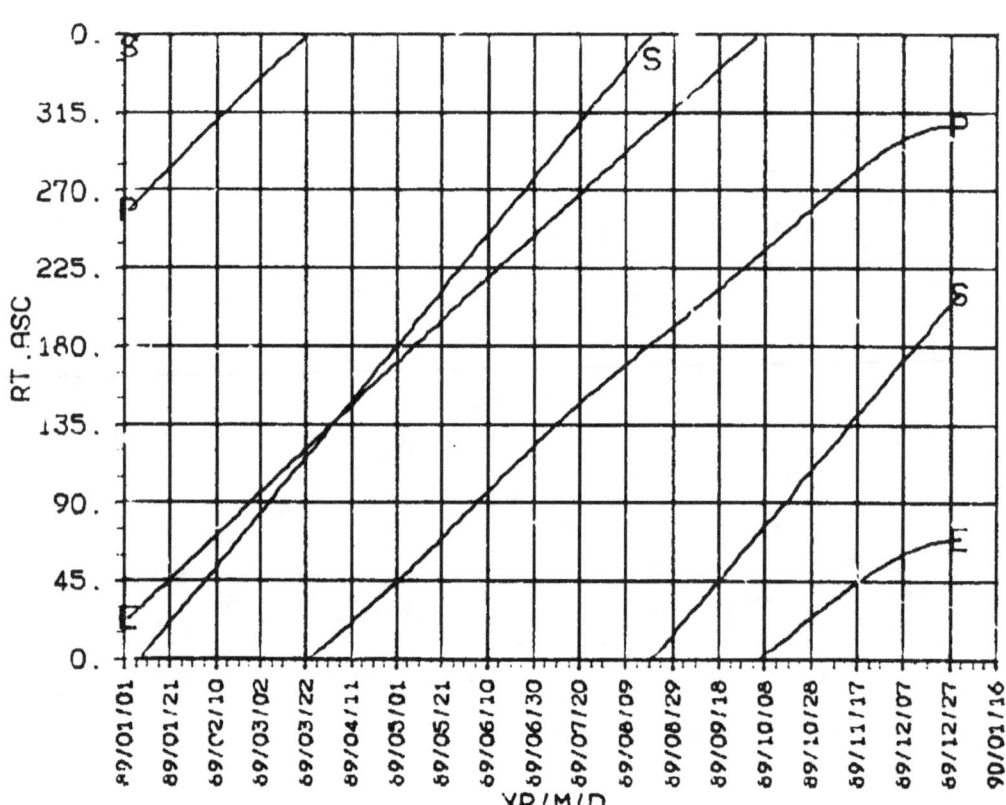

VENUS 1989

DISTANCE
EC.LON
1989

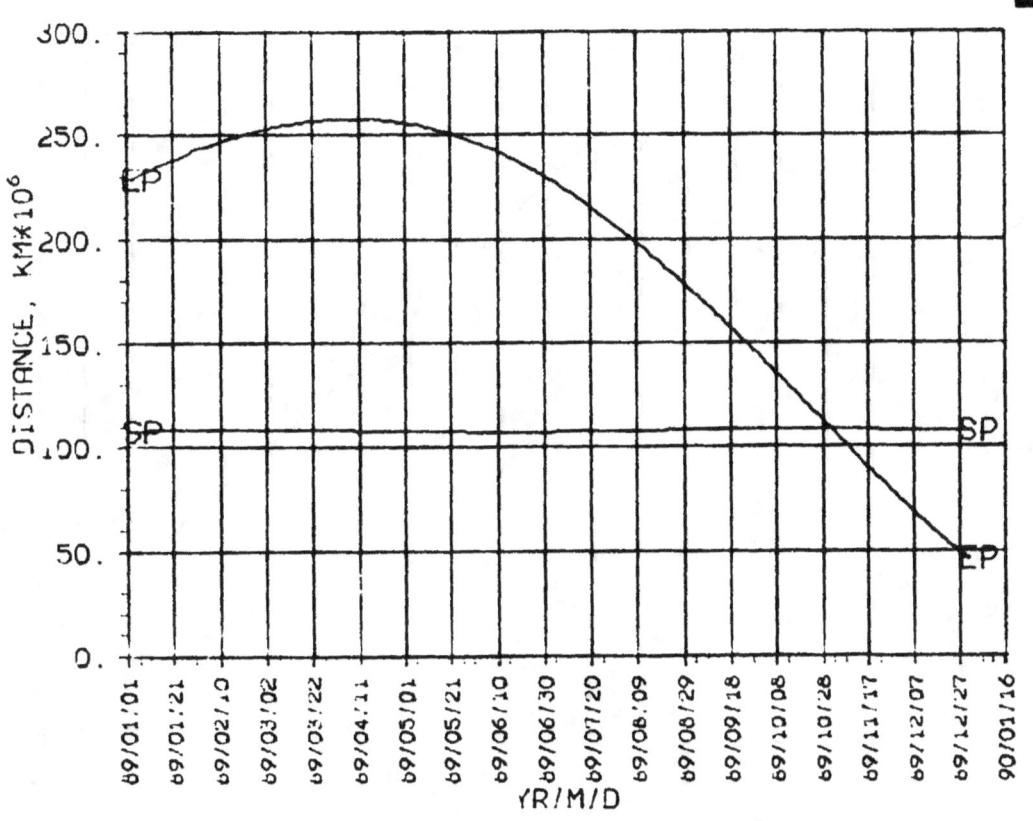

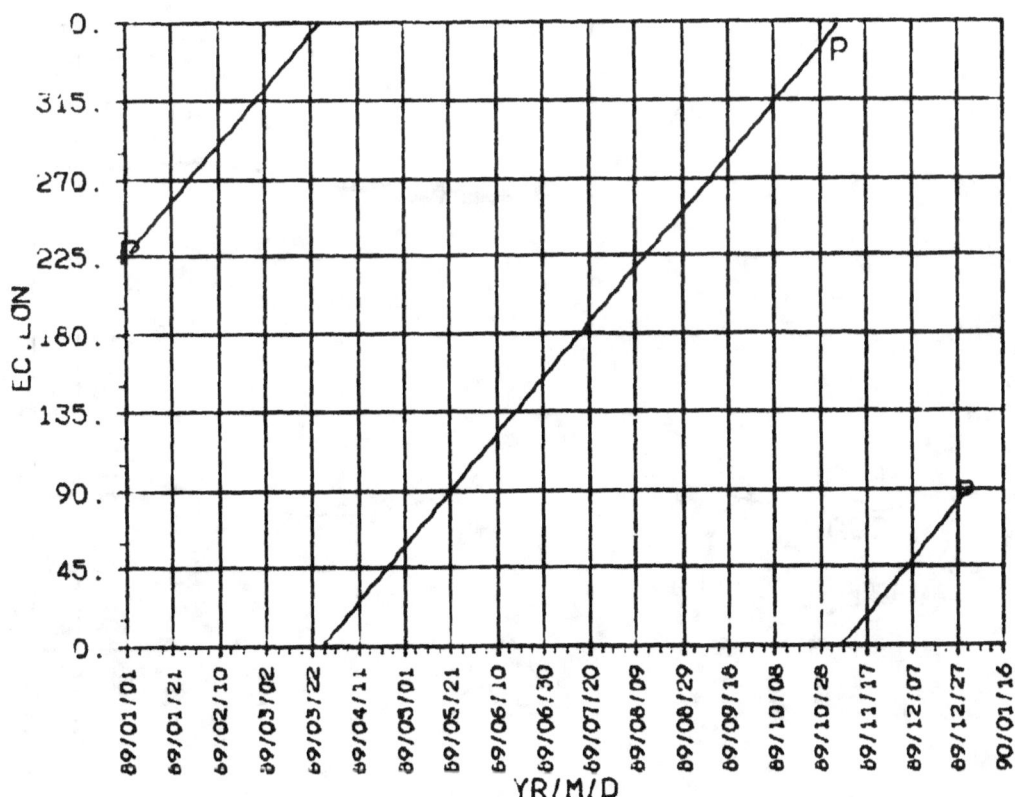

**SEP, ESP
CA, KA
1989**

VENUS 1989

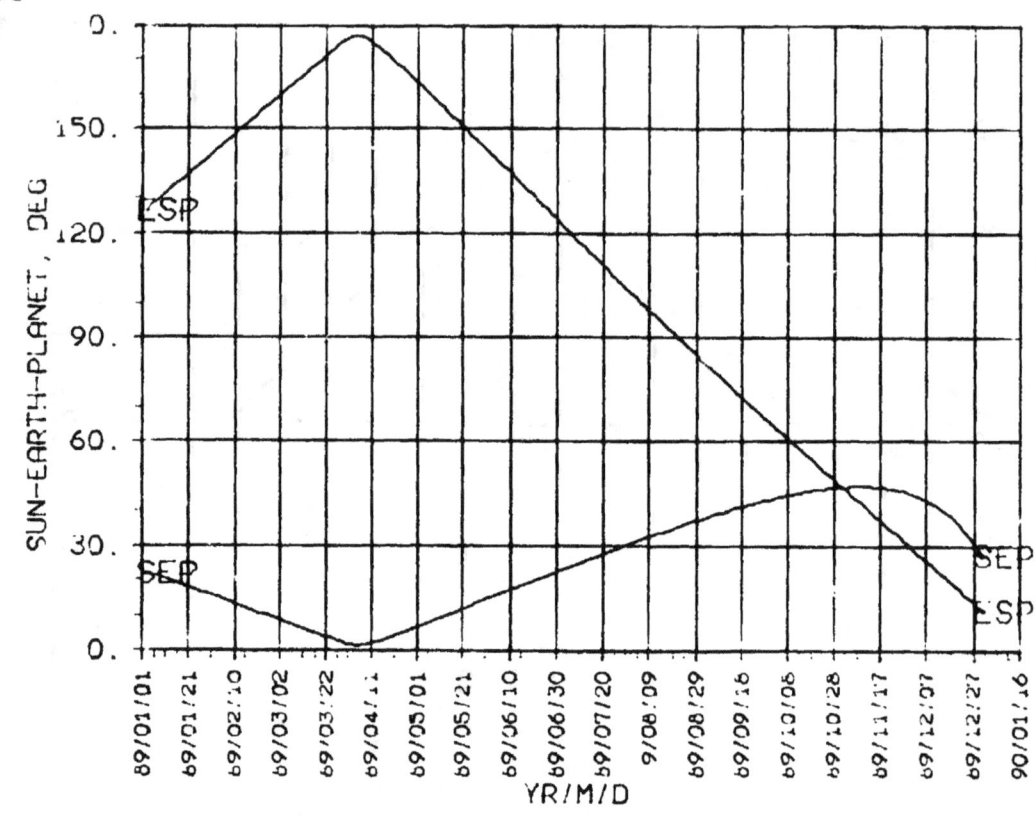

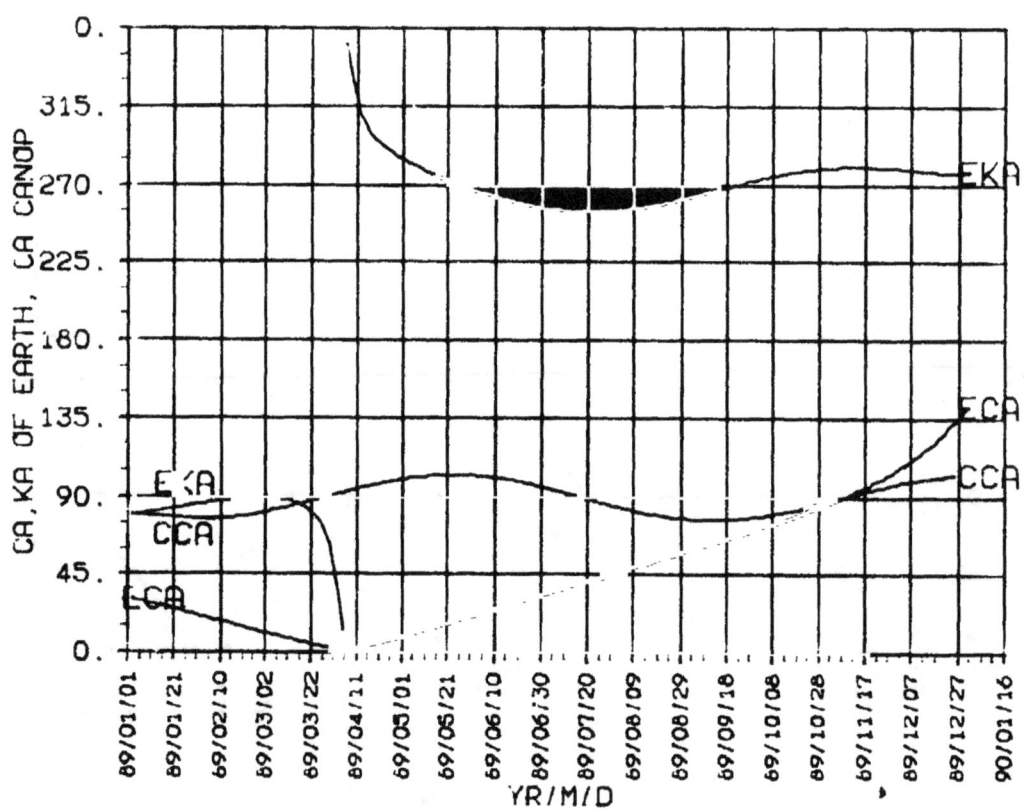

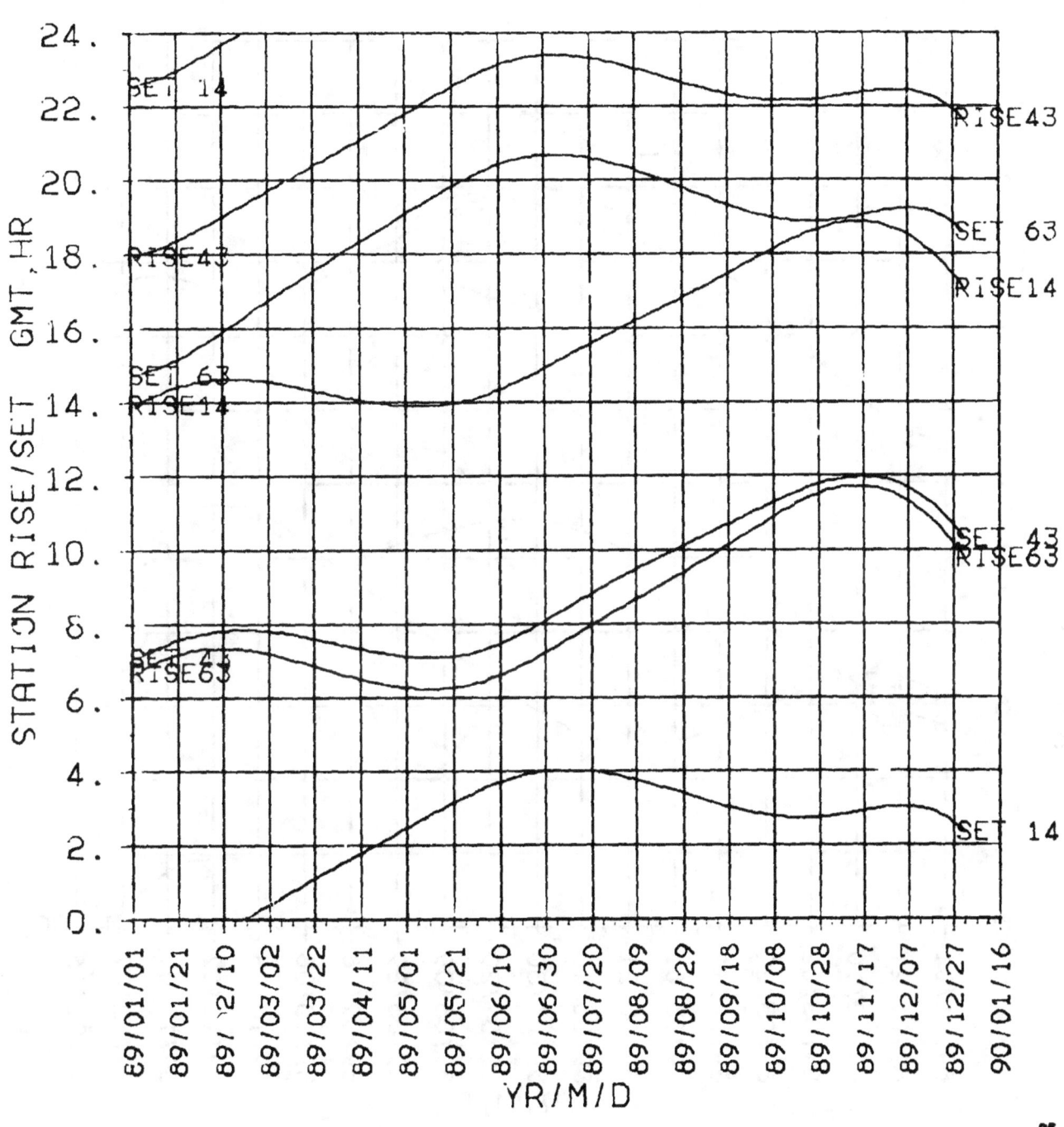

**STA R/S
NON-DSN
1989**

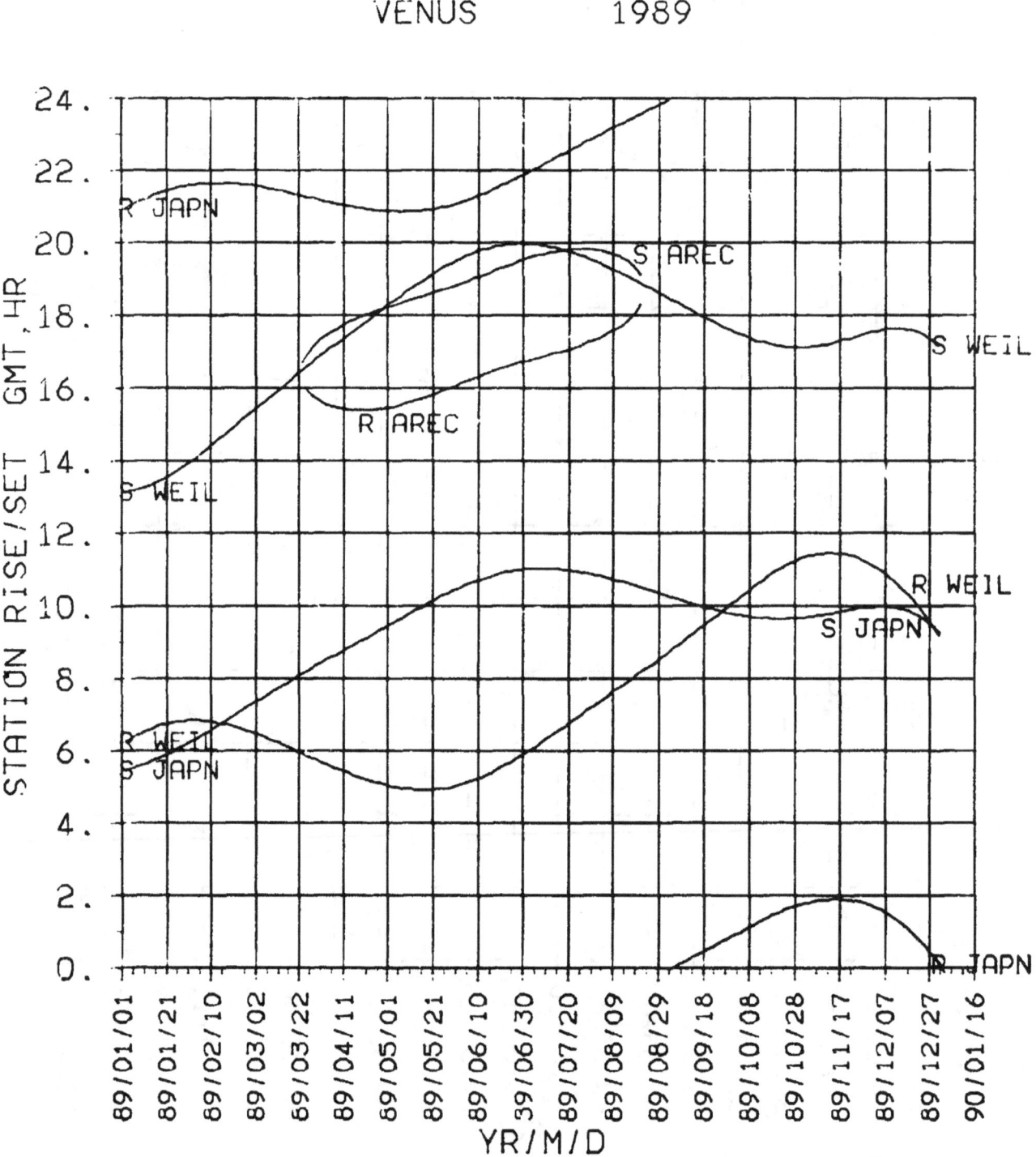

Venus

1990

DECLIN RT.ASC 1990

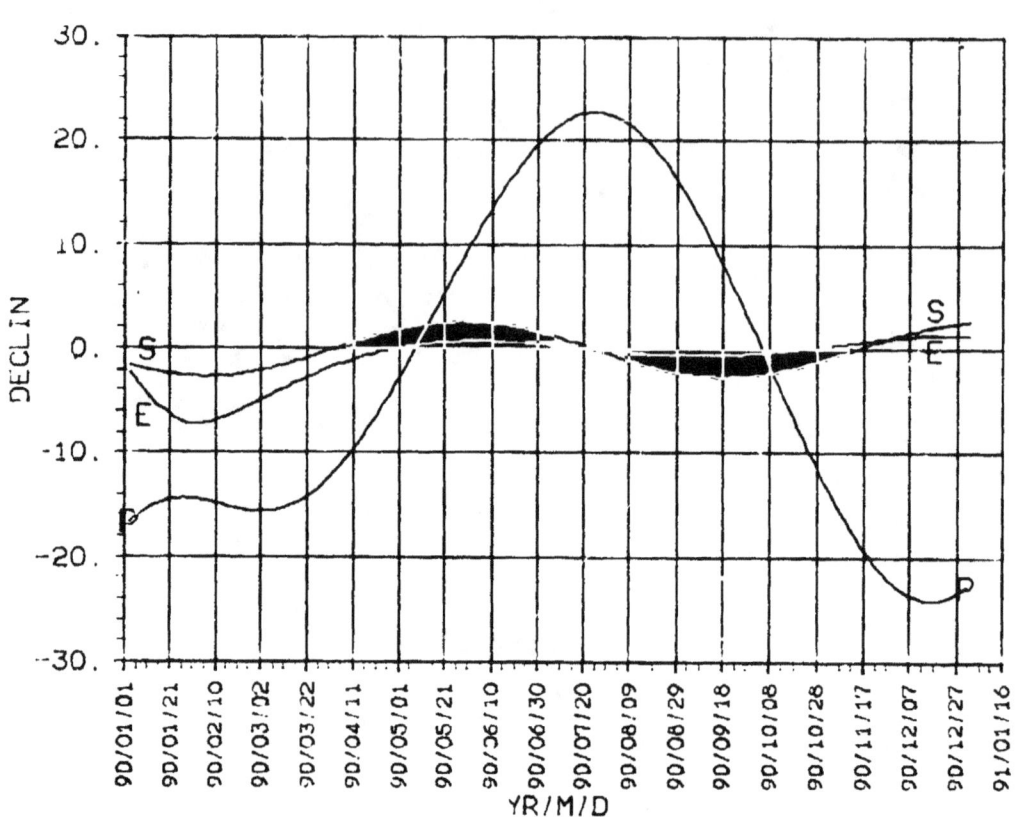

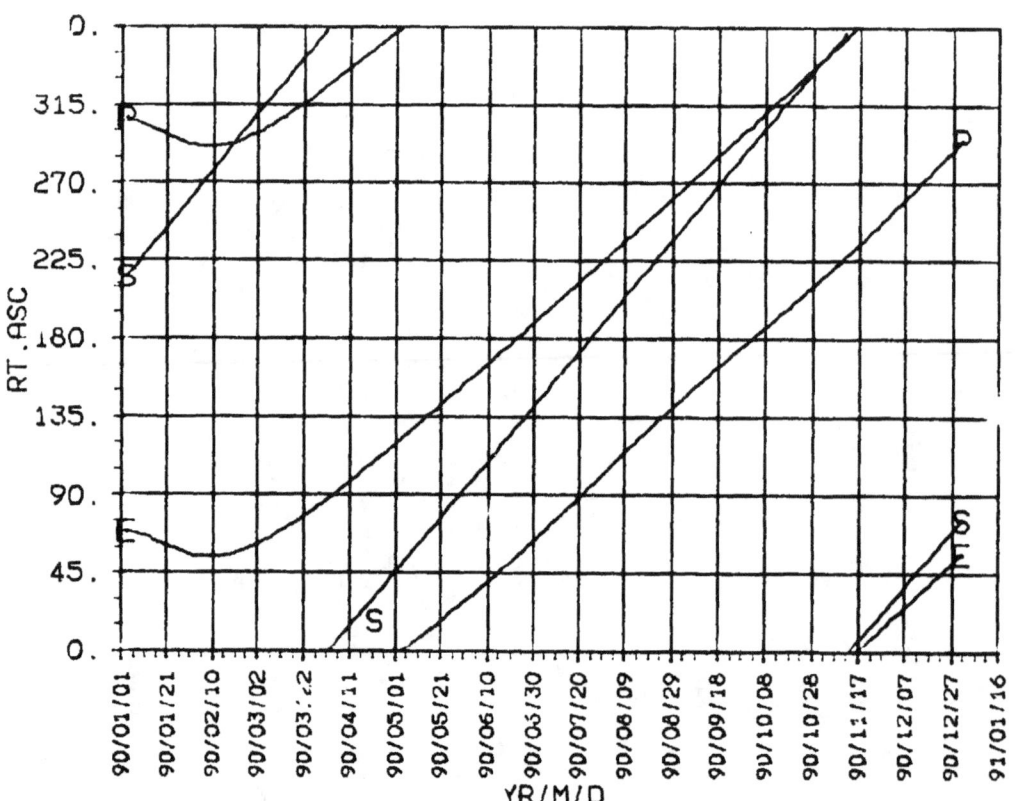

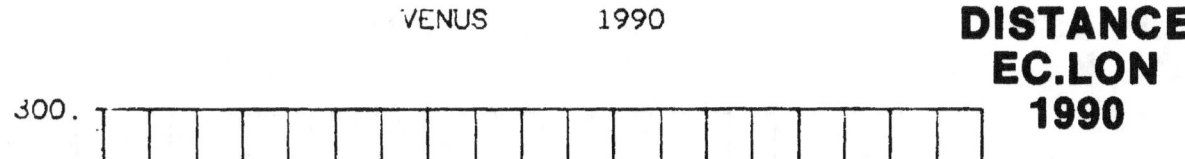

DISTANCE EC.LON 1990

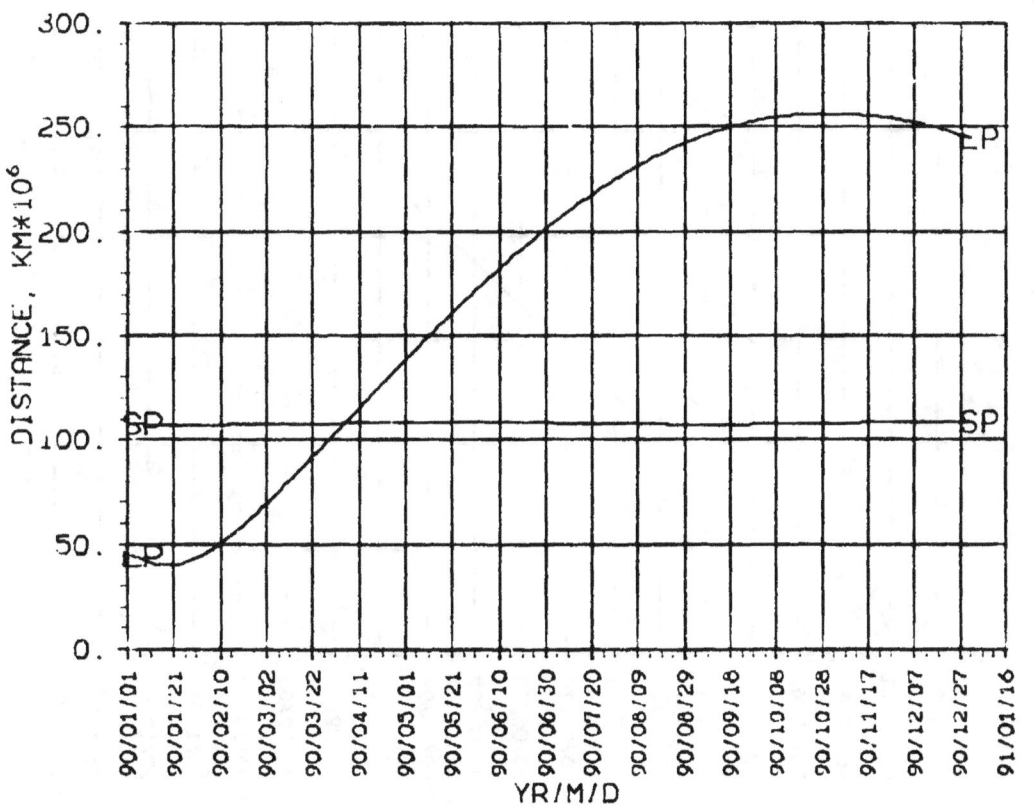

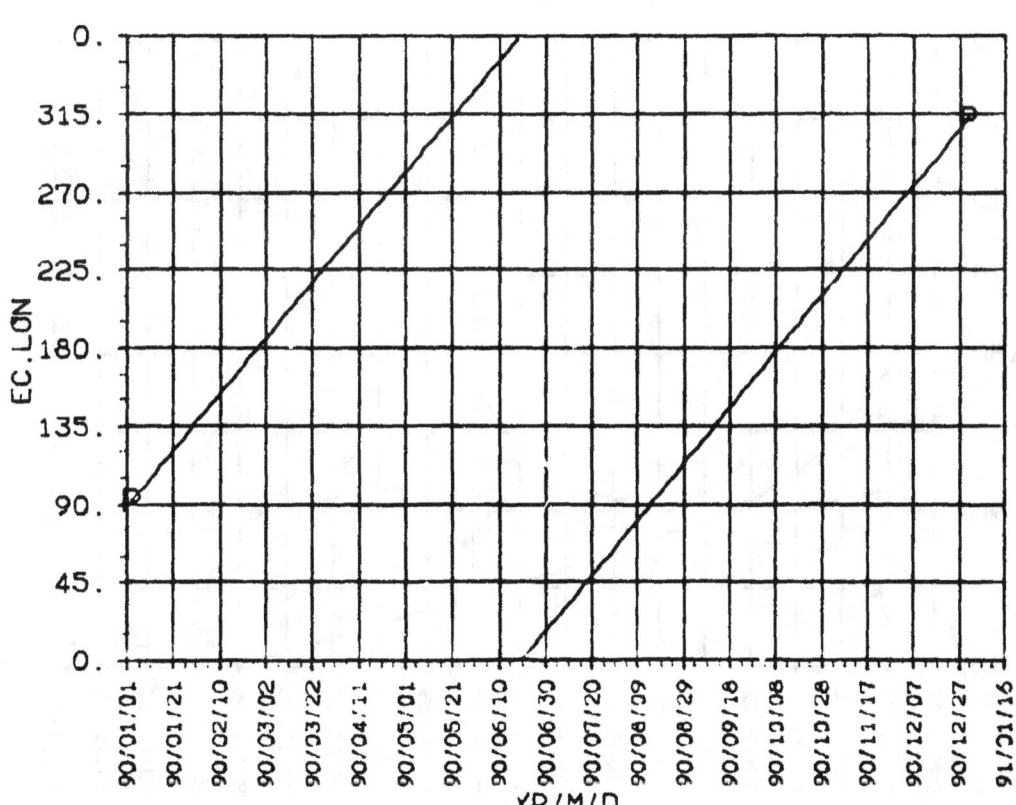

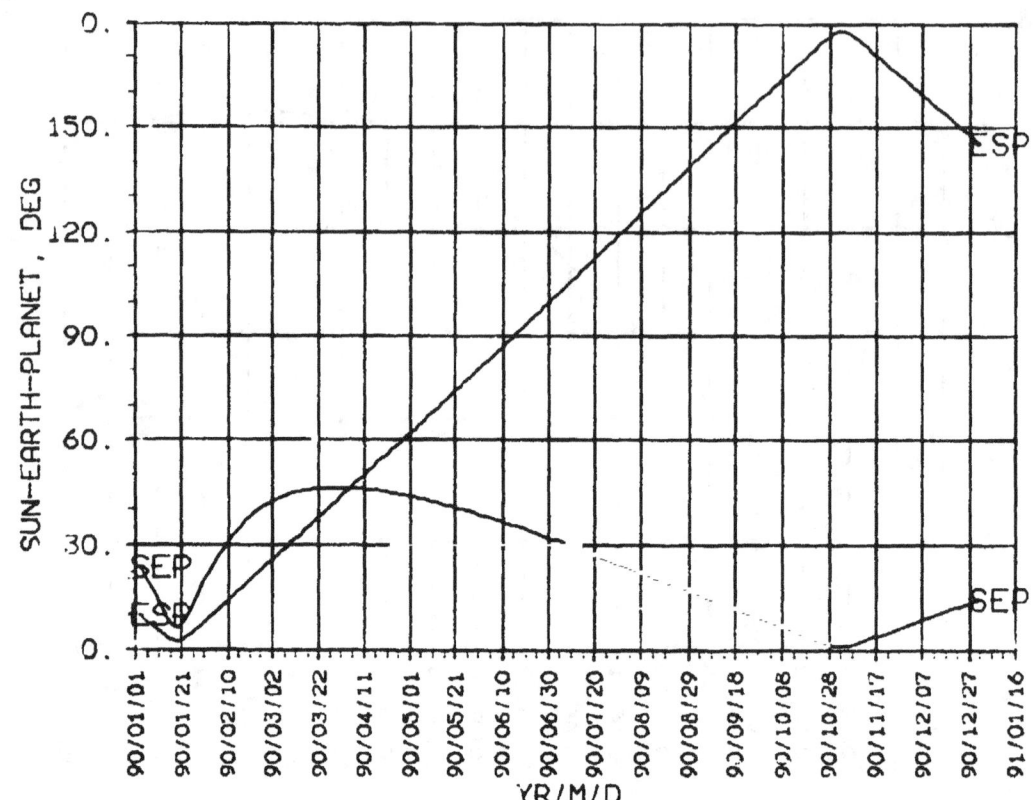
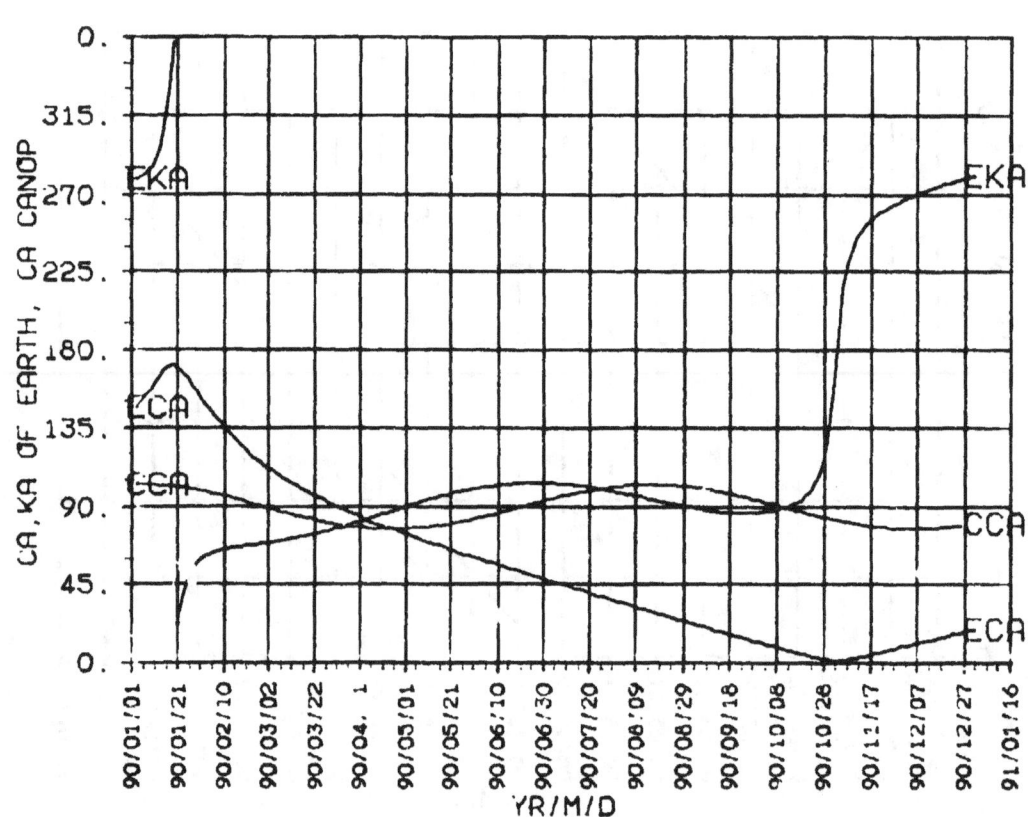

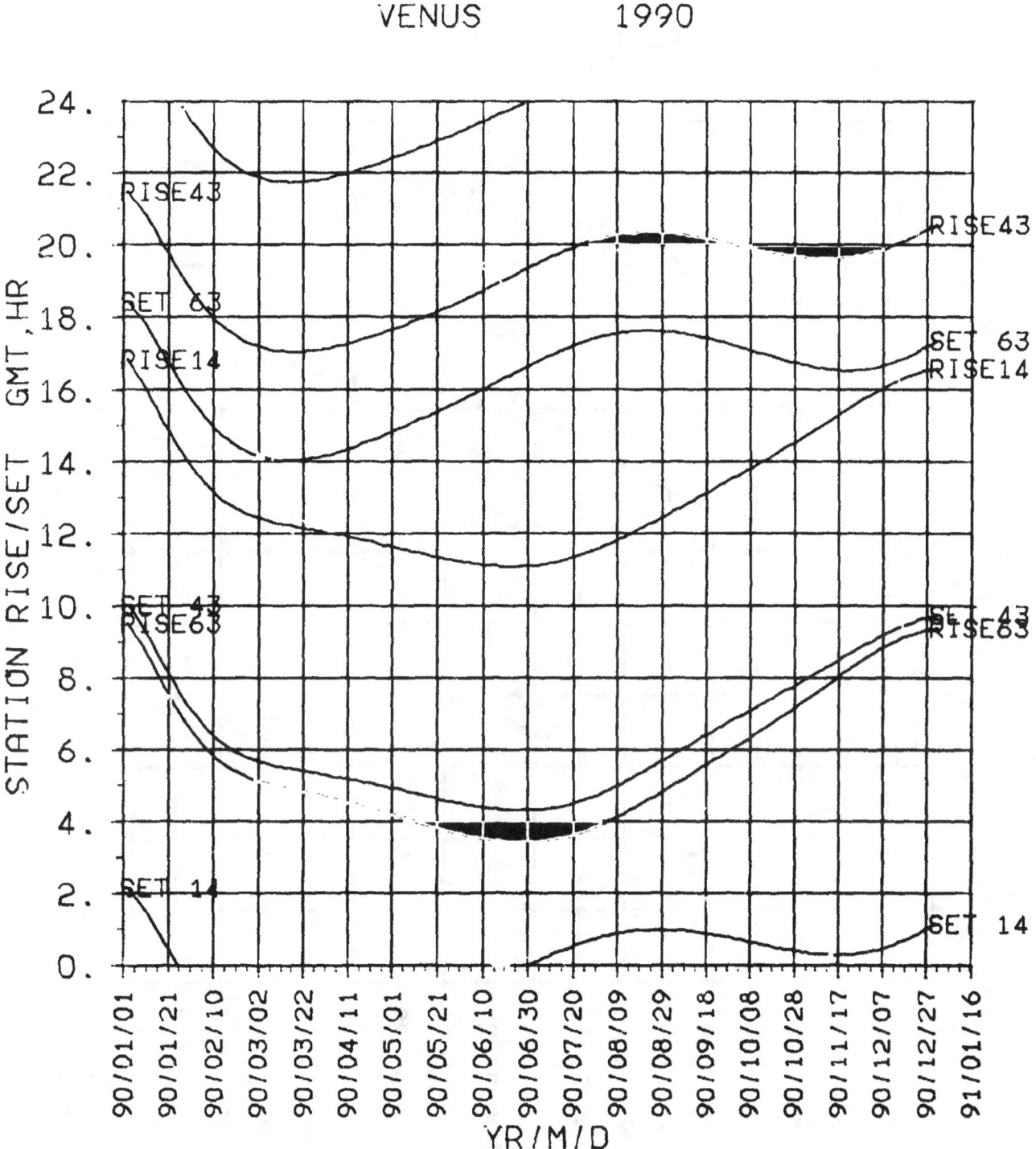

**STA R/S
NON-DSN
1990**

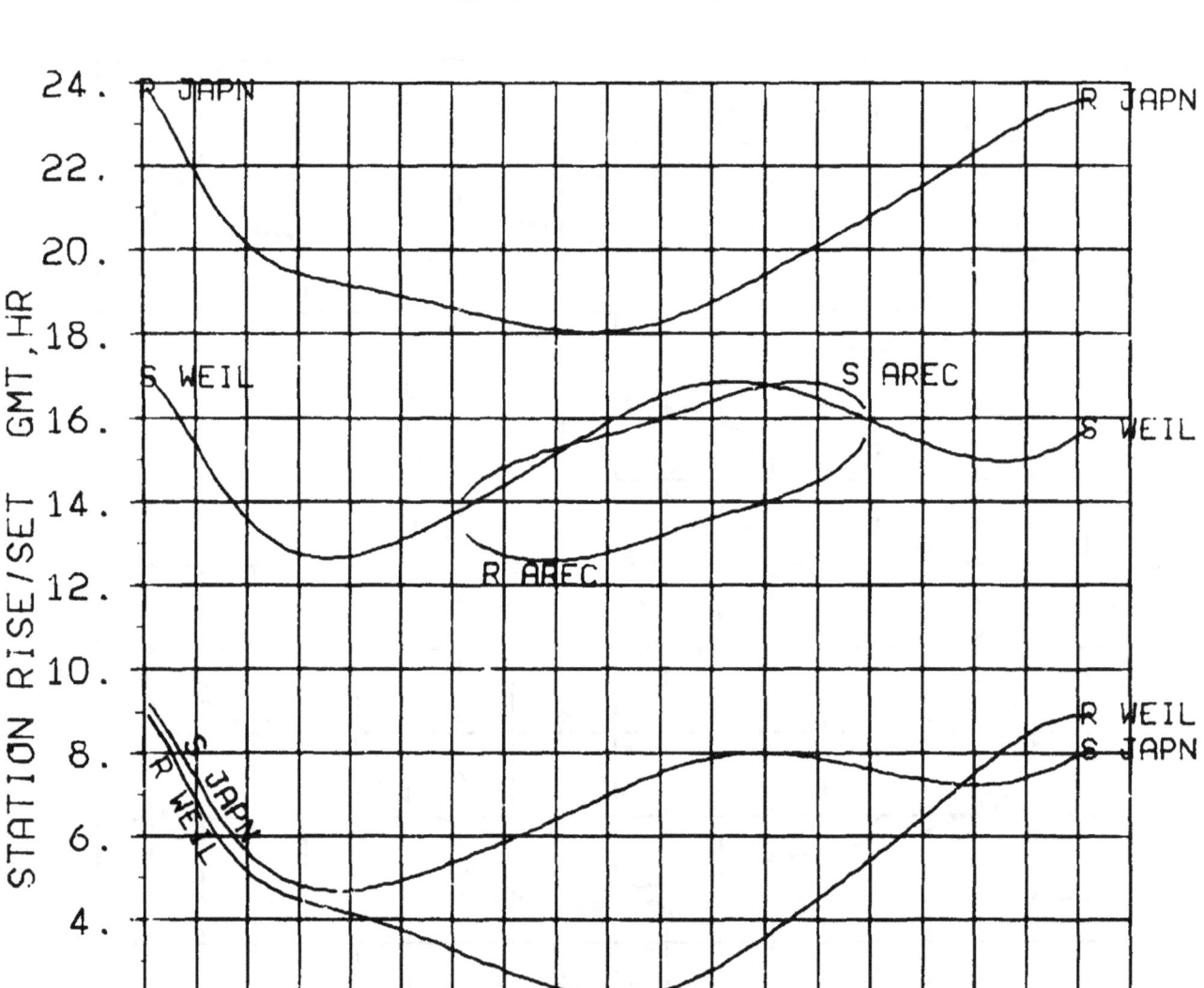

Venus

1991

DECLIN RT.ASC 1991

VENUS 1991

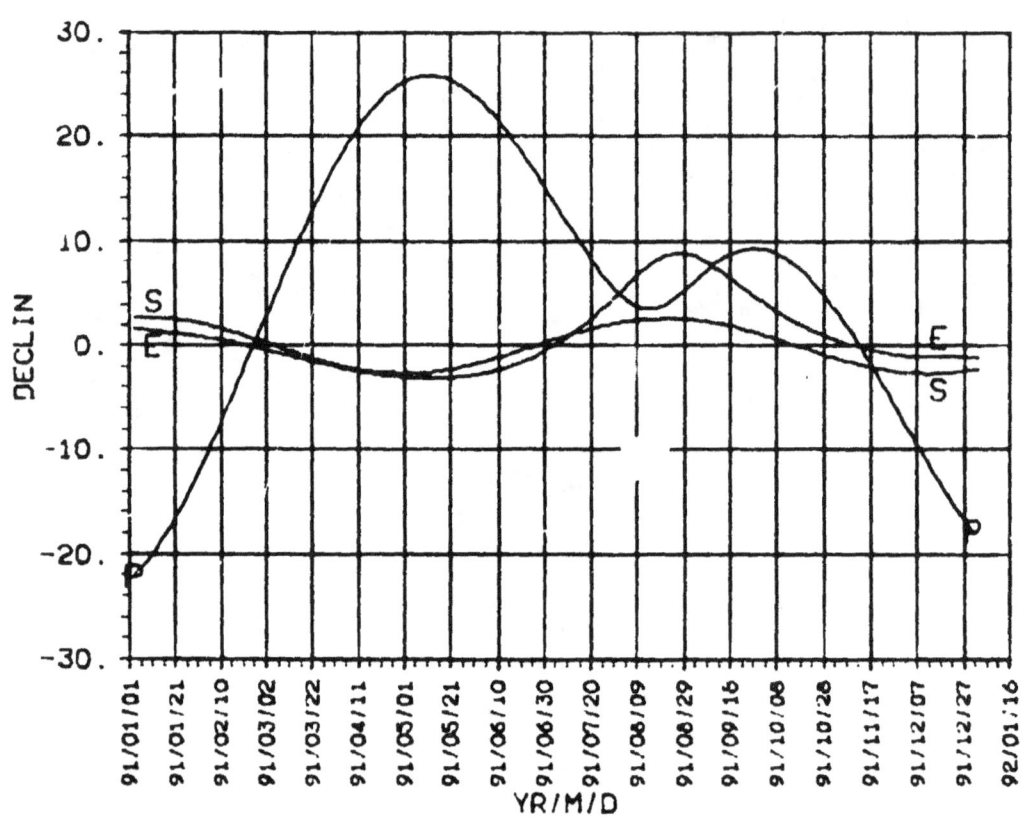

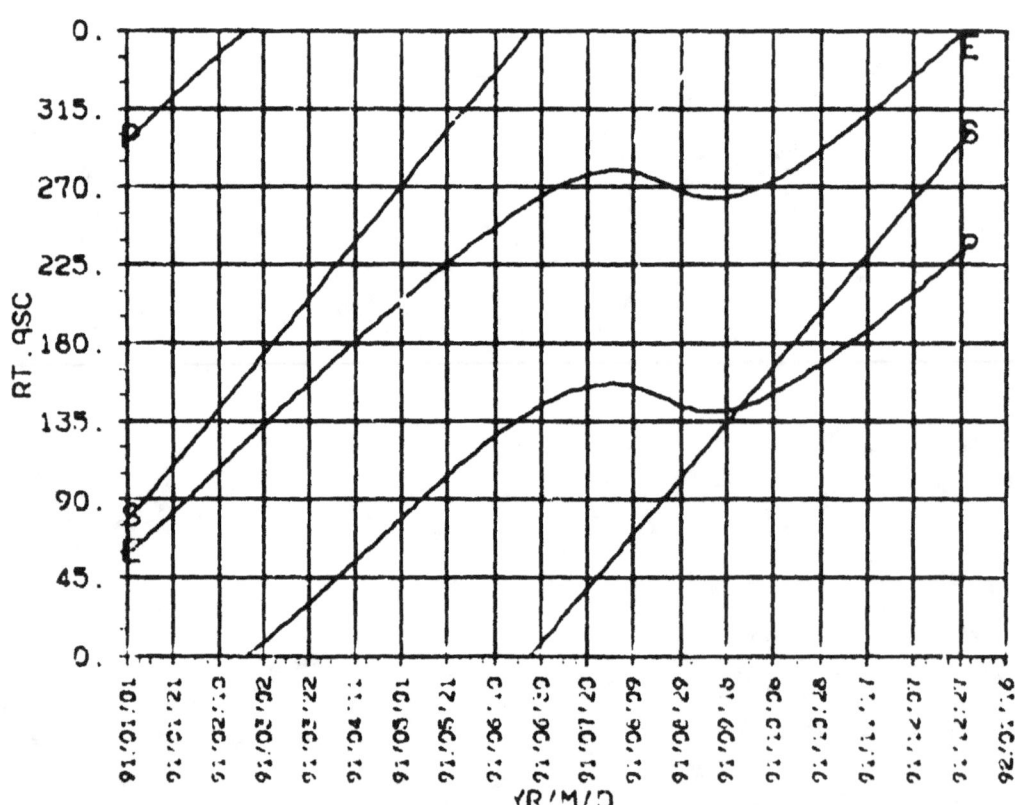

VENUS 1991

DISTANCE EC.LON 1991

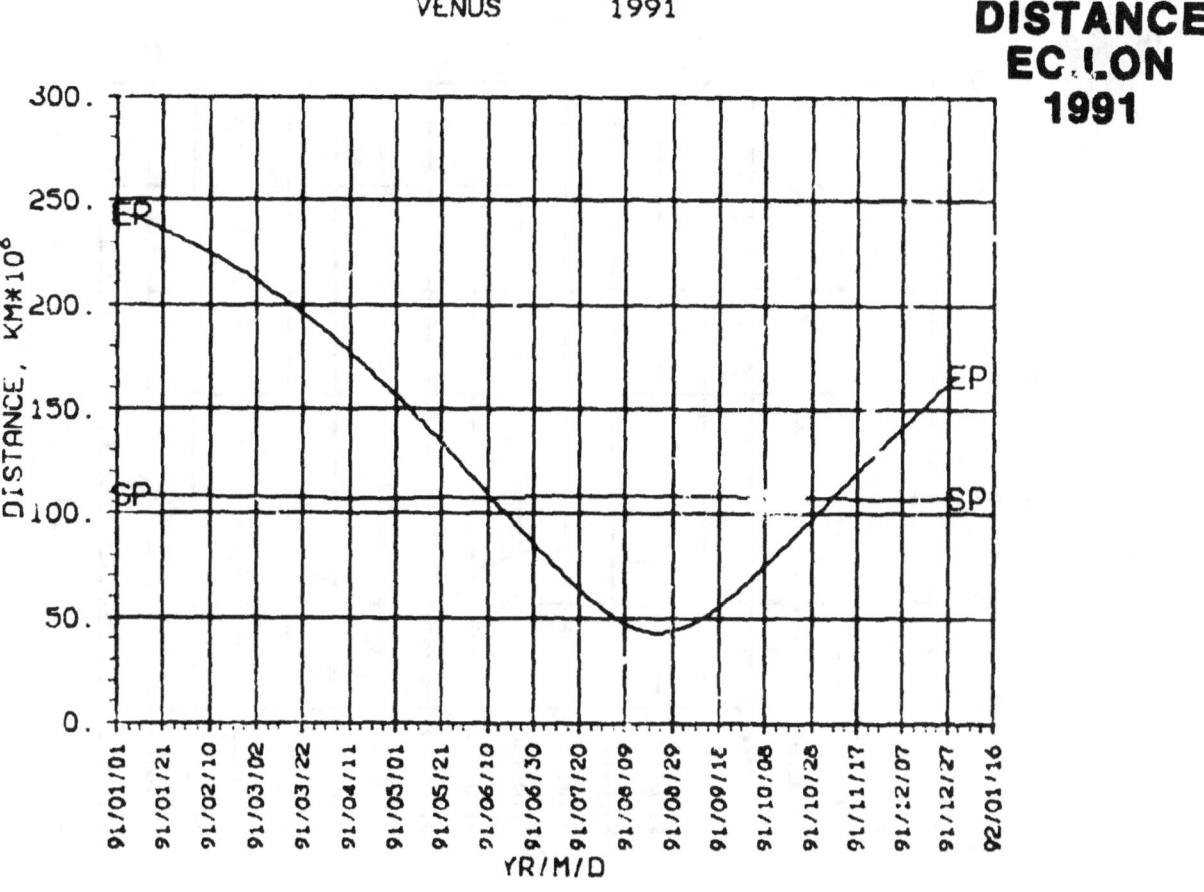

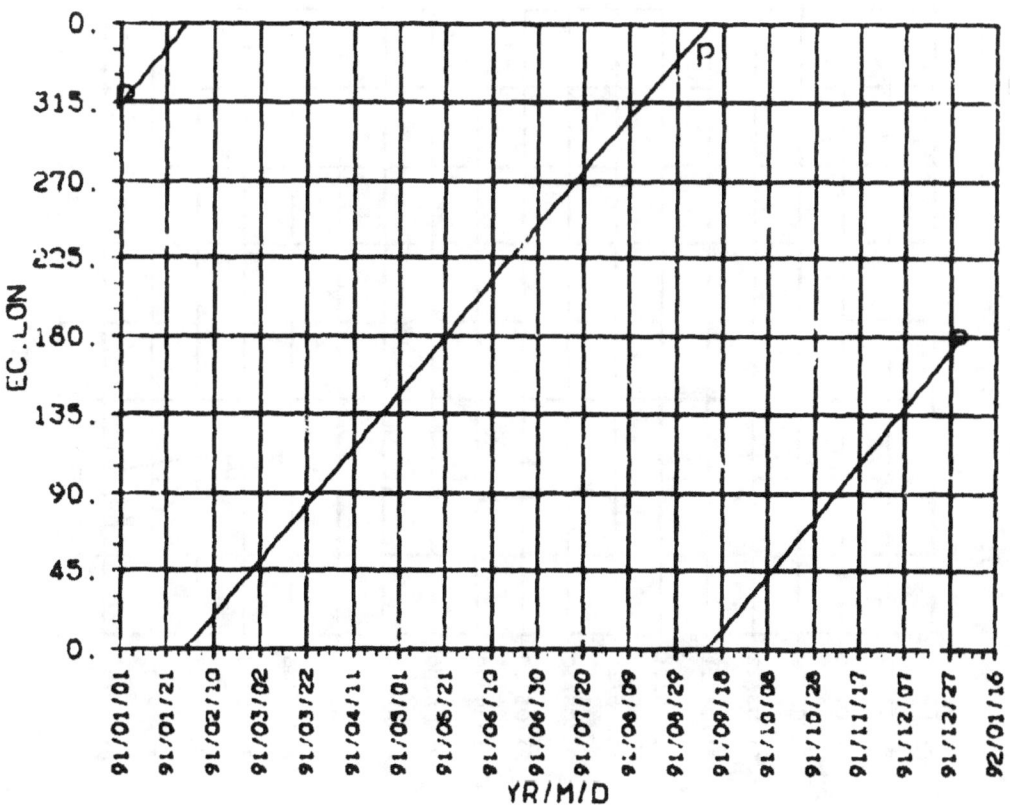

**SEP, ESP
CA, KA
1991**

VENUS 1991

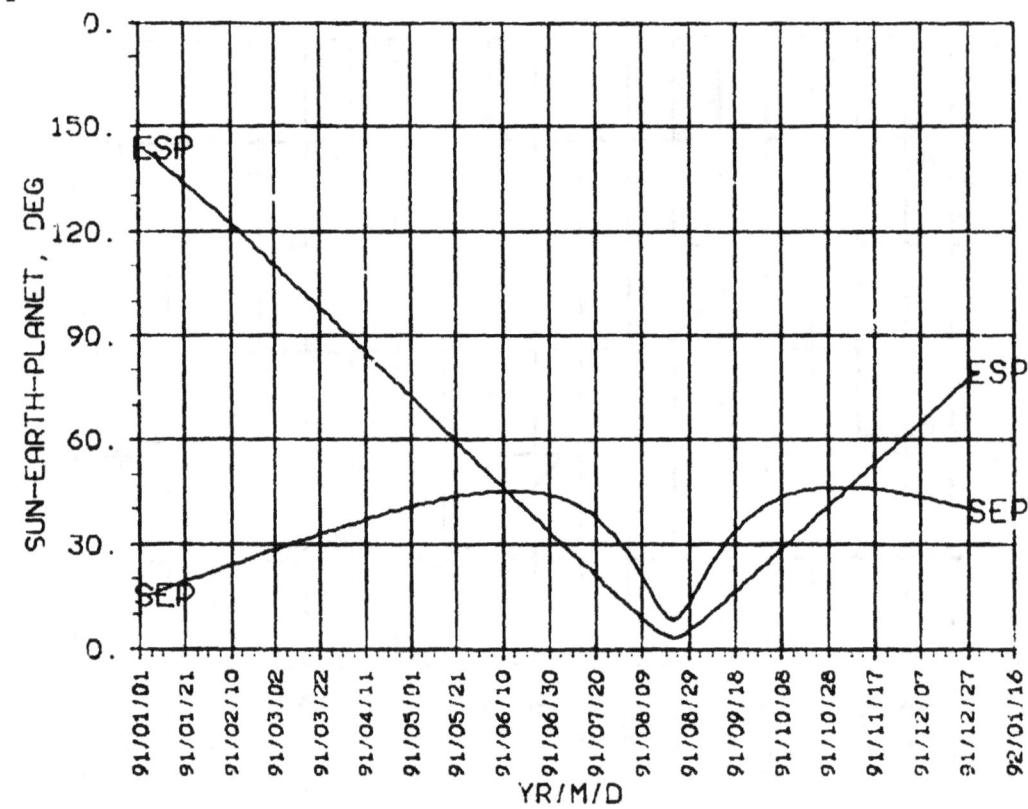

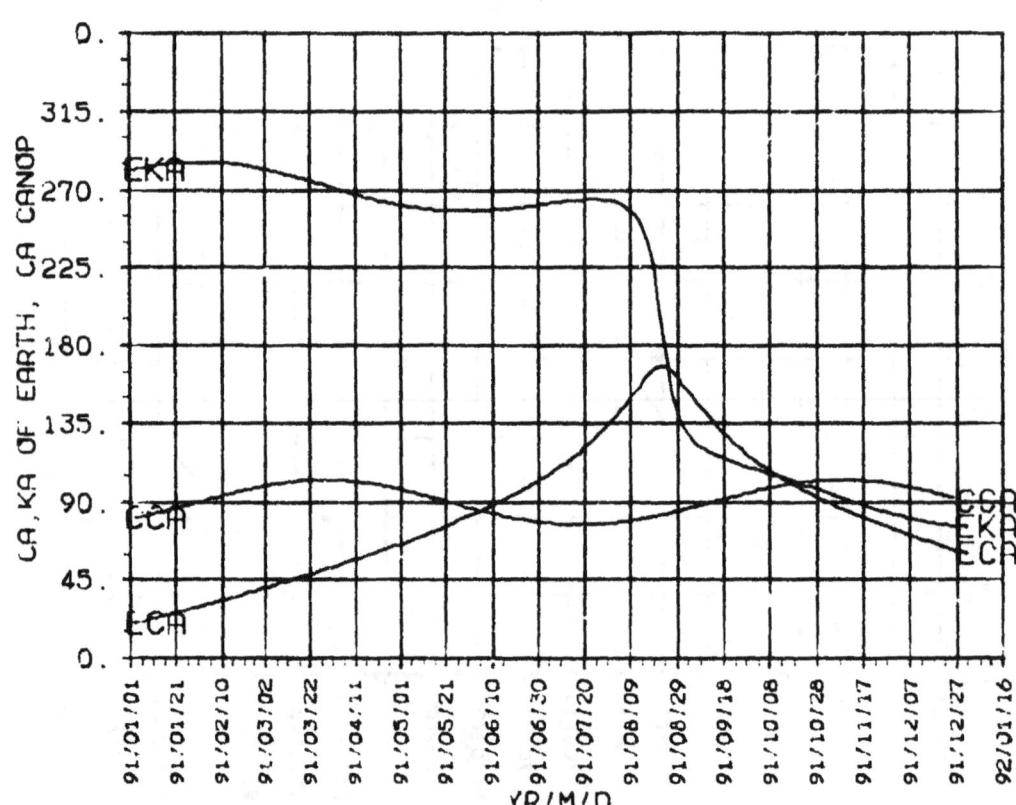

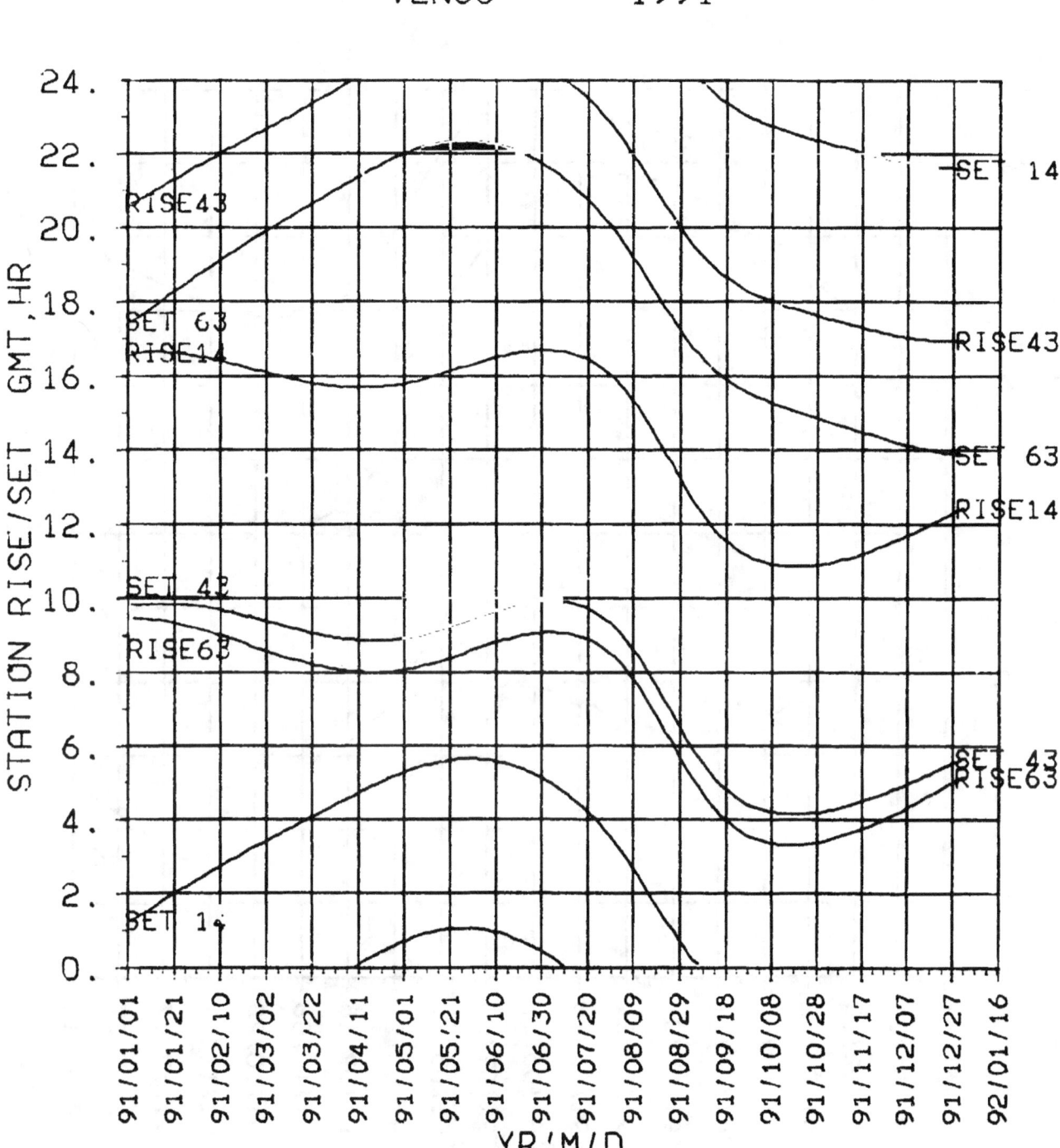

STA R/S
NON-DSN
1991

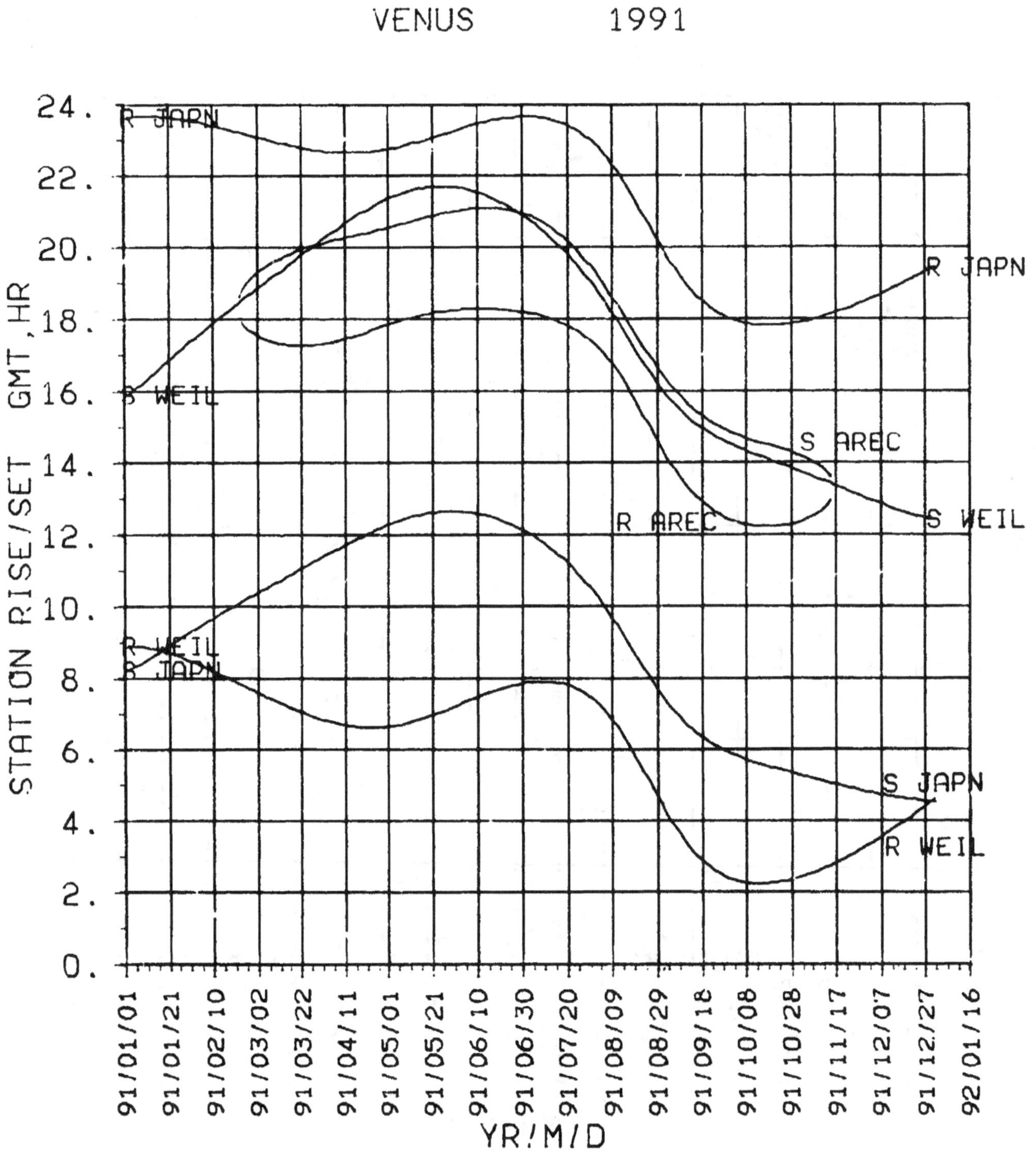

Venus

1992

DECLIN RT.ASC 1992

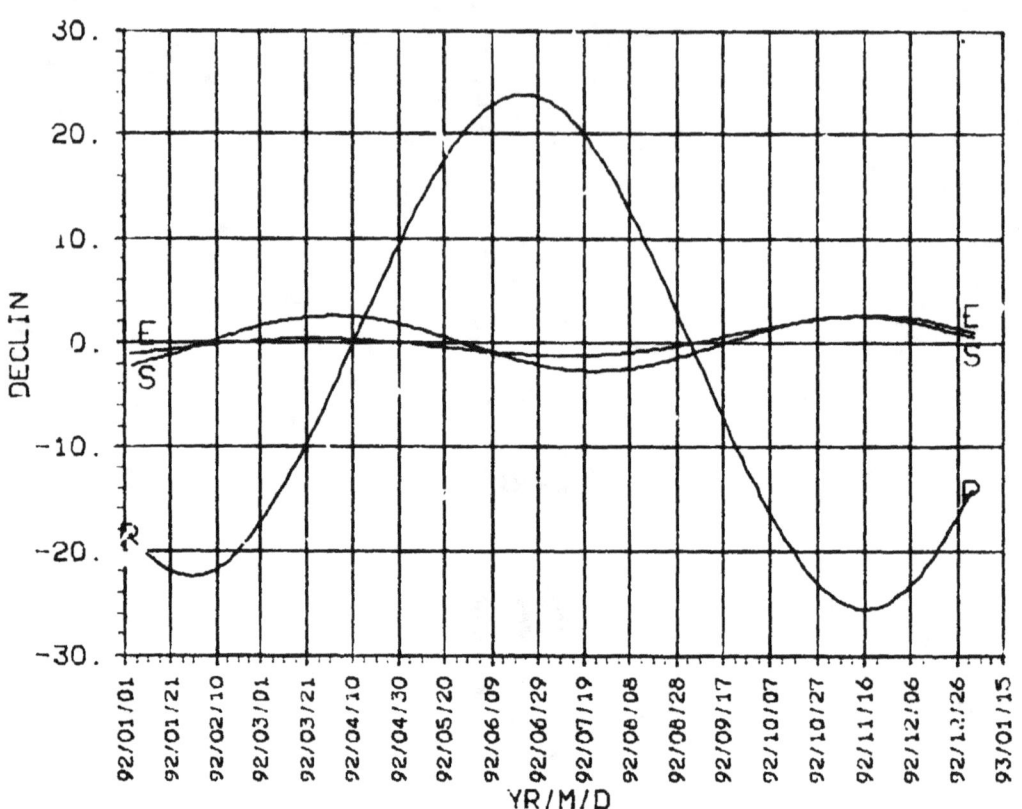

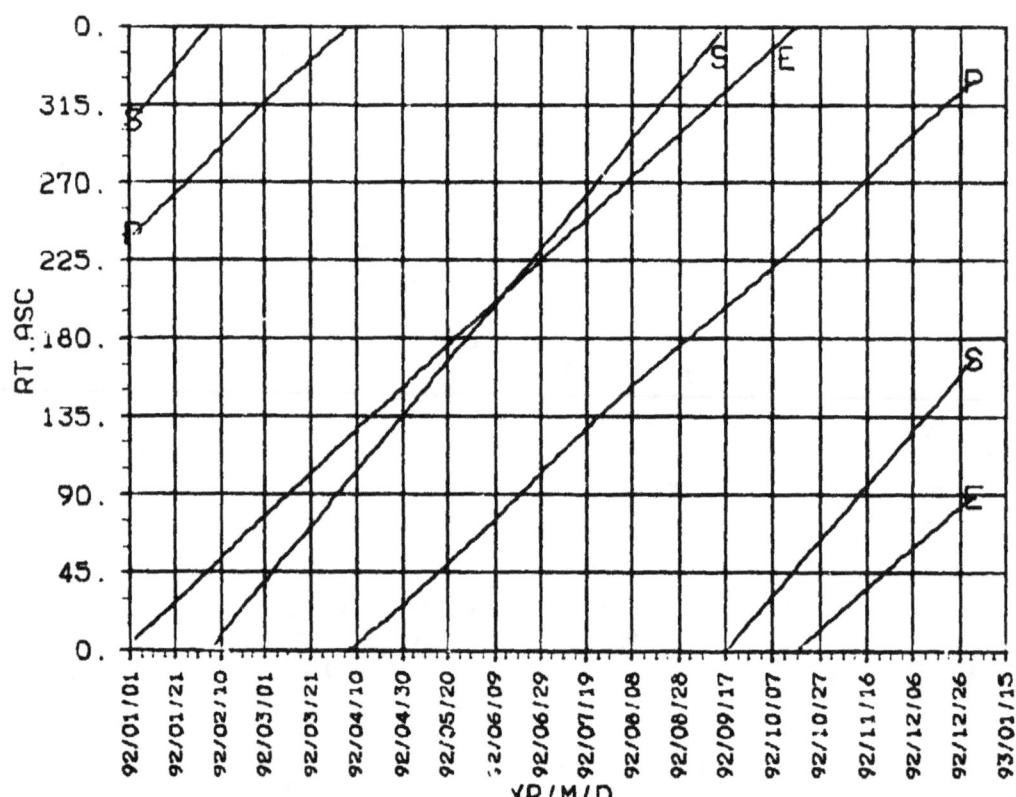

VENUS 1992

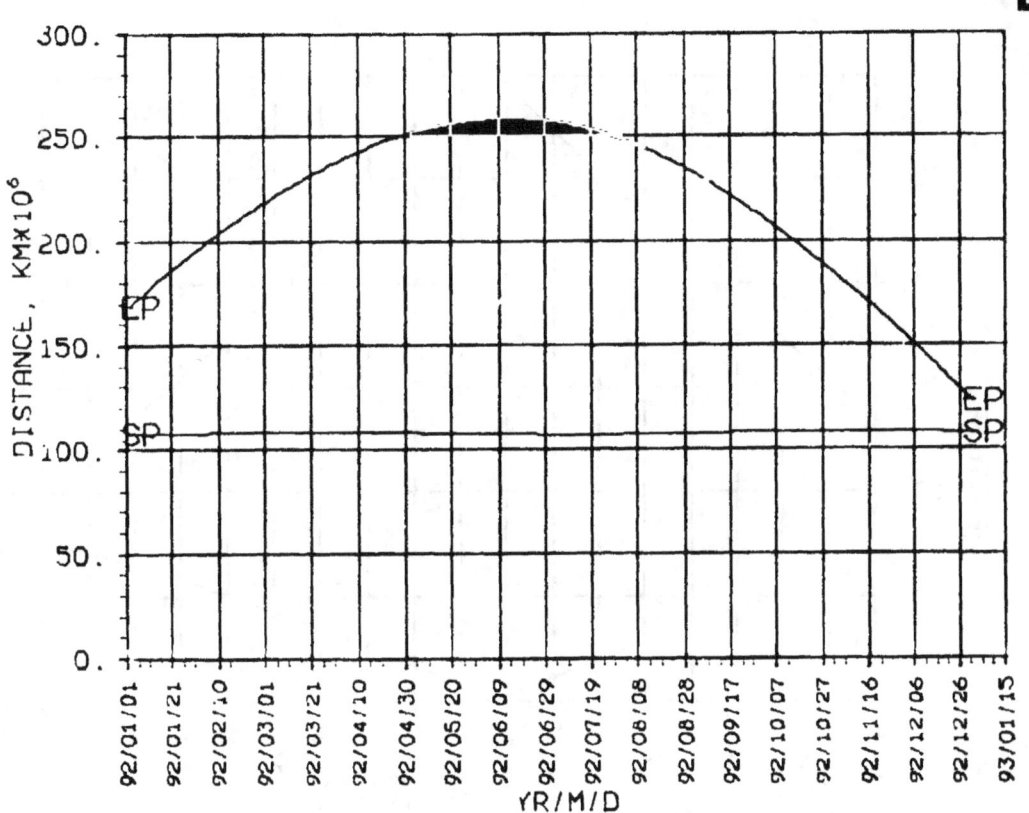

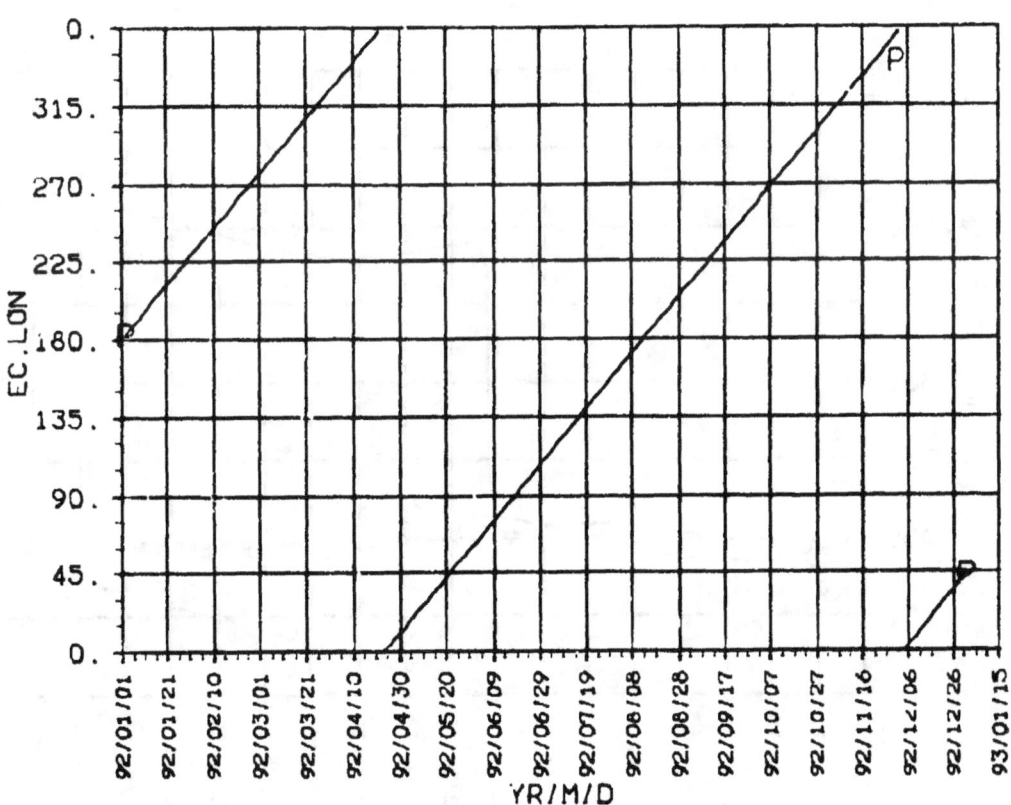

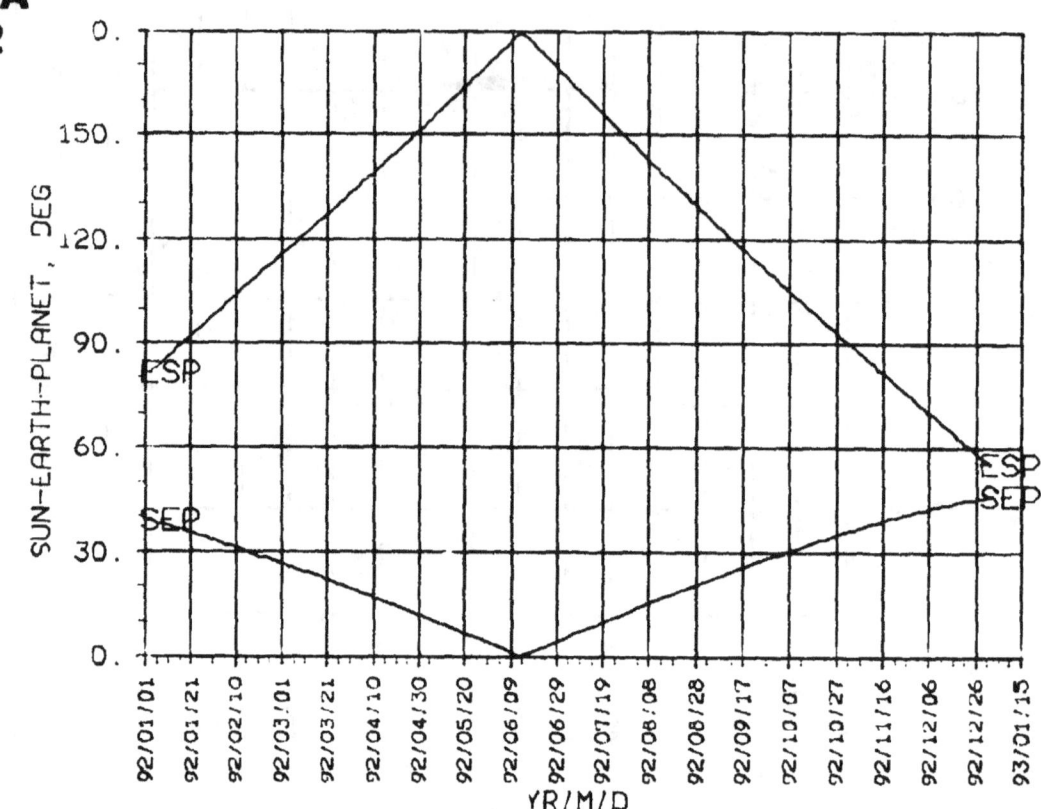
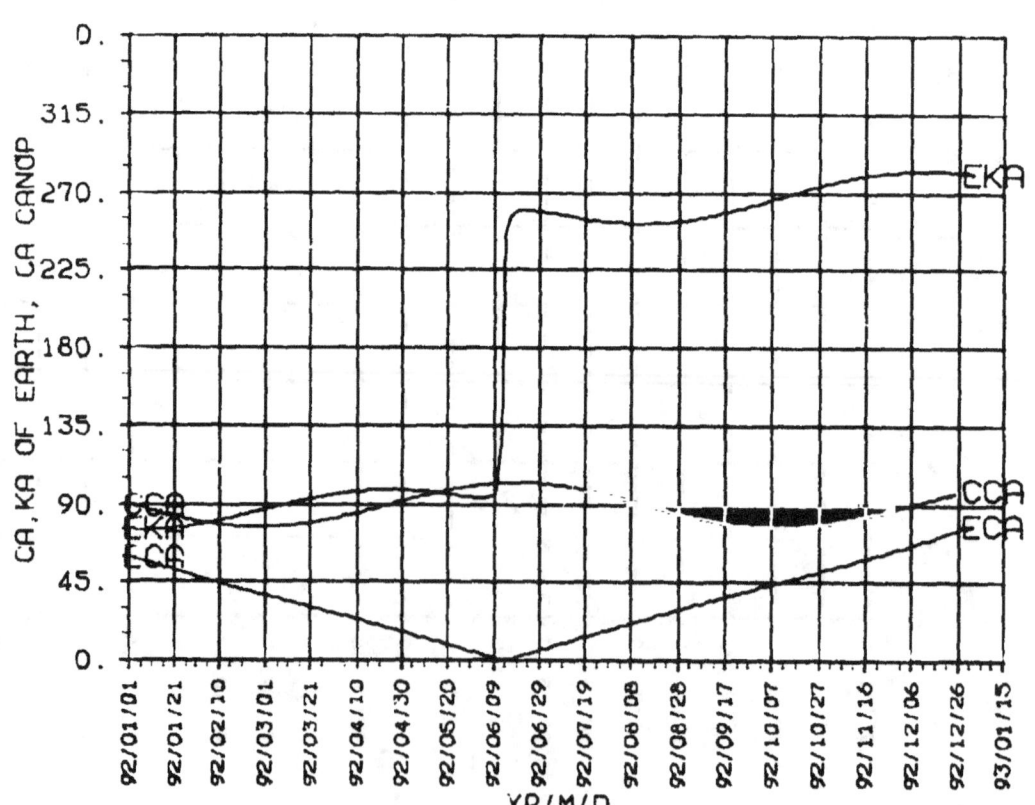

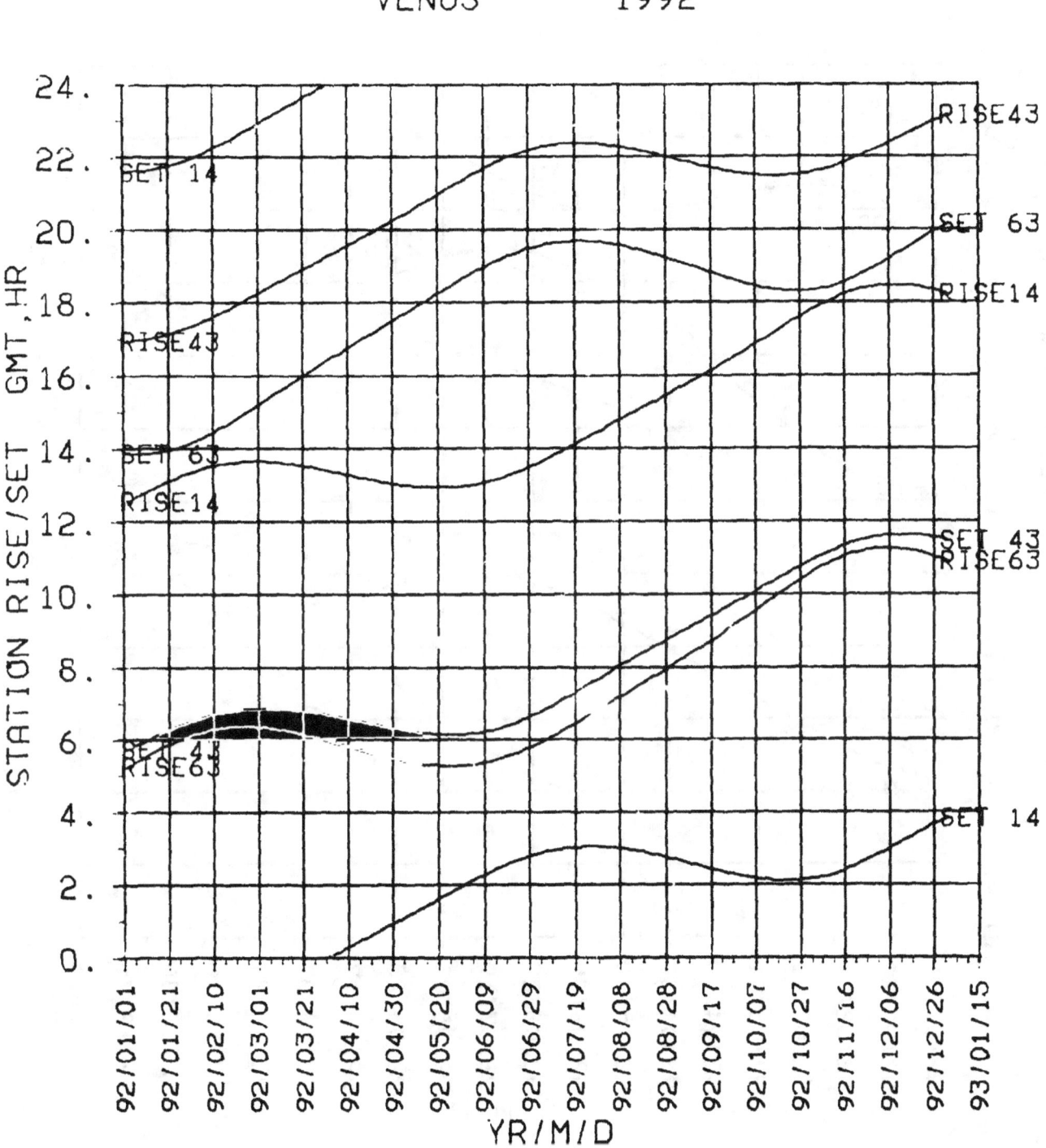

**STA R/S
NON-DSN
1992**

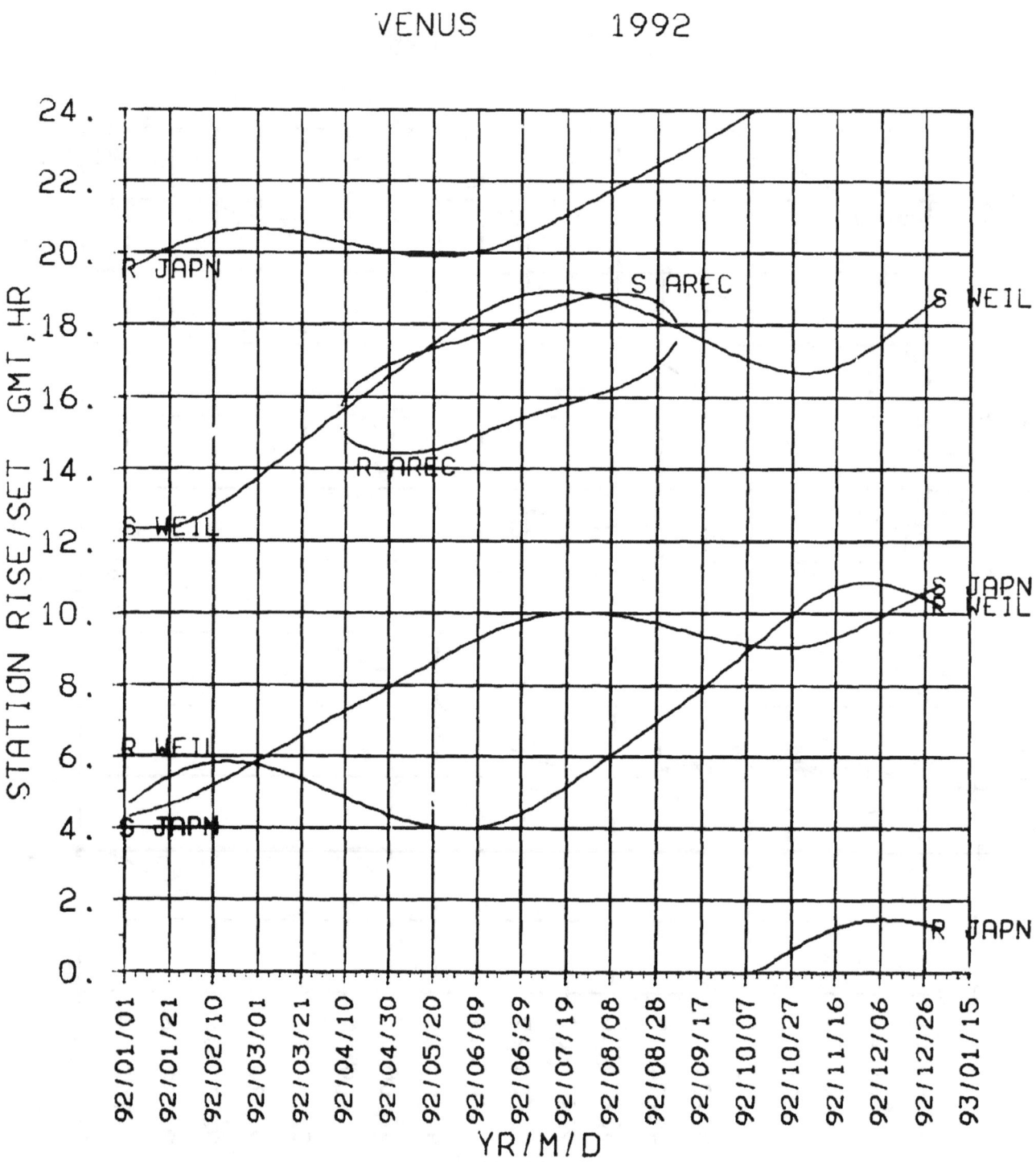

Venus

1993

DECLIN RT.ASC 1993

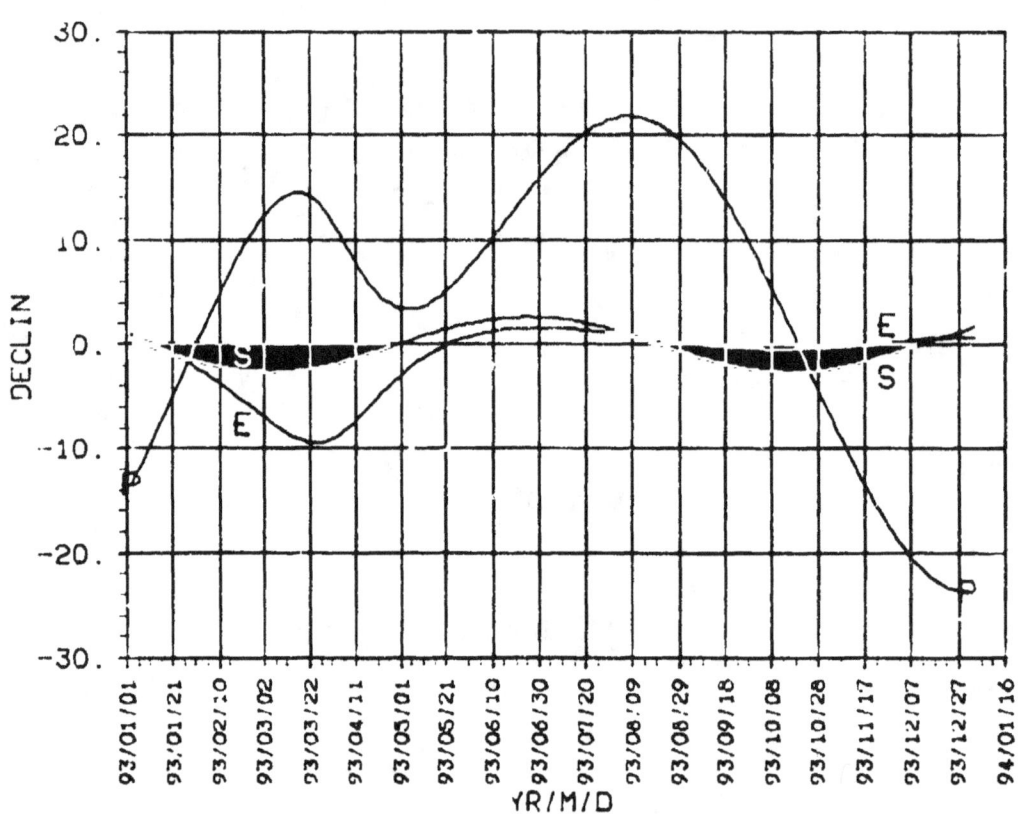

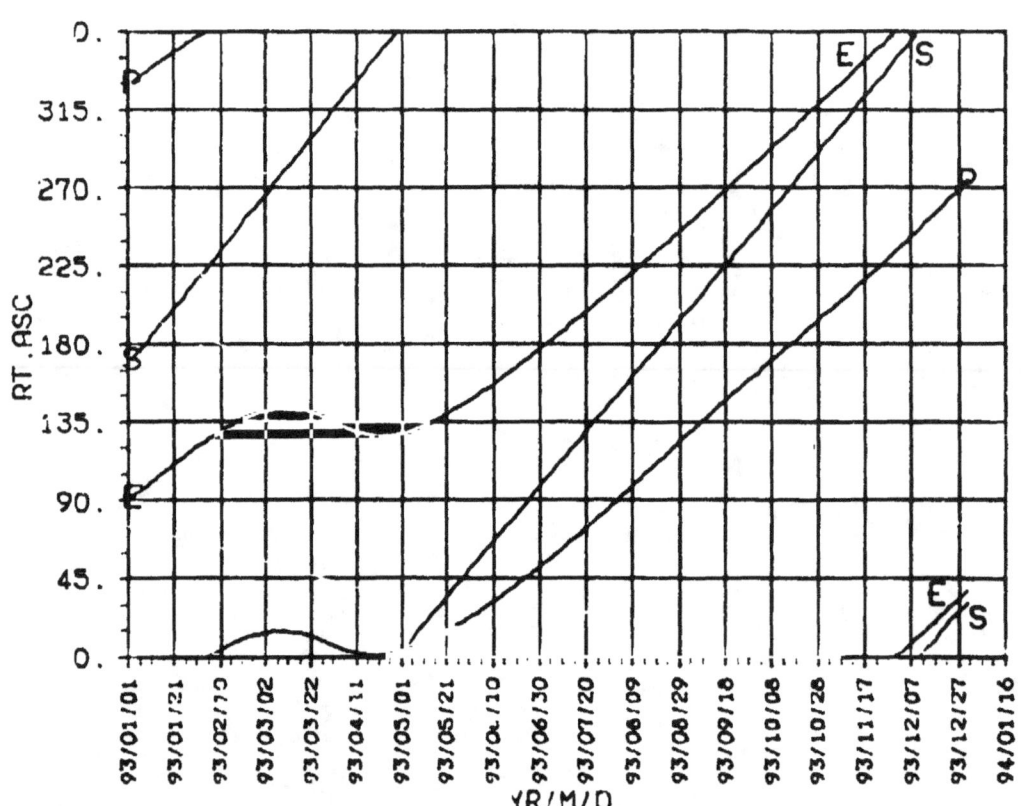

VENUS 1993

DISTANCE
EC.LON
1993

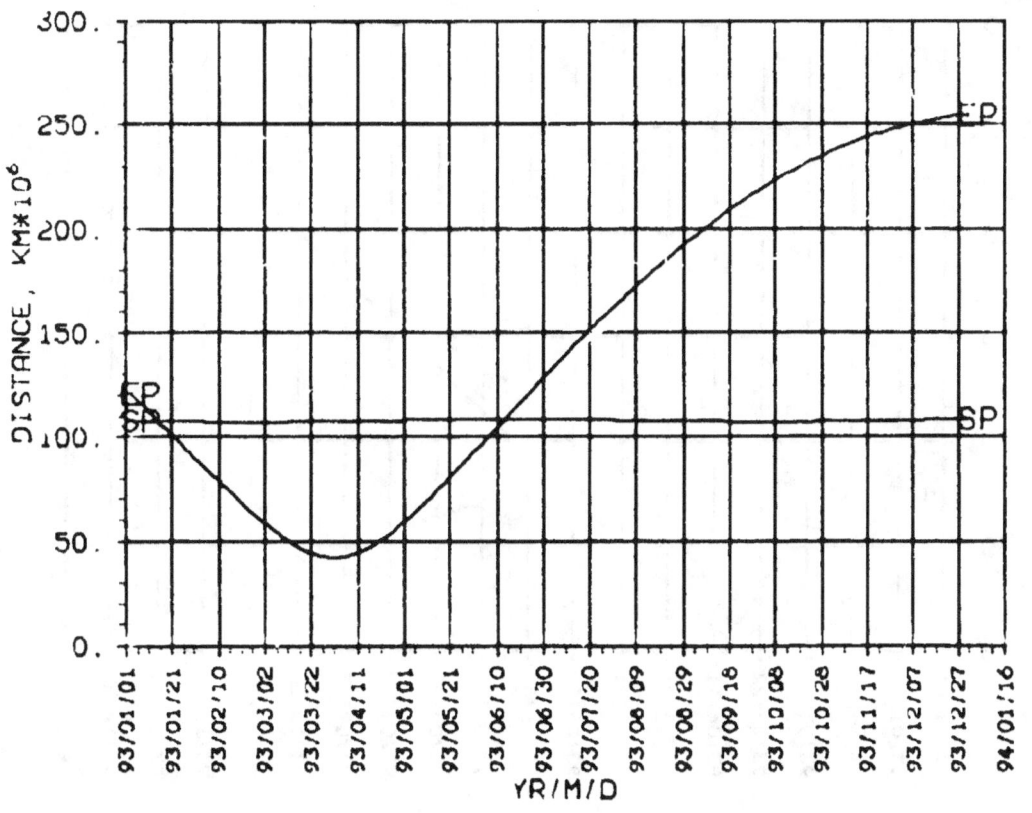

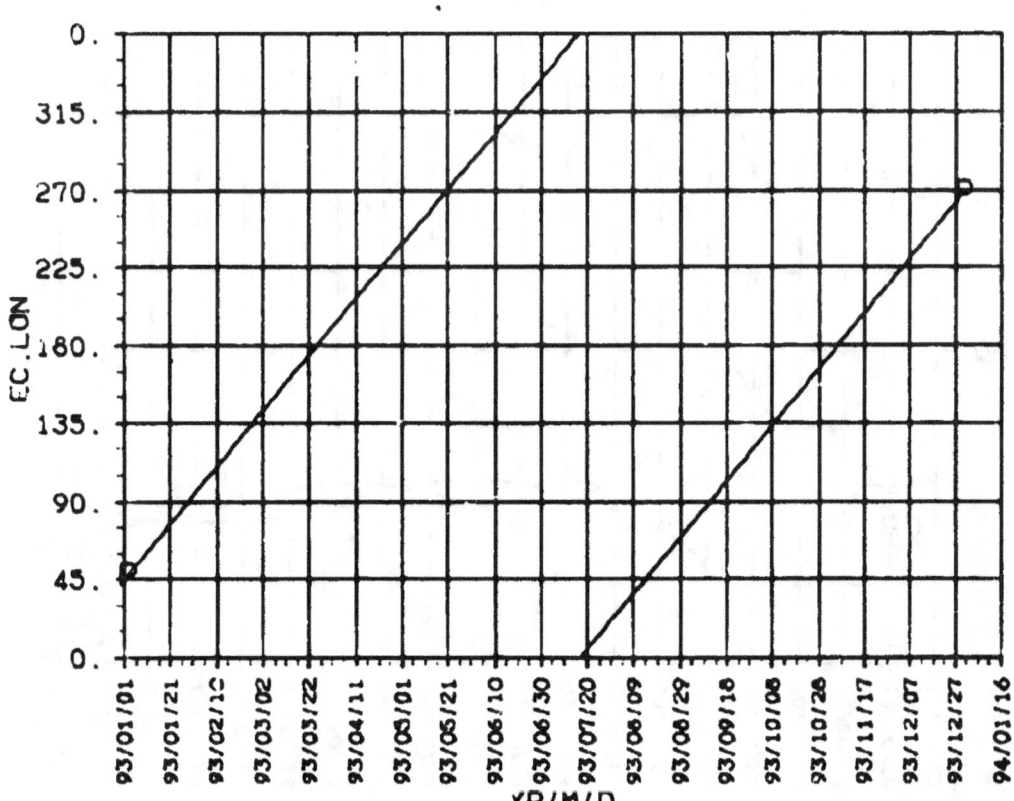

**SEP, ESP
CA, KA
1993**

VENUS 1993

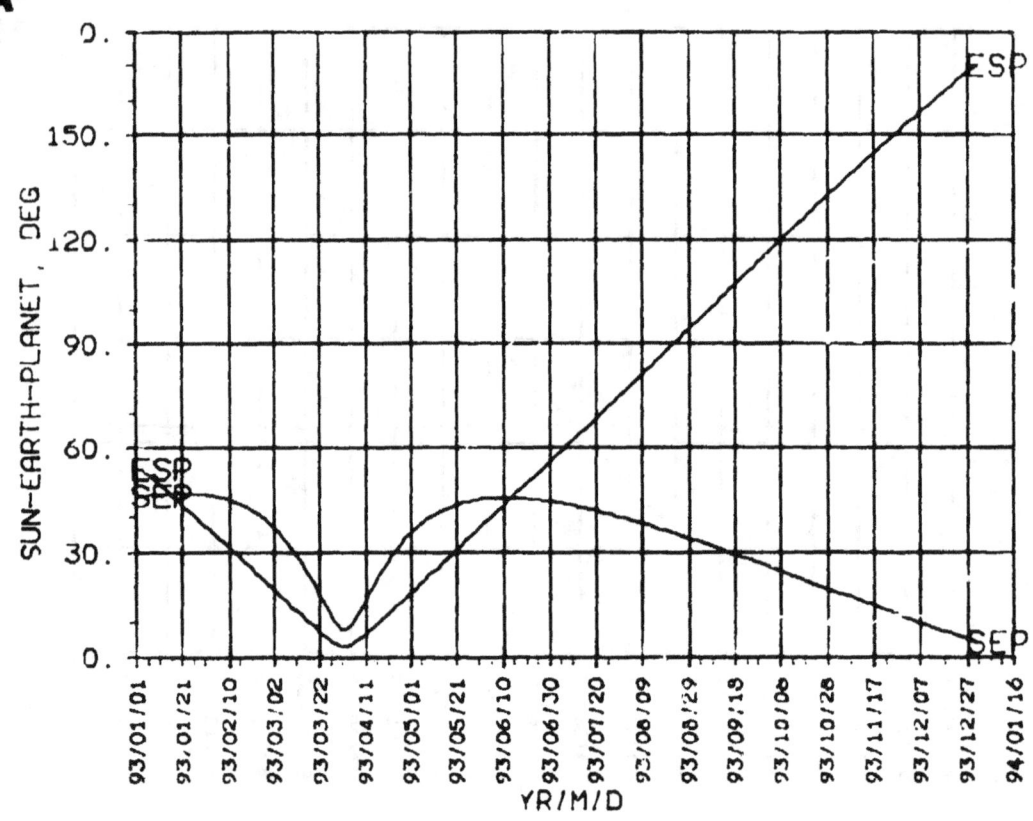

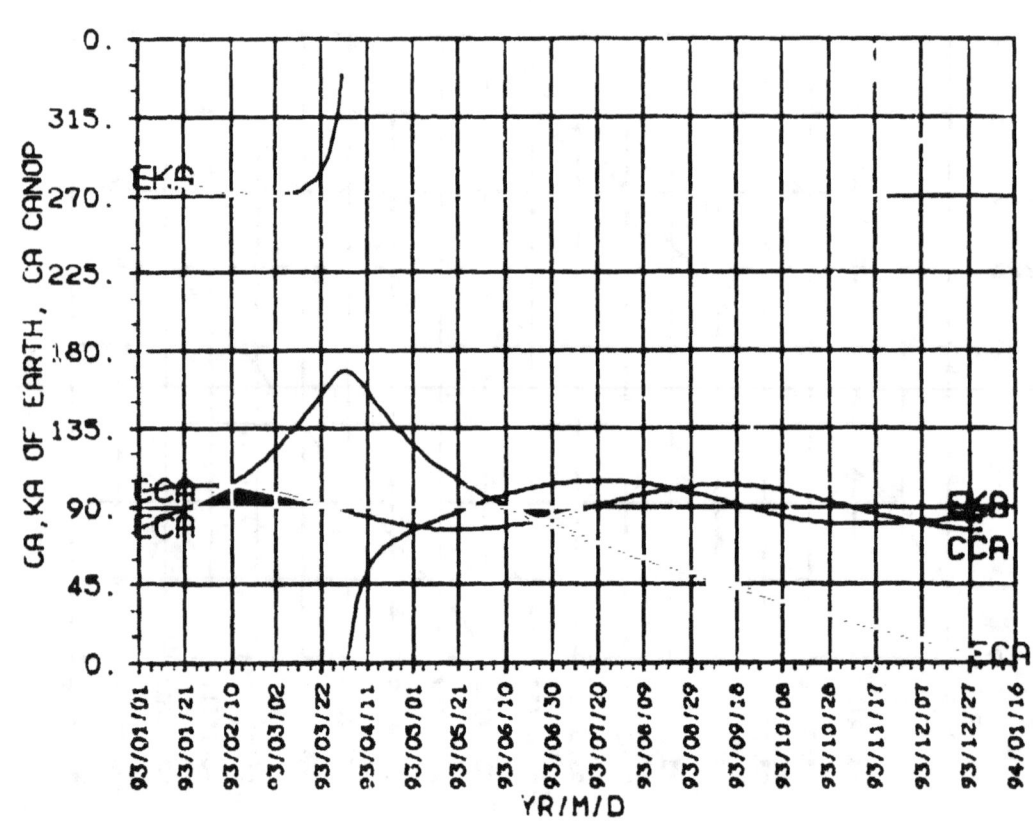

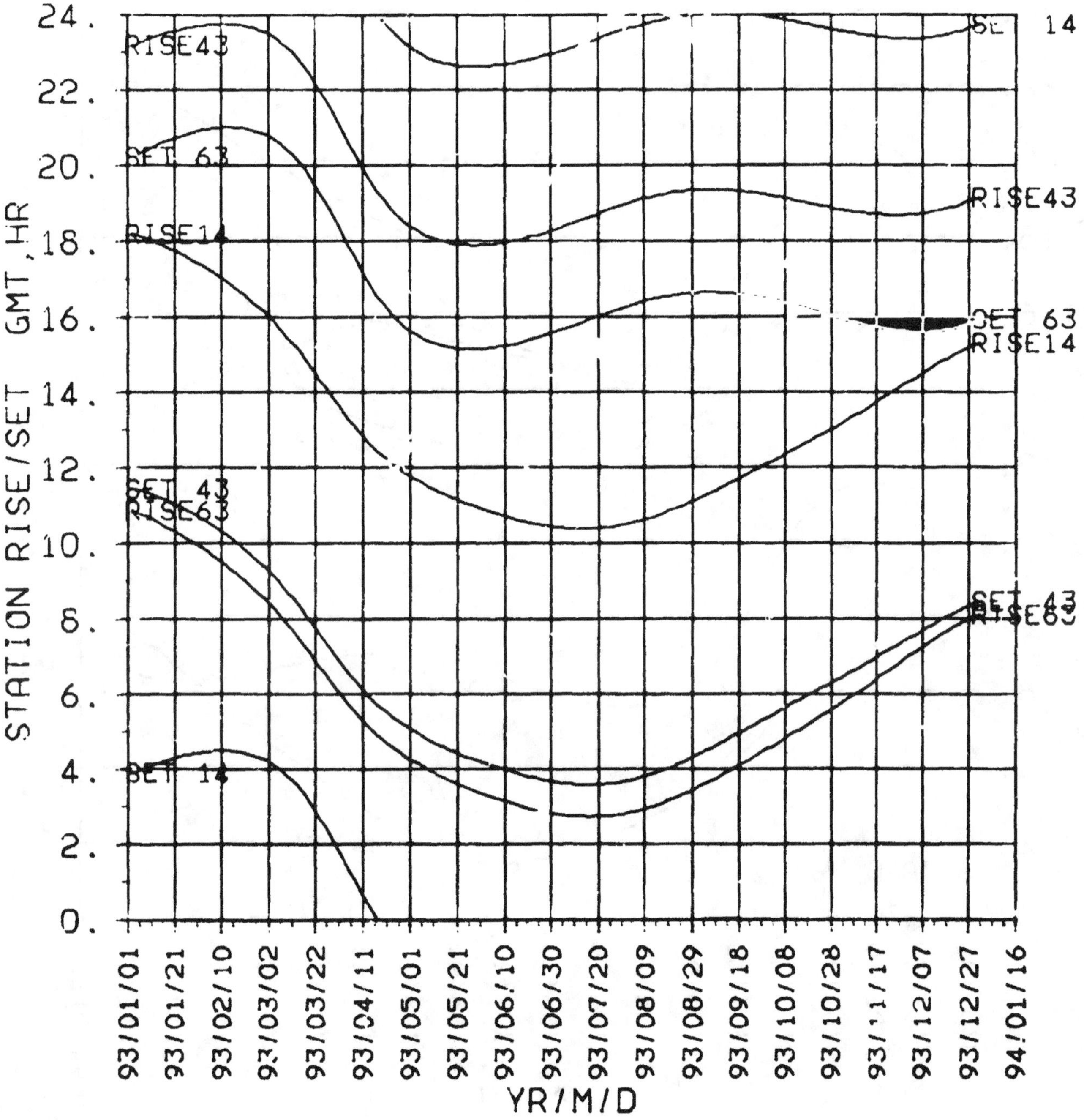

STA R/S NON-DSN 1993

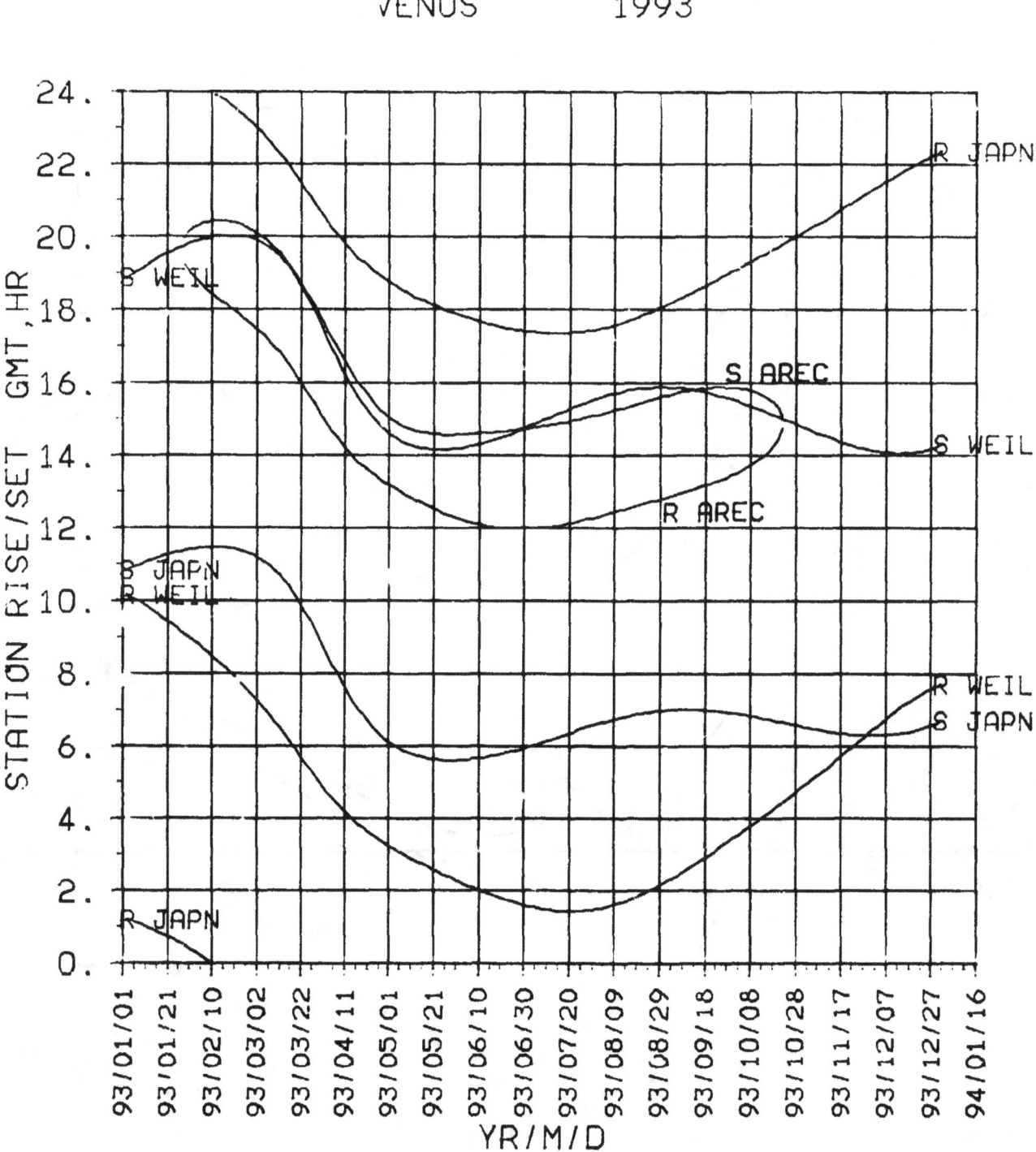

Venus

1994

DECLIN RT.ASC 1994

VENUS 1994

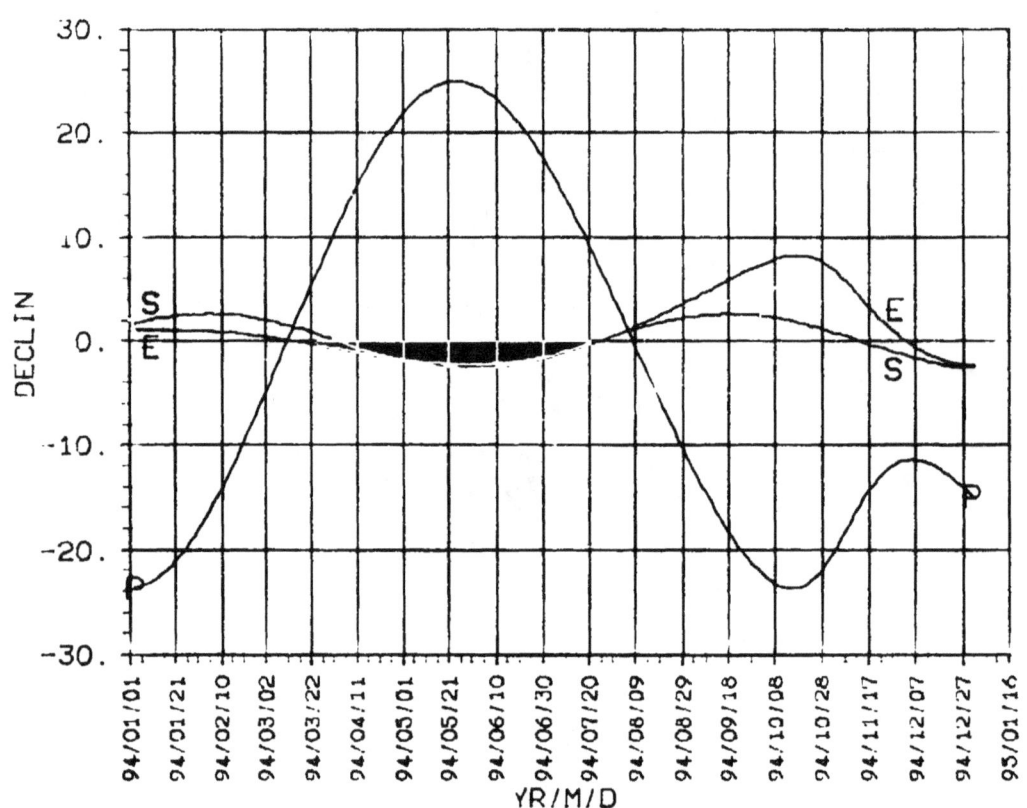

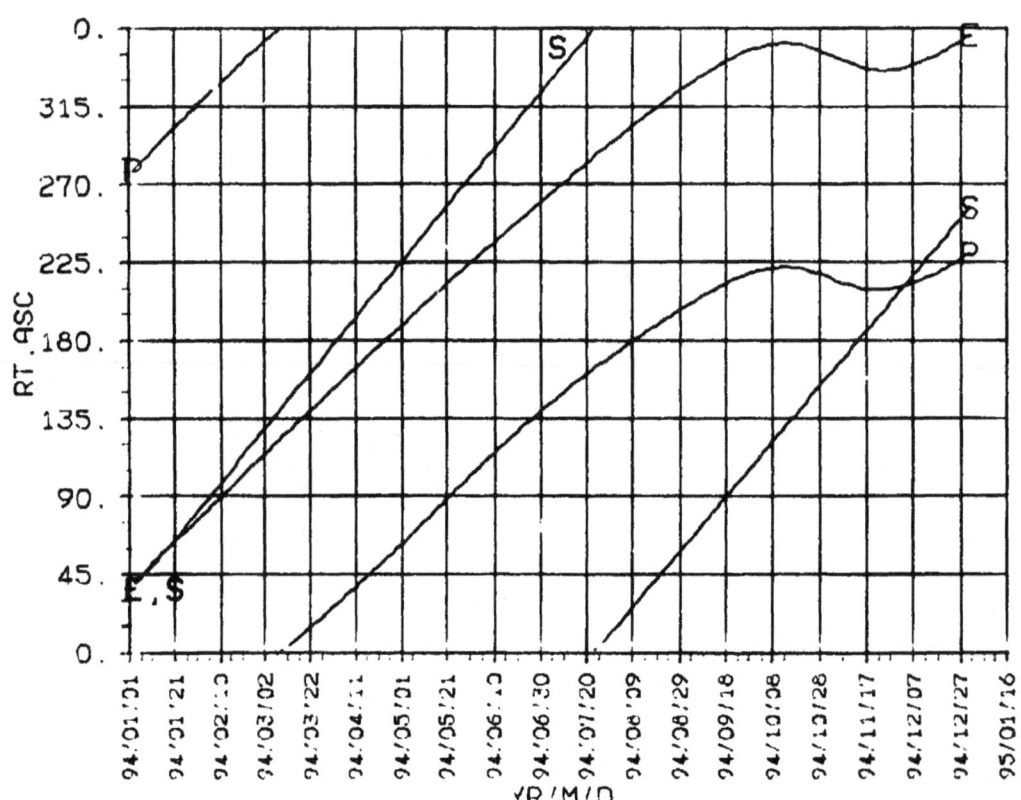

VENUS 1994

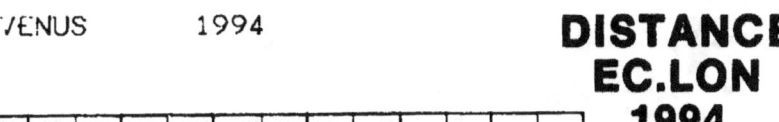

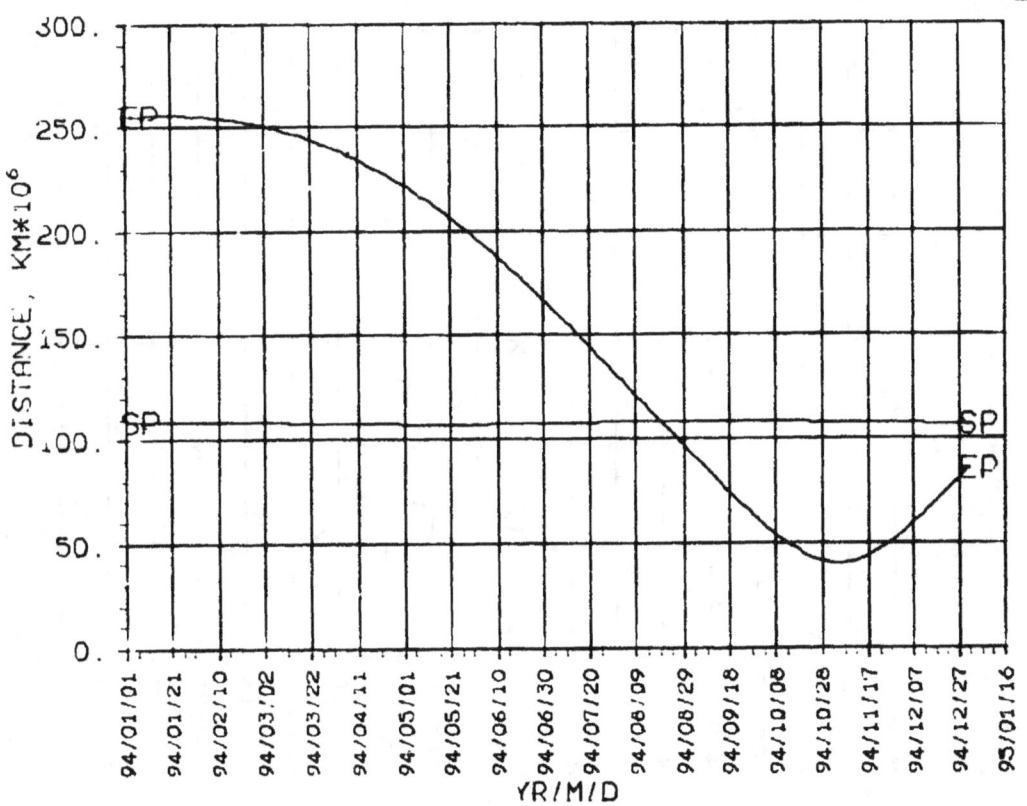

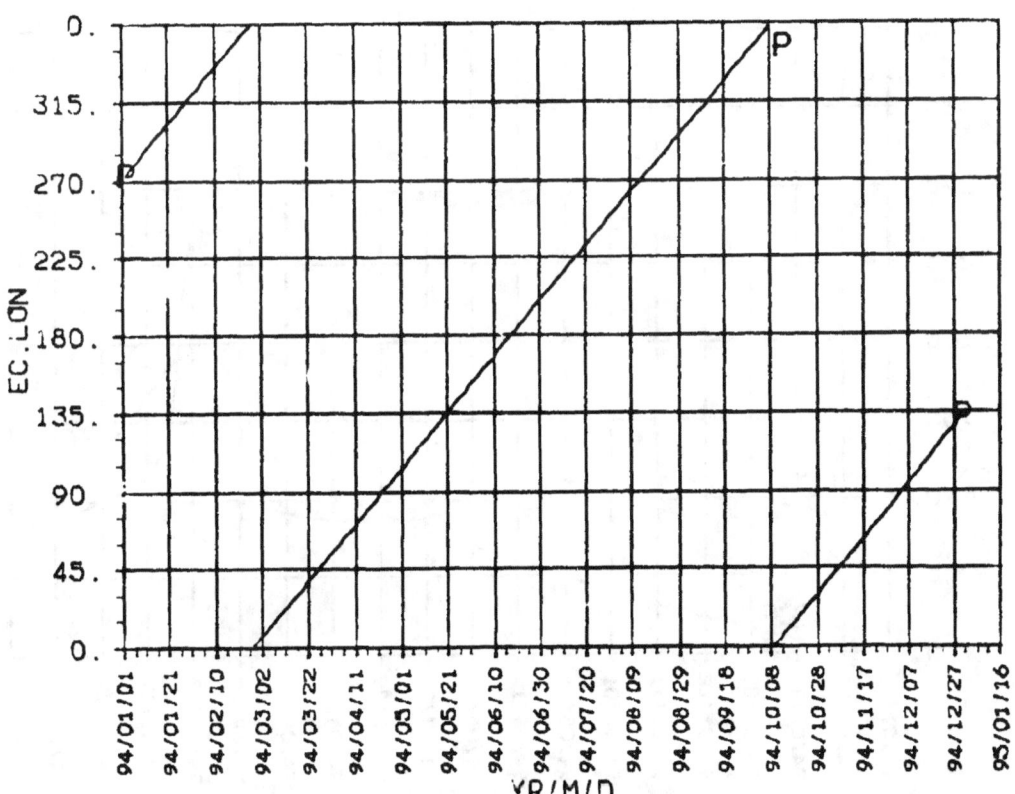

SEP, ESP CA, KA 1994

VENUS 1994

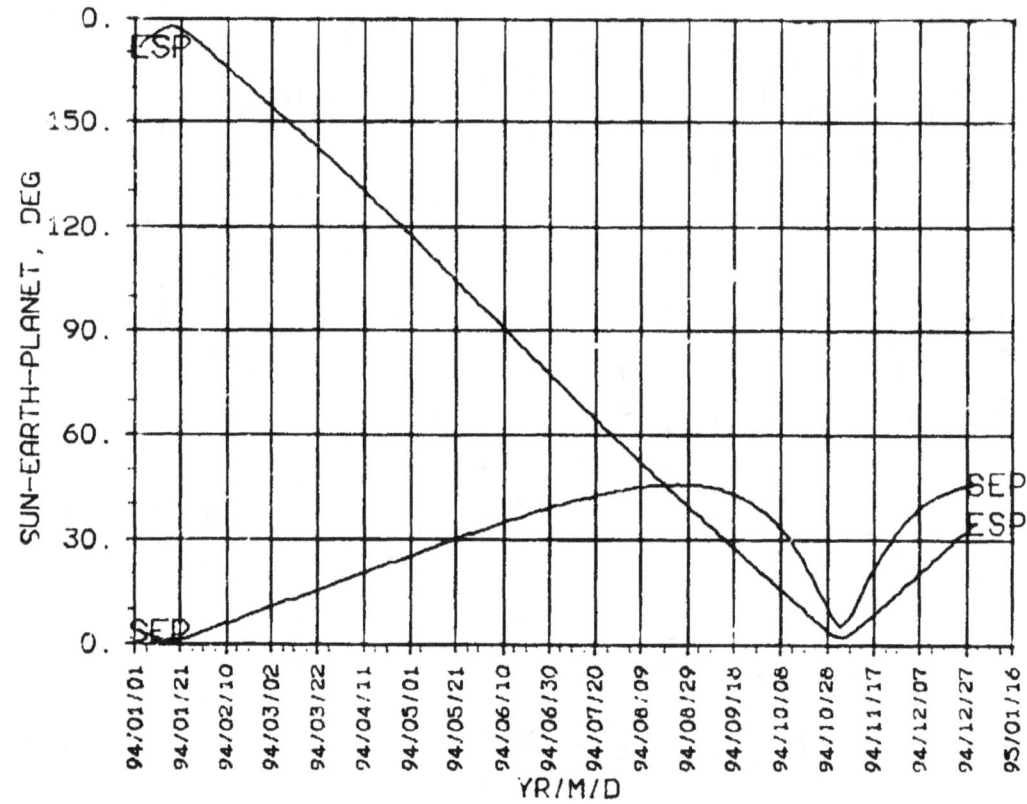

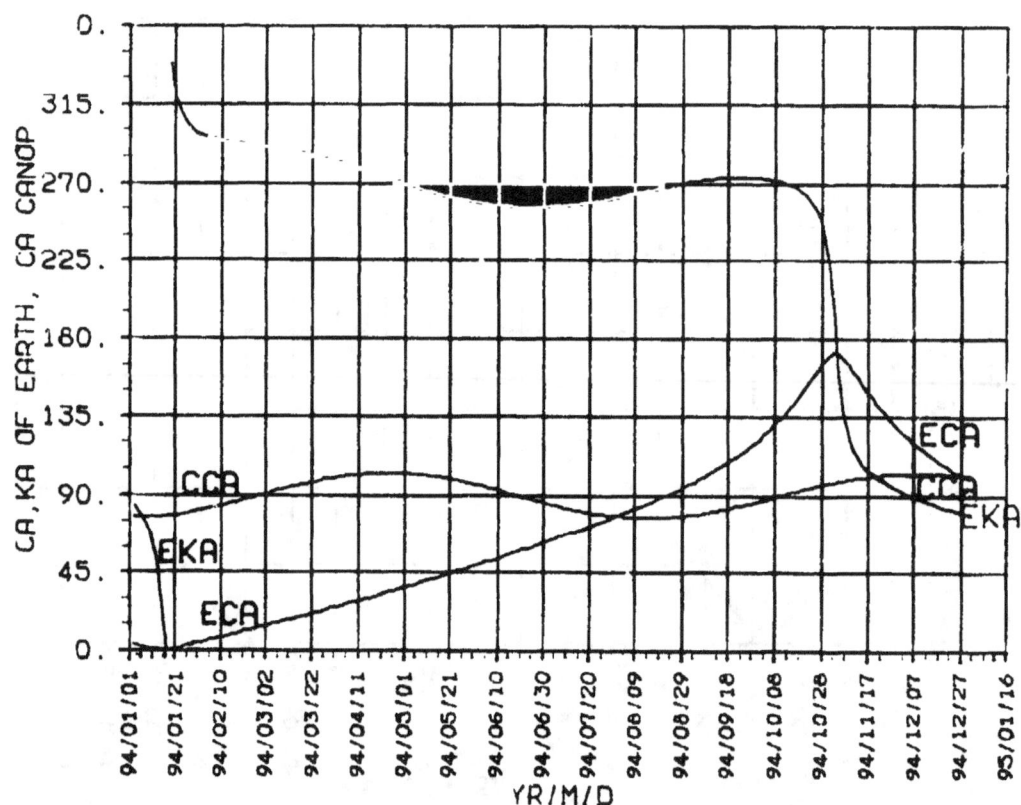

VENUS 1994

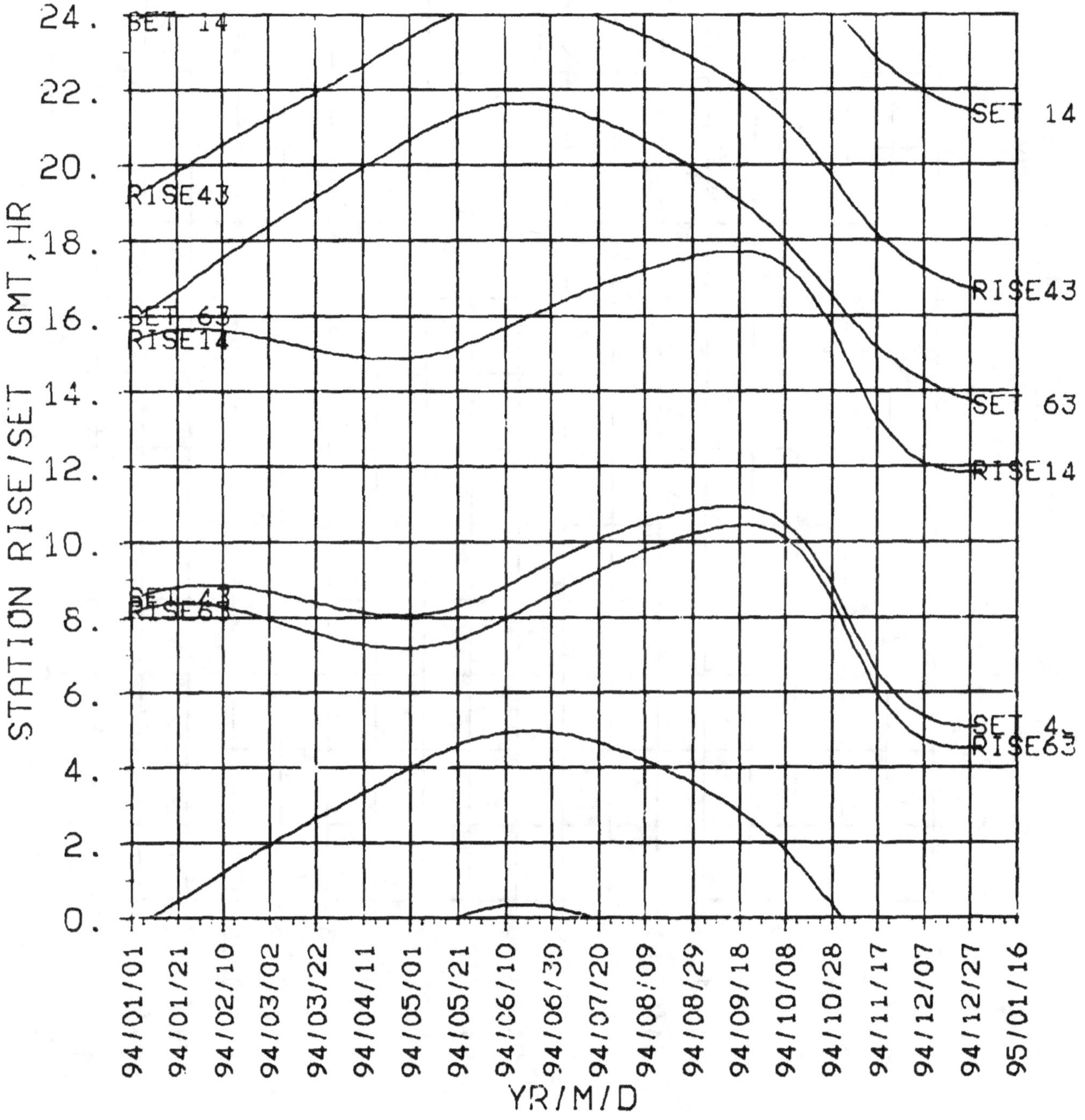
SET 14
RISE43
SET 63
RISE14
SET 43
RISE63
SET 14
RISE43
SET 63
RISE14
SET 43
RISE63
24.
22.
20.
18.
16.
14.
12.
10.
8.
6.
4.
2.
0.
STATION RISE/SET GMT,HR
94/01/01
94/01/21
94/02/10
94/03/02
94/03/22
94/04/11
94/05/01
94/05/21
94/06/10
94/06/30
94/07/20
94/08/09
94/08/29
94/09/18
94/10/08
94/10/28
94/11/17
94/12/07
94/12/27
95/01/16
YR/M/D

STA R/S NON-DSN 1994

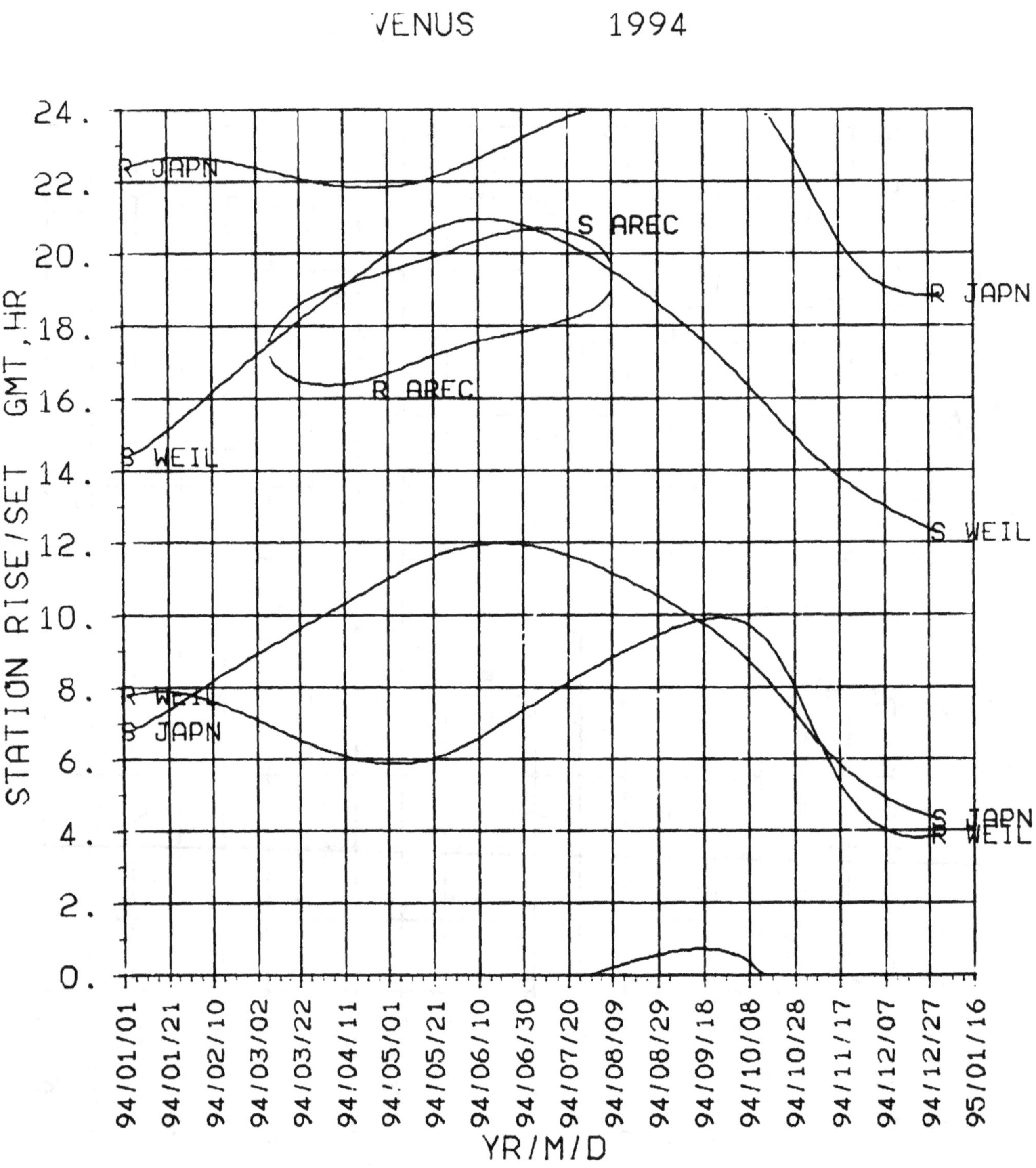

Venus

1995

DECLIN RT.ASC 1995

VENUS 1995

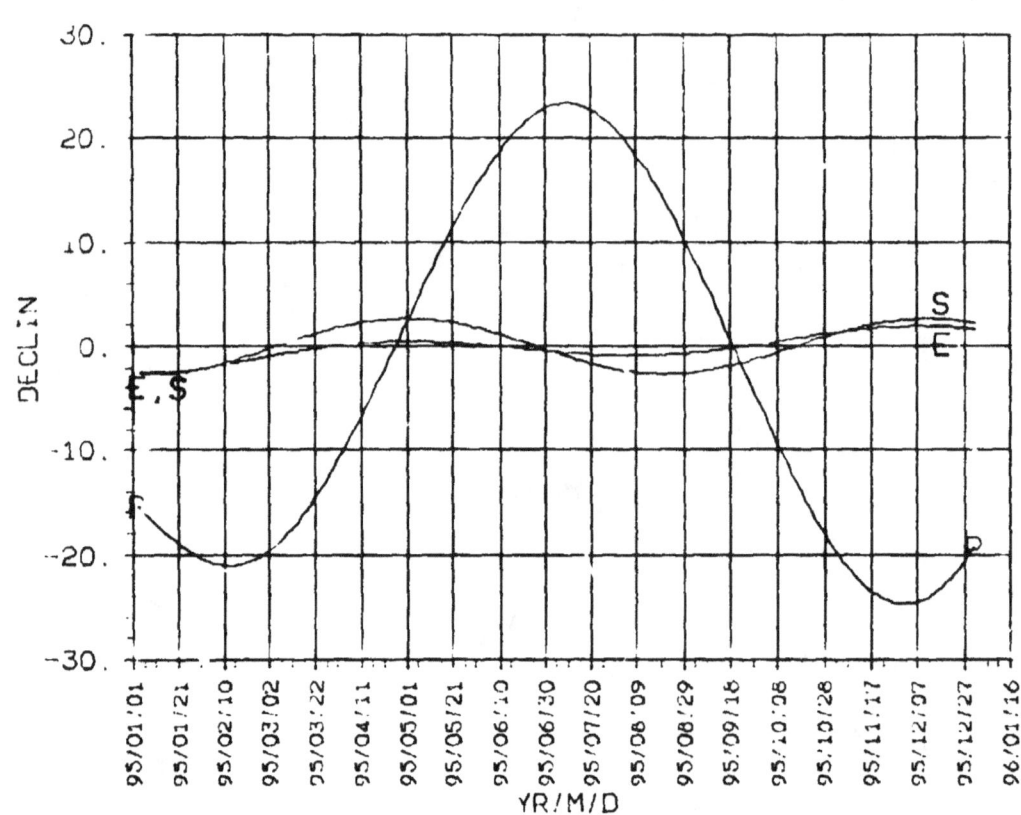

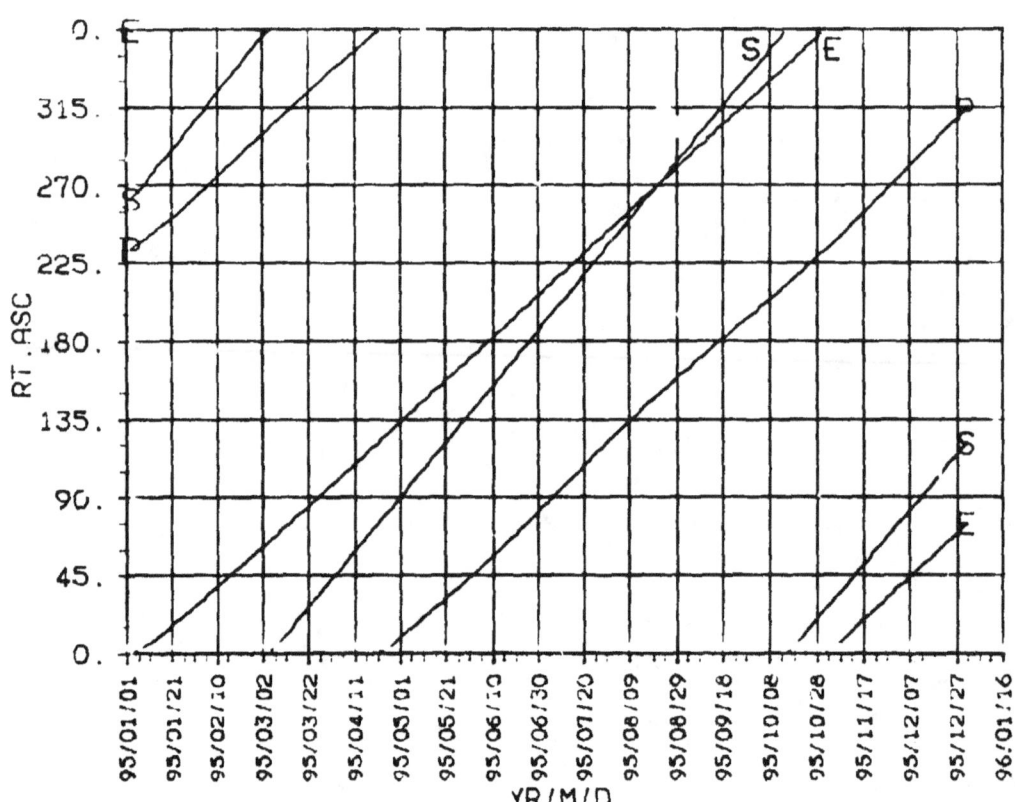

VENUS 1995

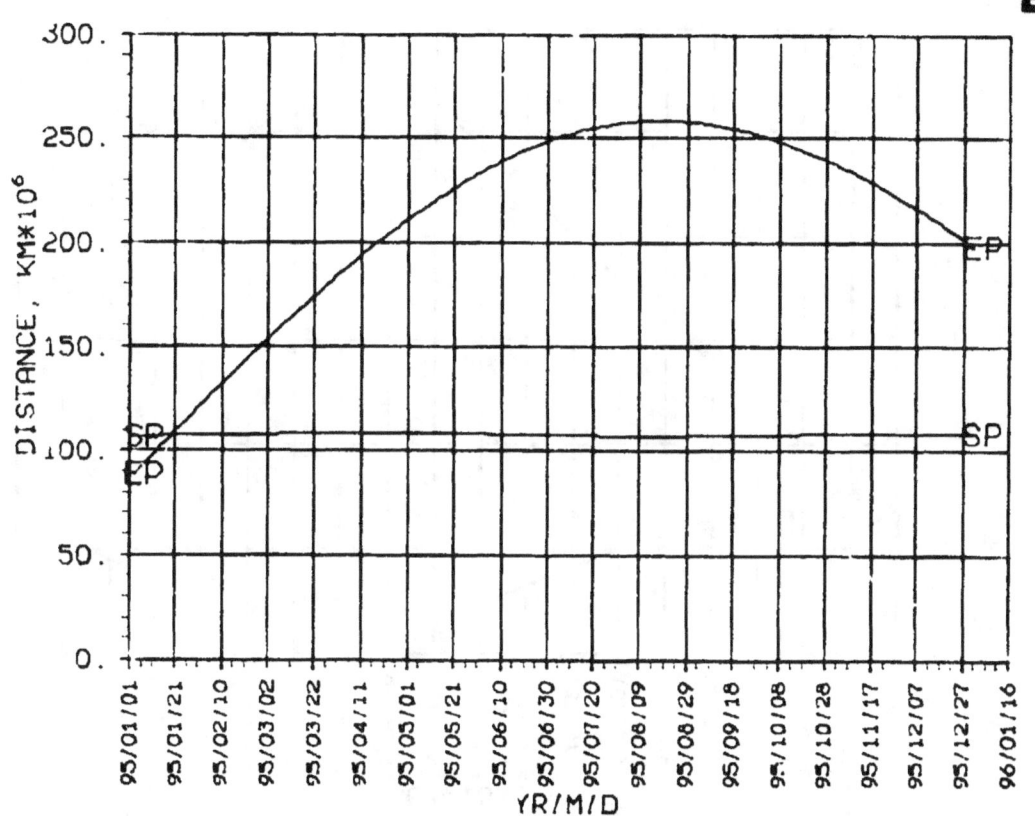

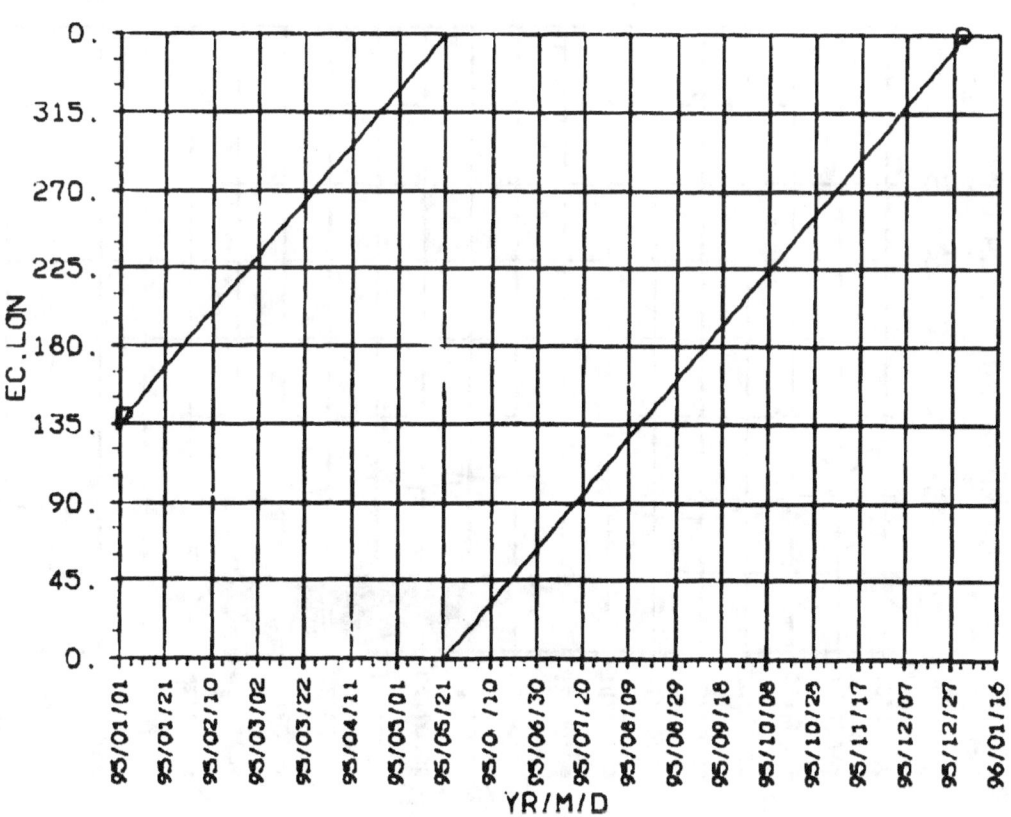

VENUS 1995

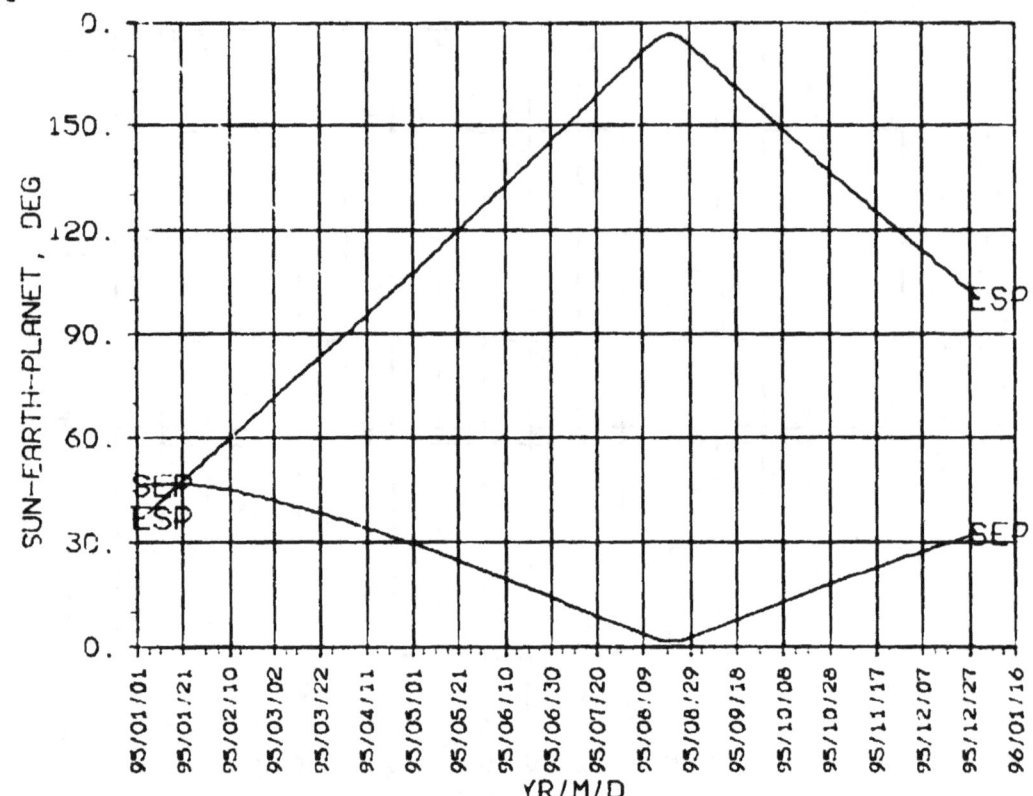

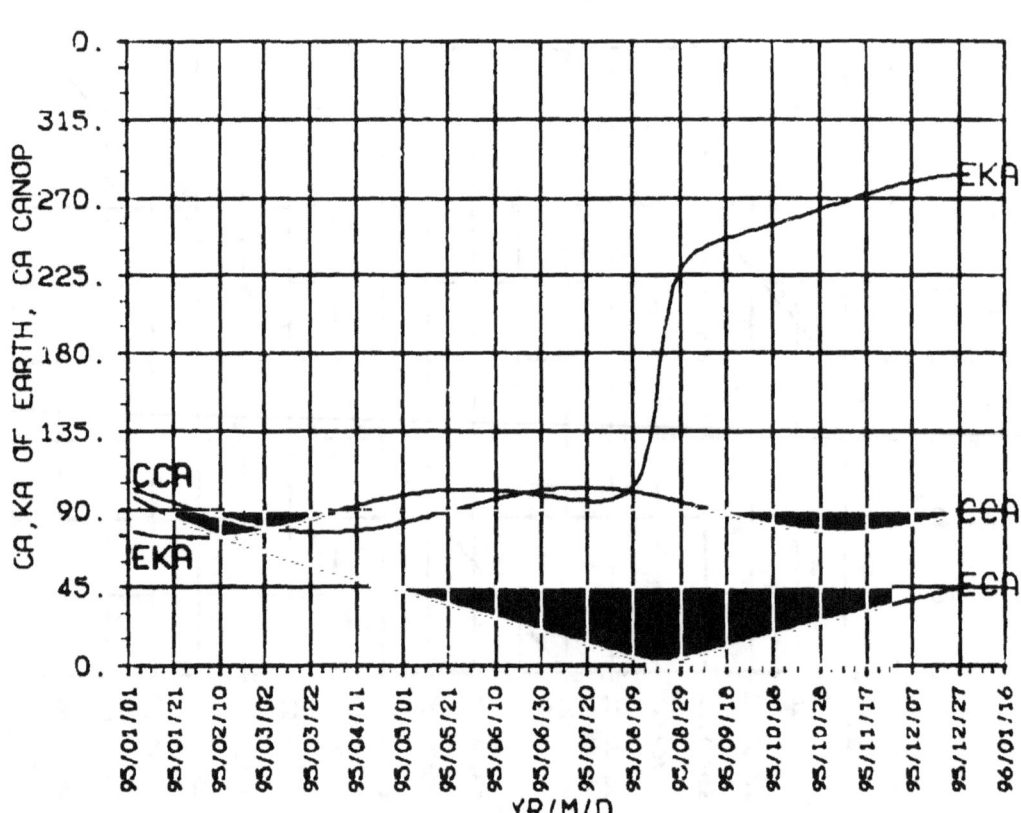

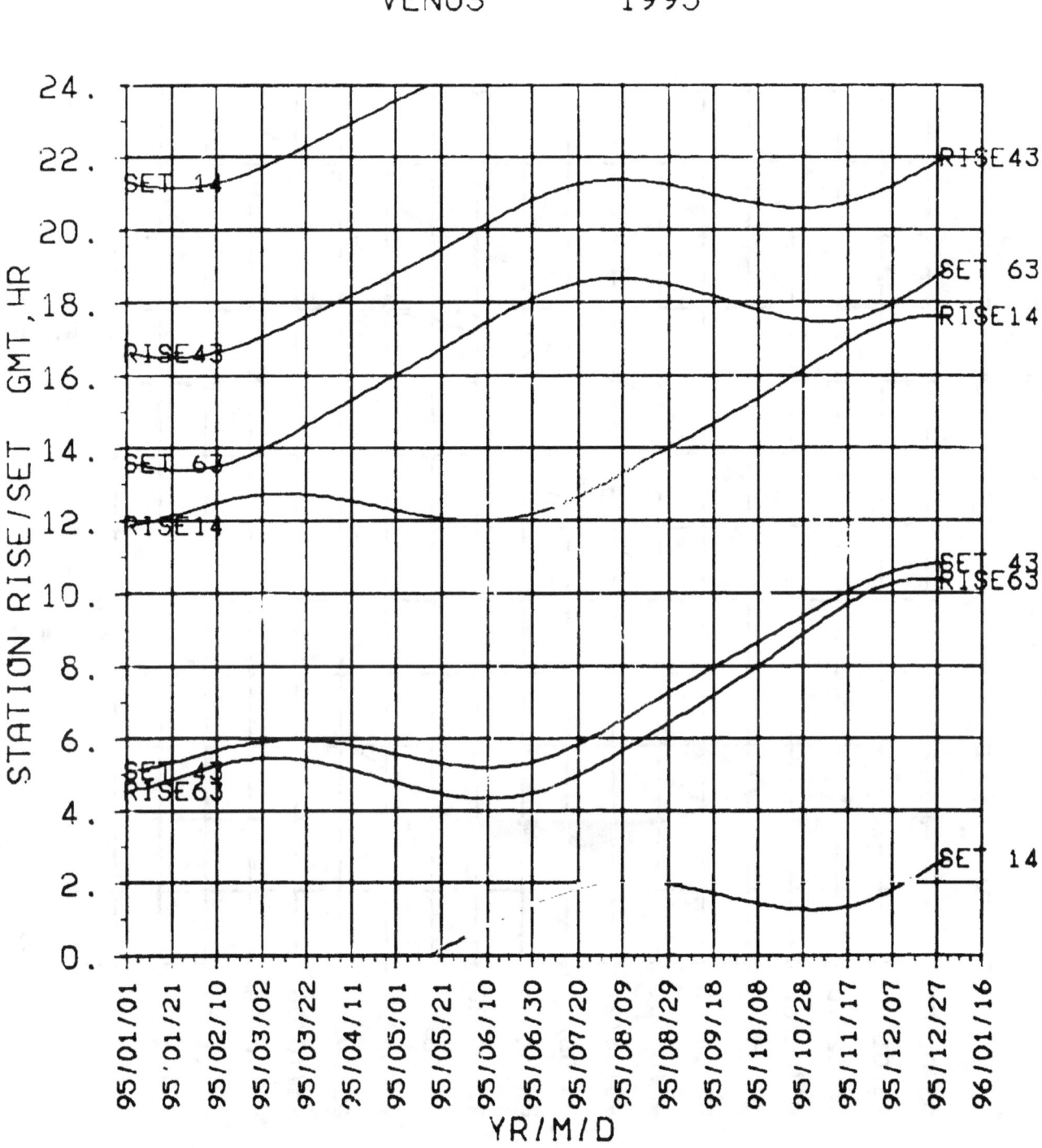

**STA R/S
NON-DSN
1995**

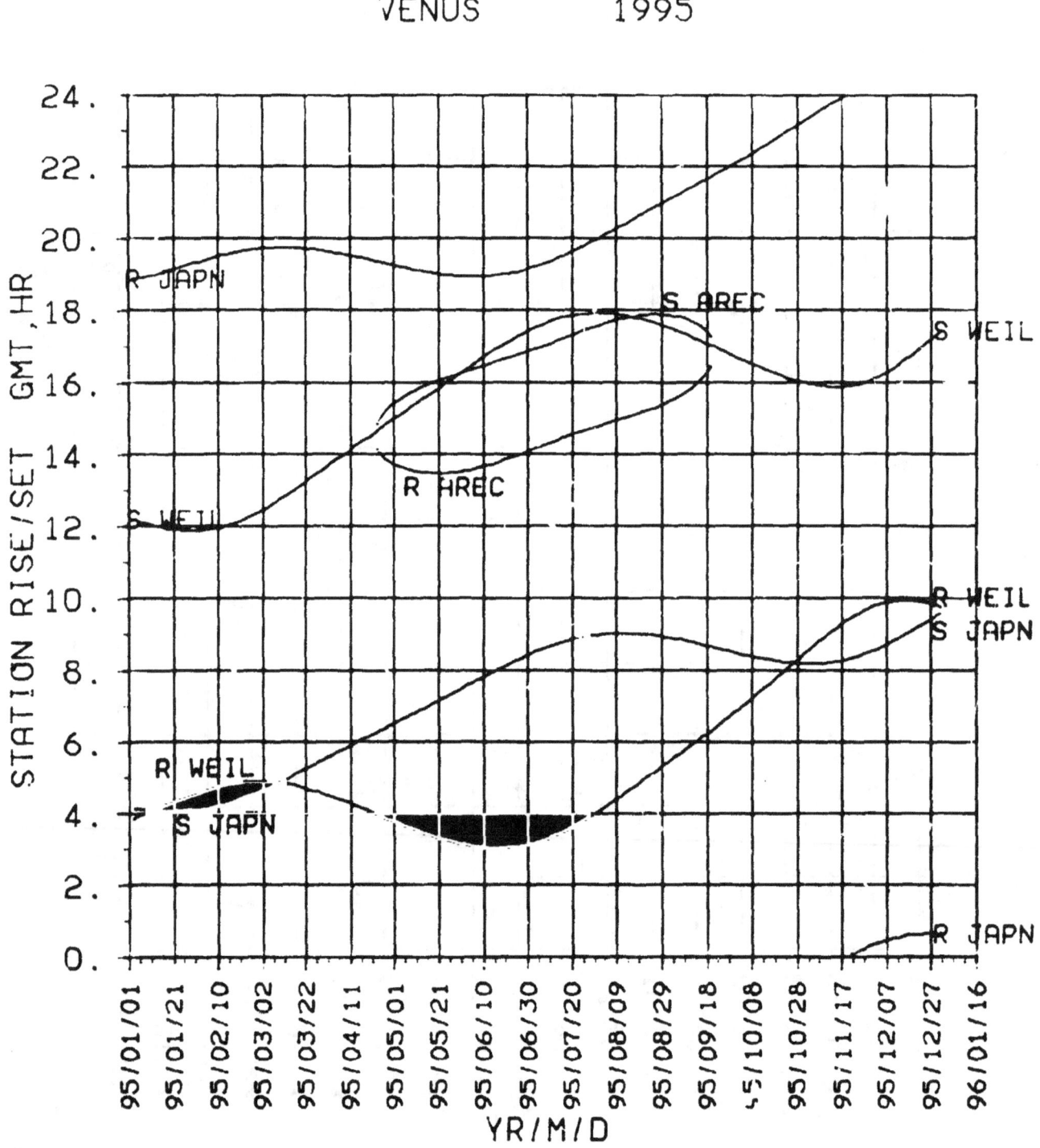

Venus

1996

DECLIN RT.ASC 1996

VENUS 1996

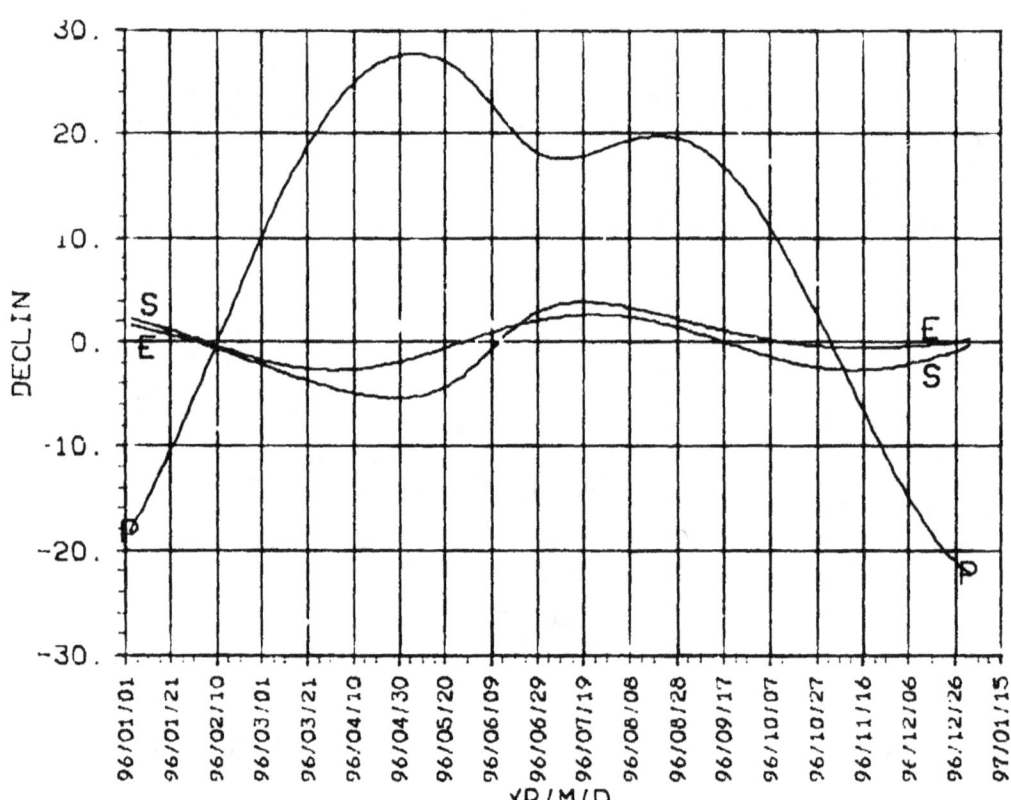

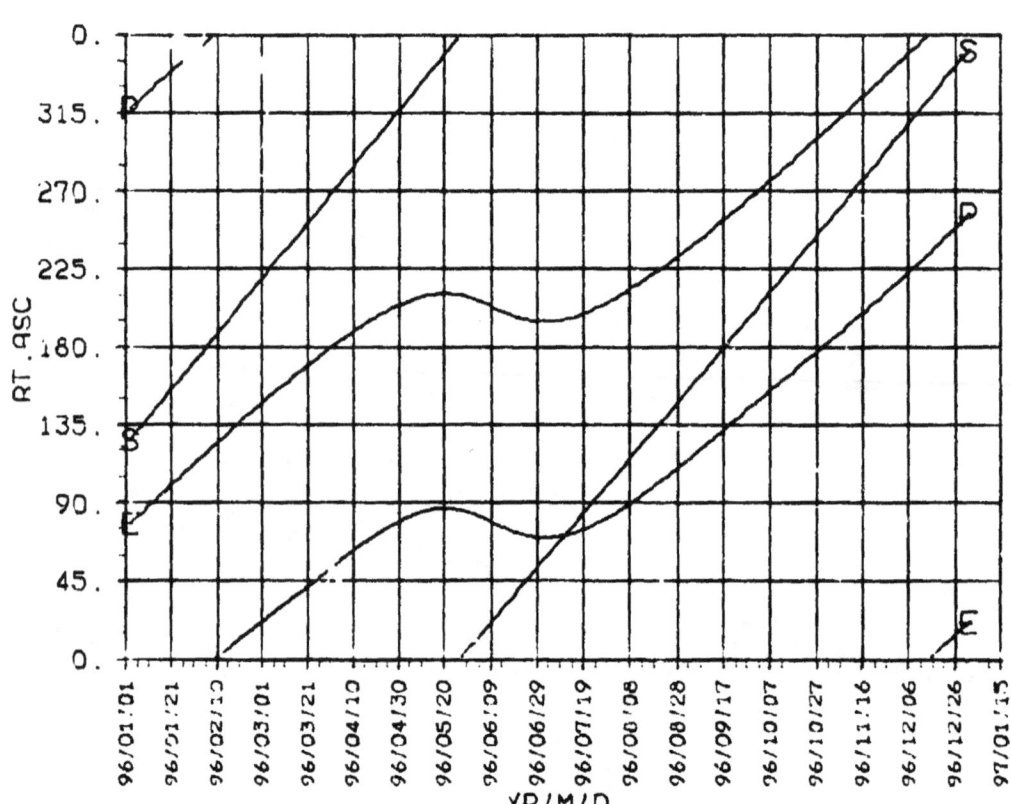

VENUS 1996

DISTANCE EC.LON 1996

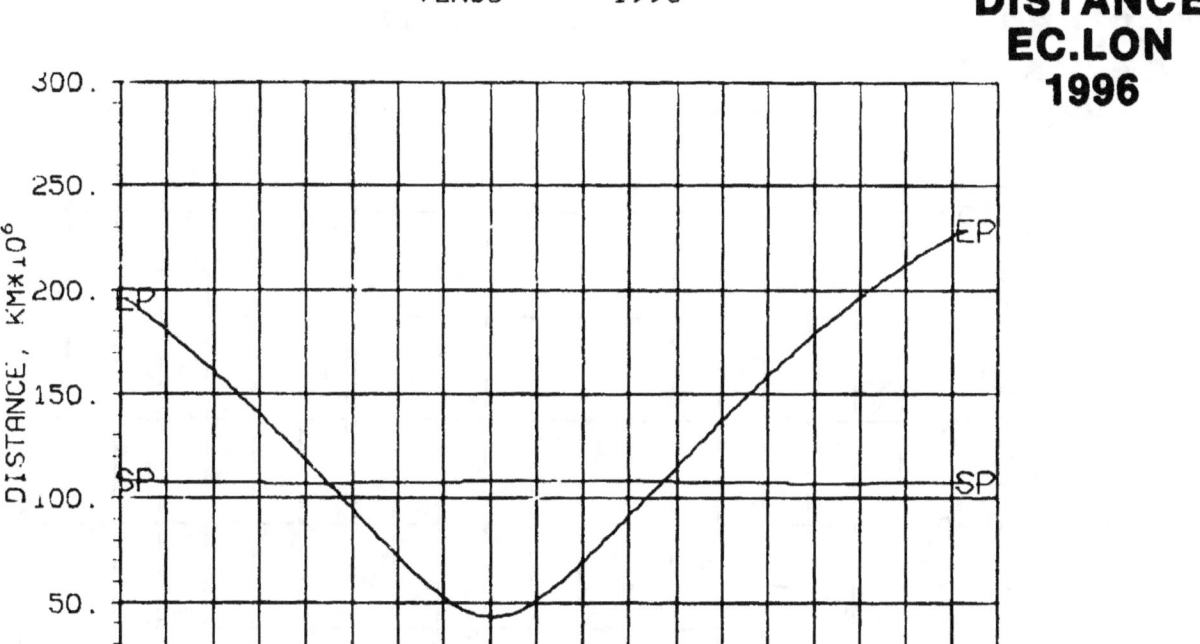

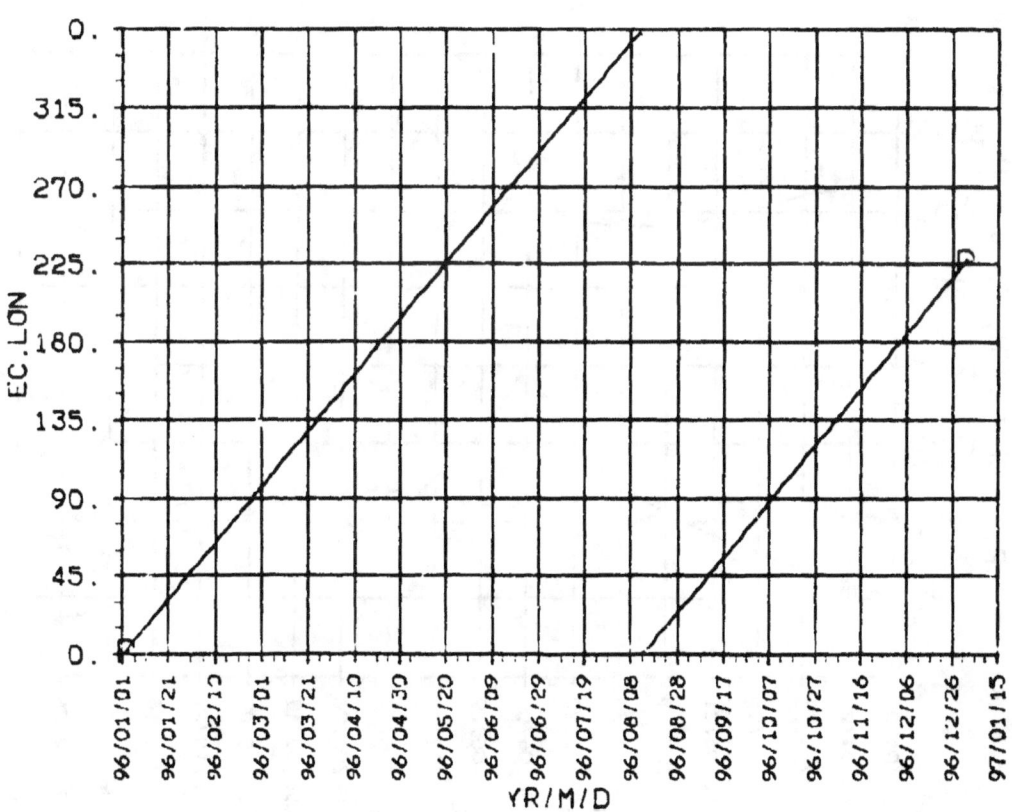

SEP, ESP CA, KA 1996

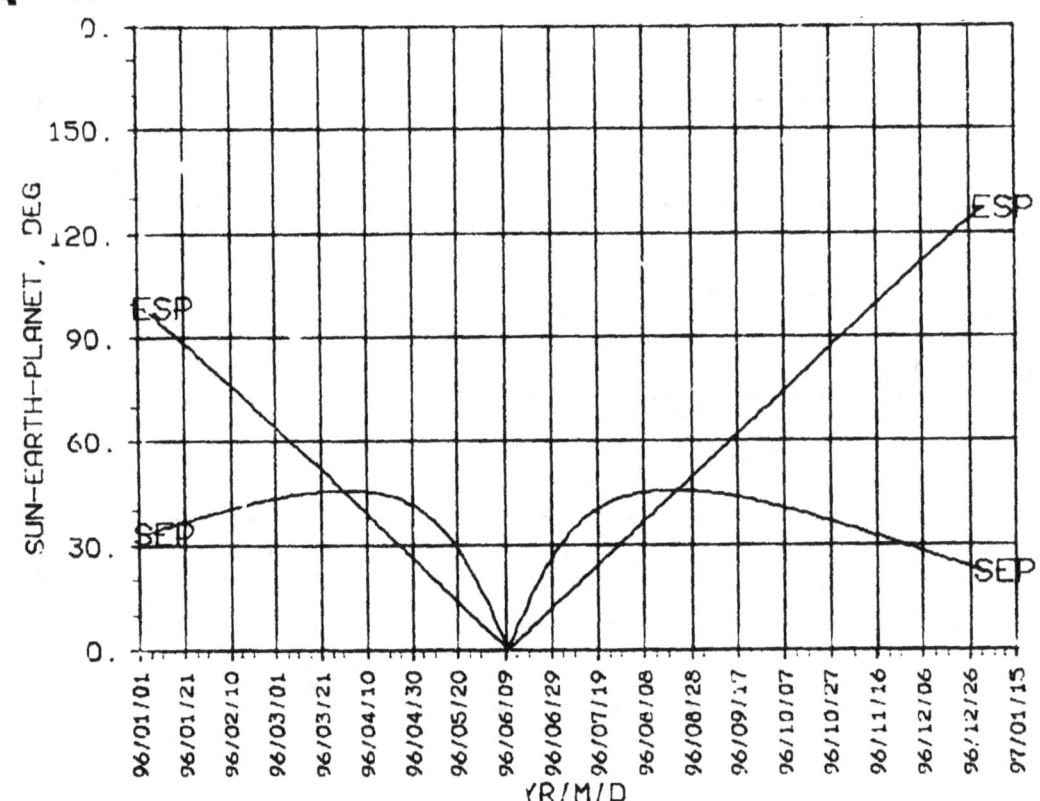

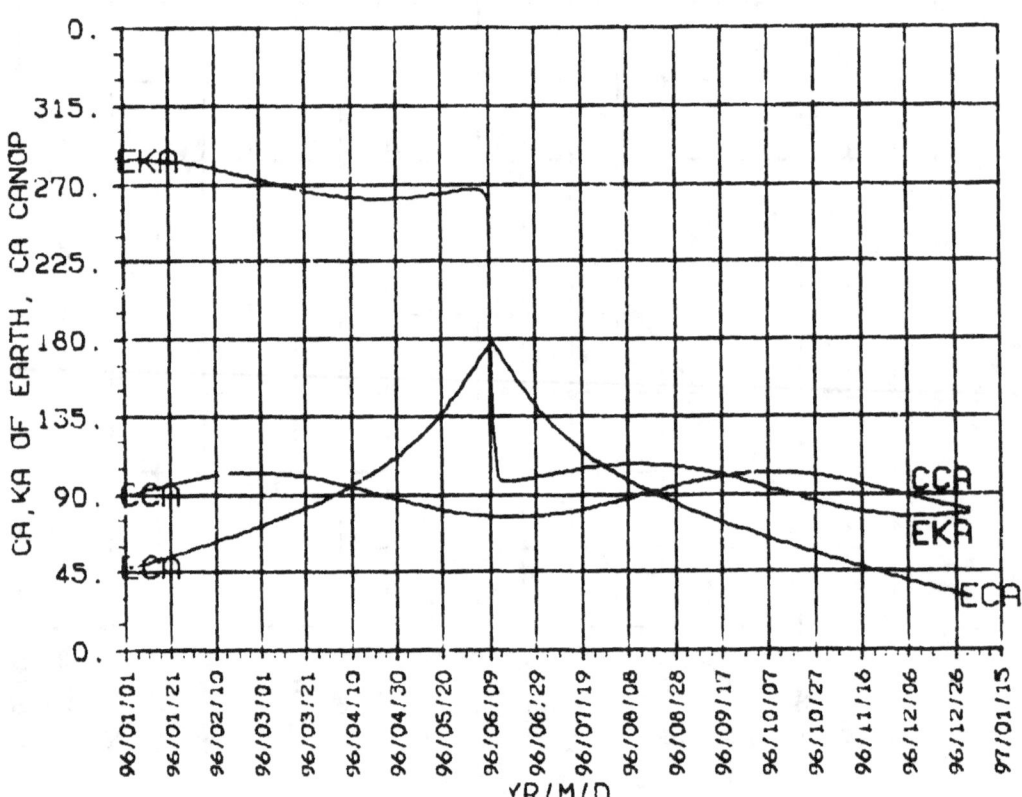

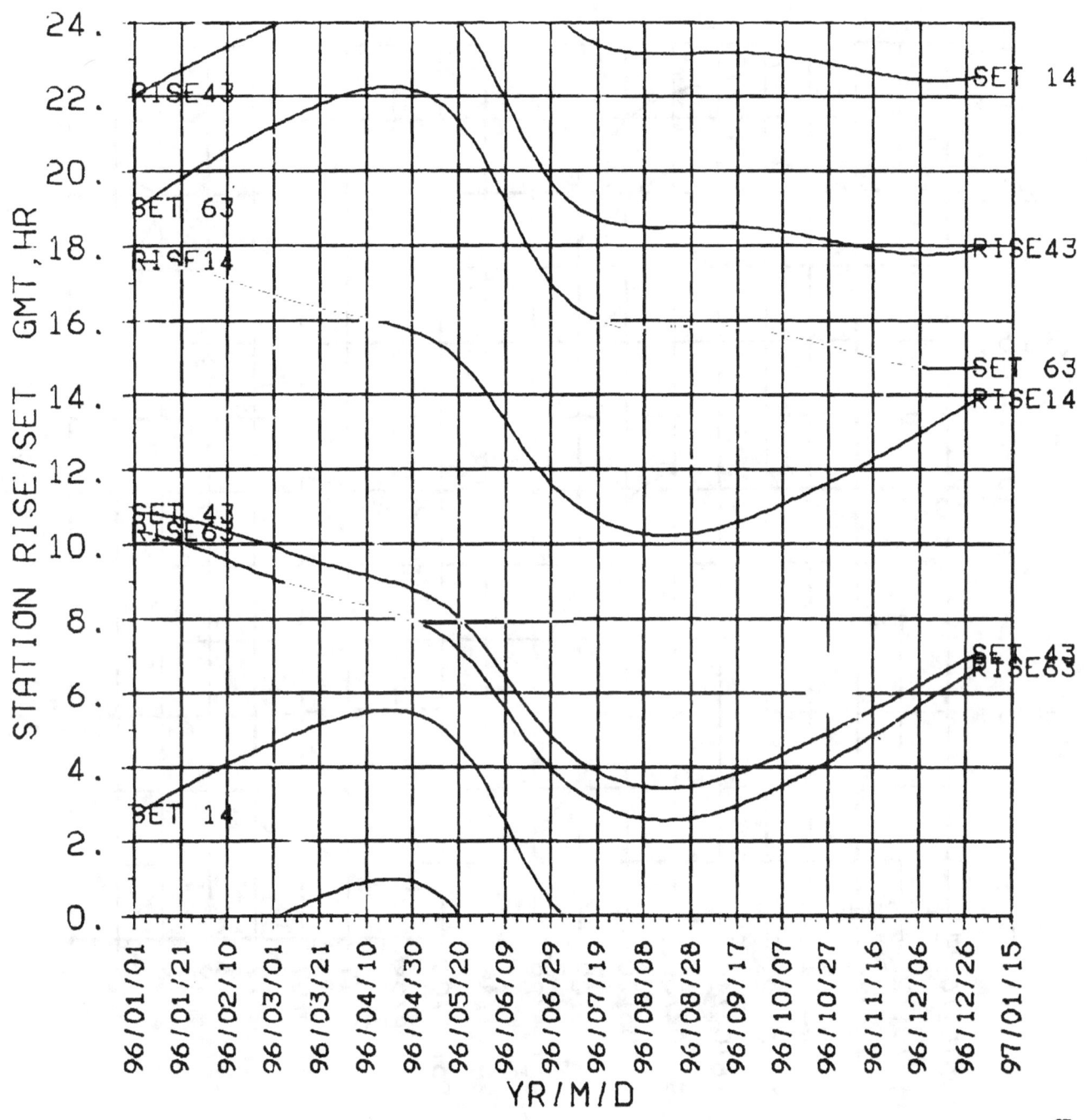

STA R/S
NON-DSN
1996

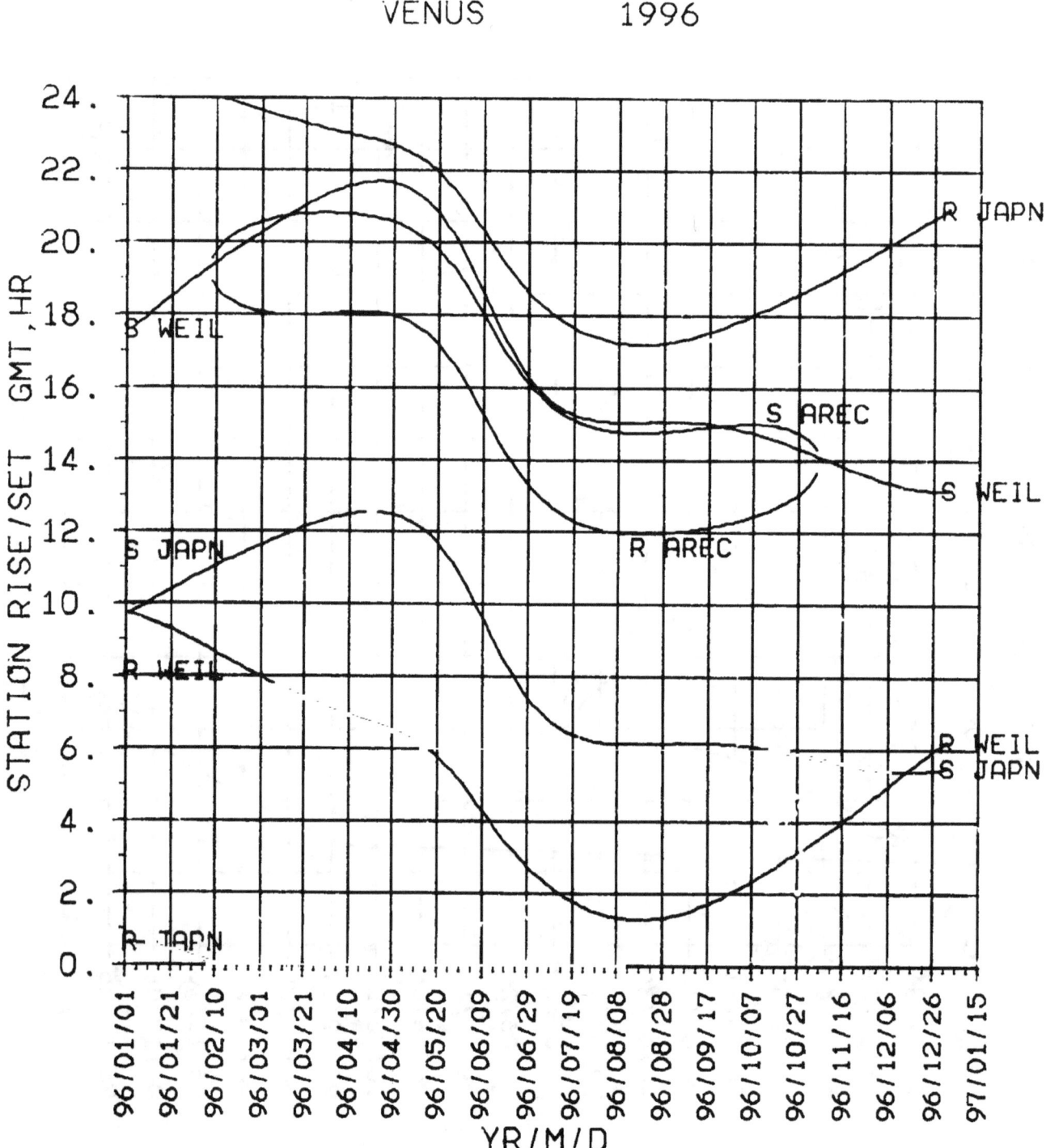

Venus

1997

DECLIN RT.ASC 1997

VENUS 1997

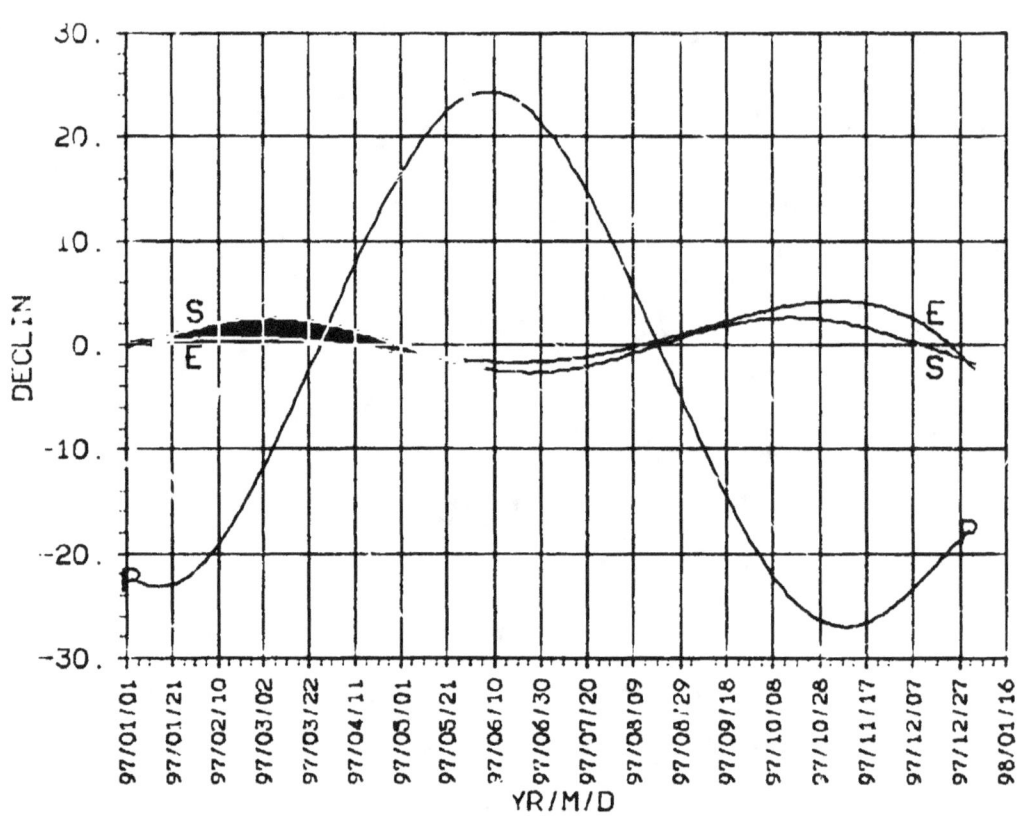

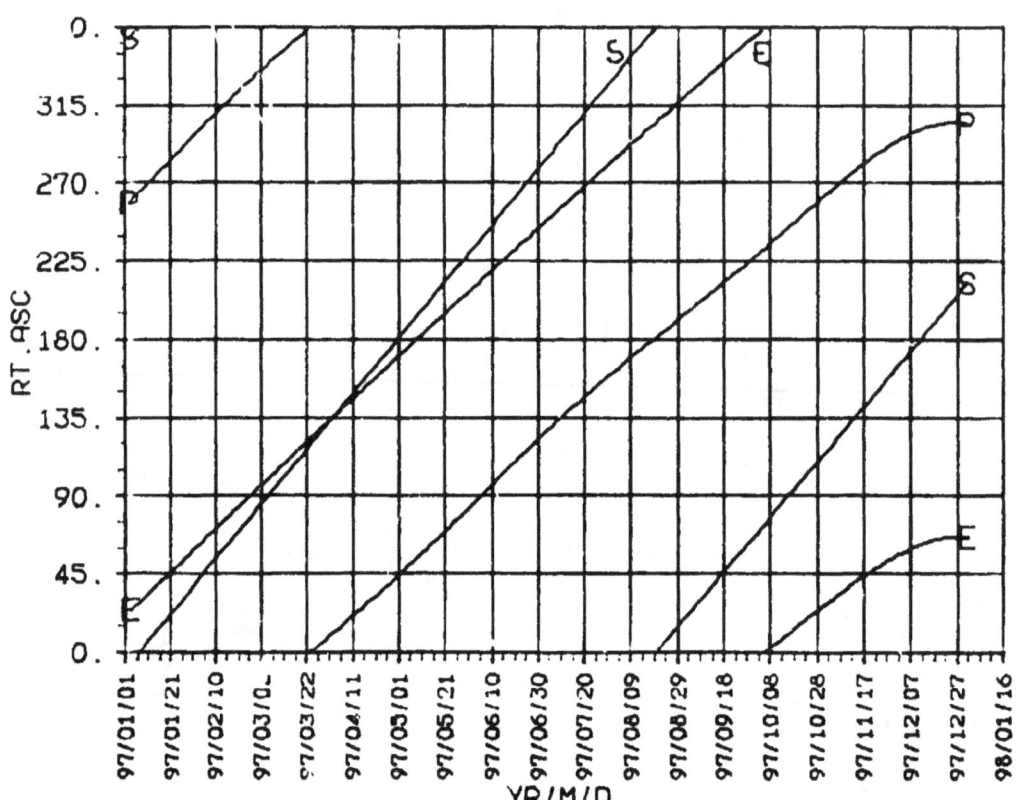

VENUS 1997

DISTANCE EC.LON 1997

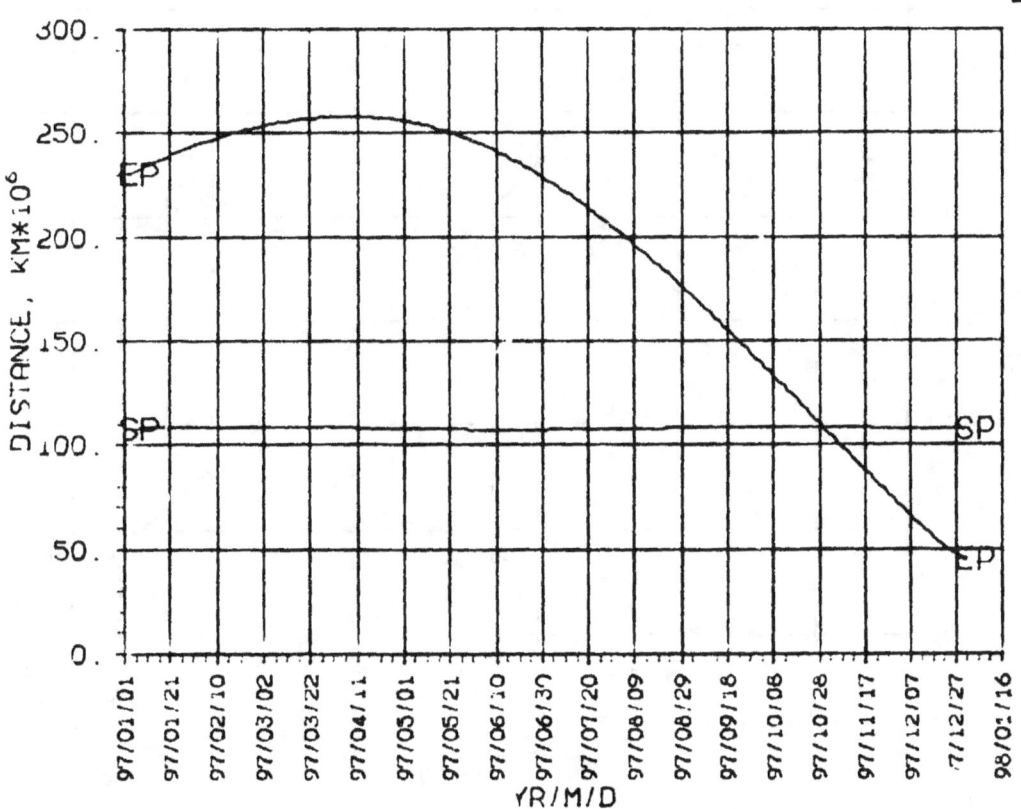

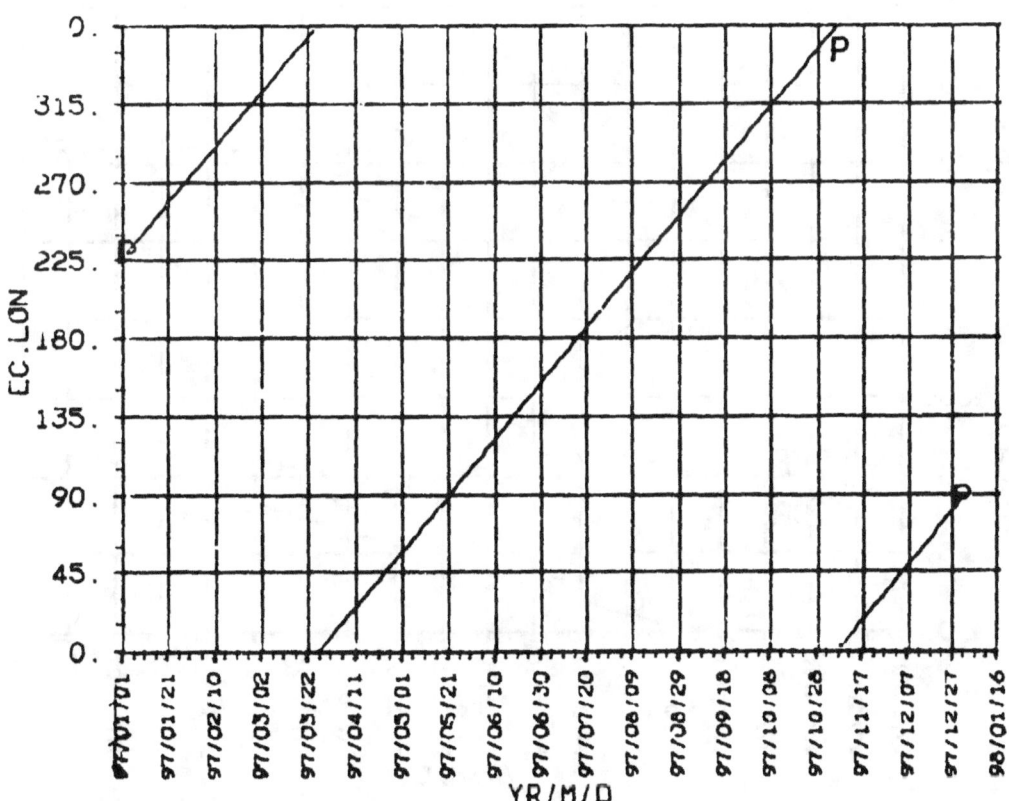

SEP, ESP CA, KA 1997

VENUS 1997

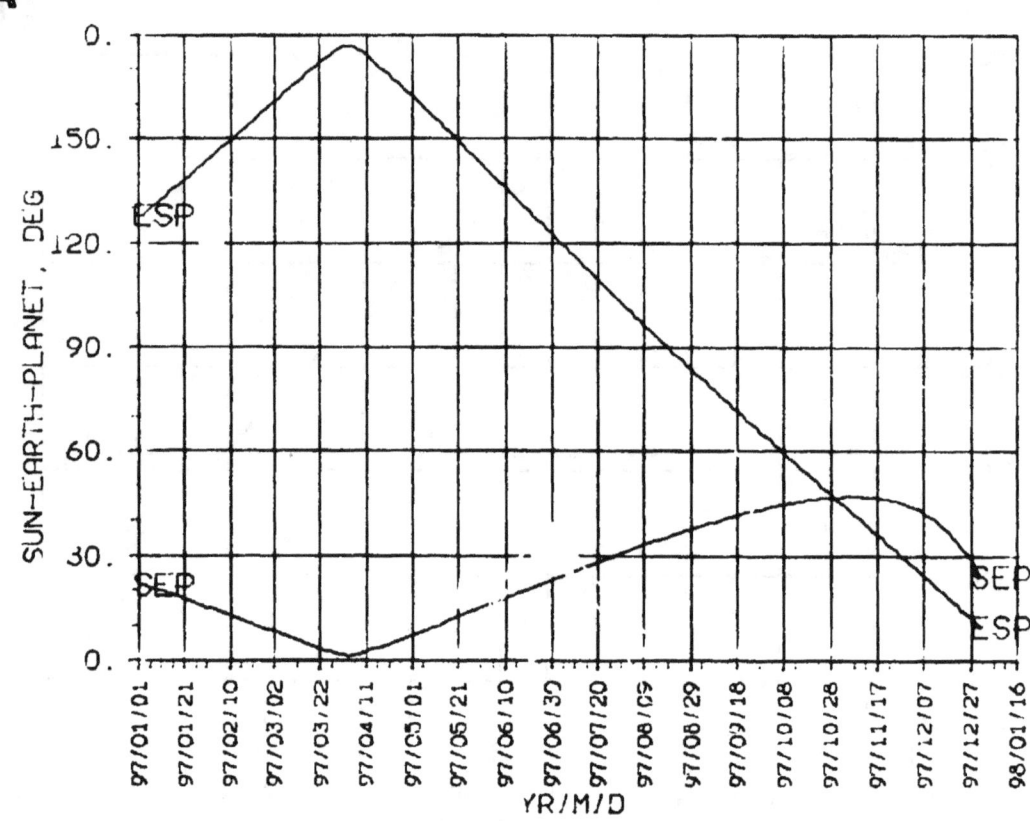

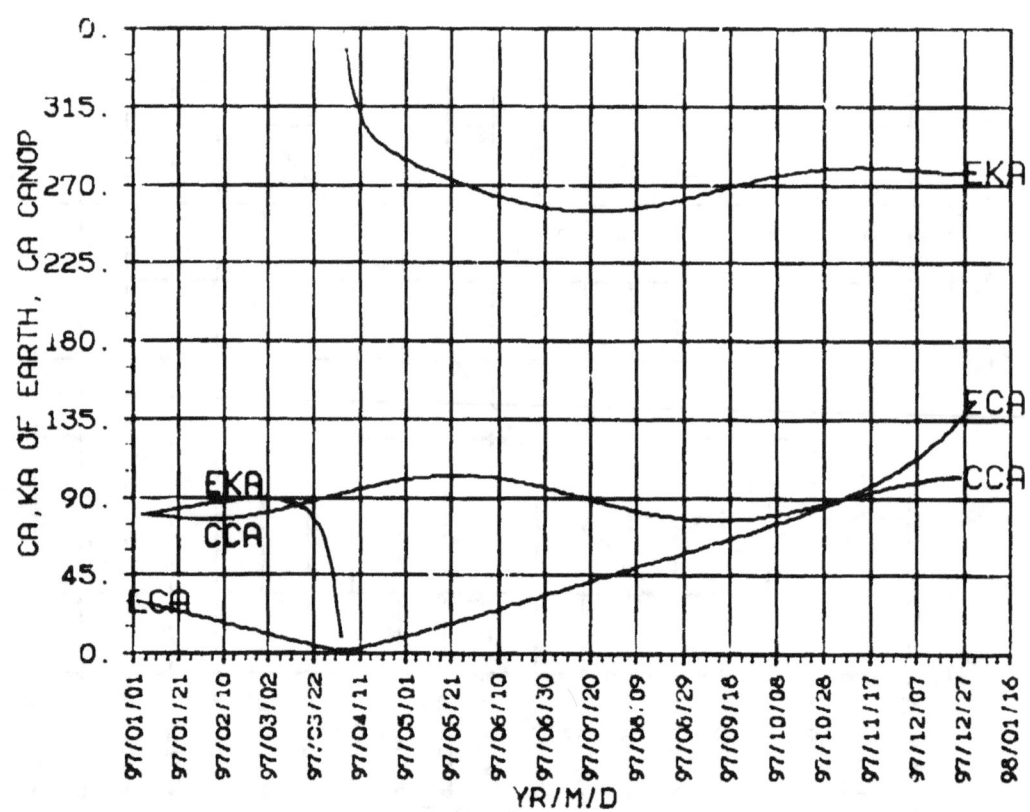

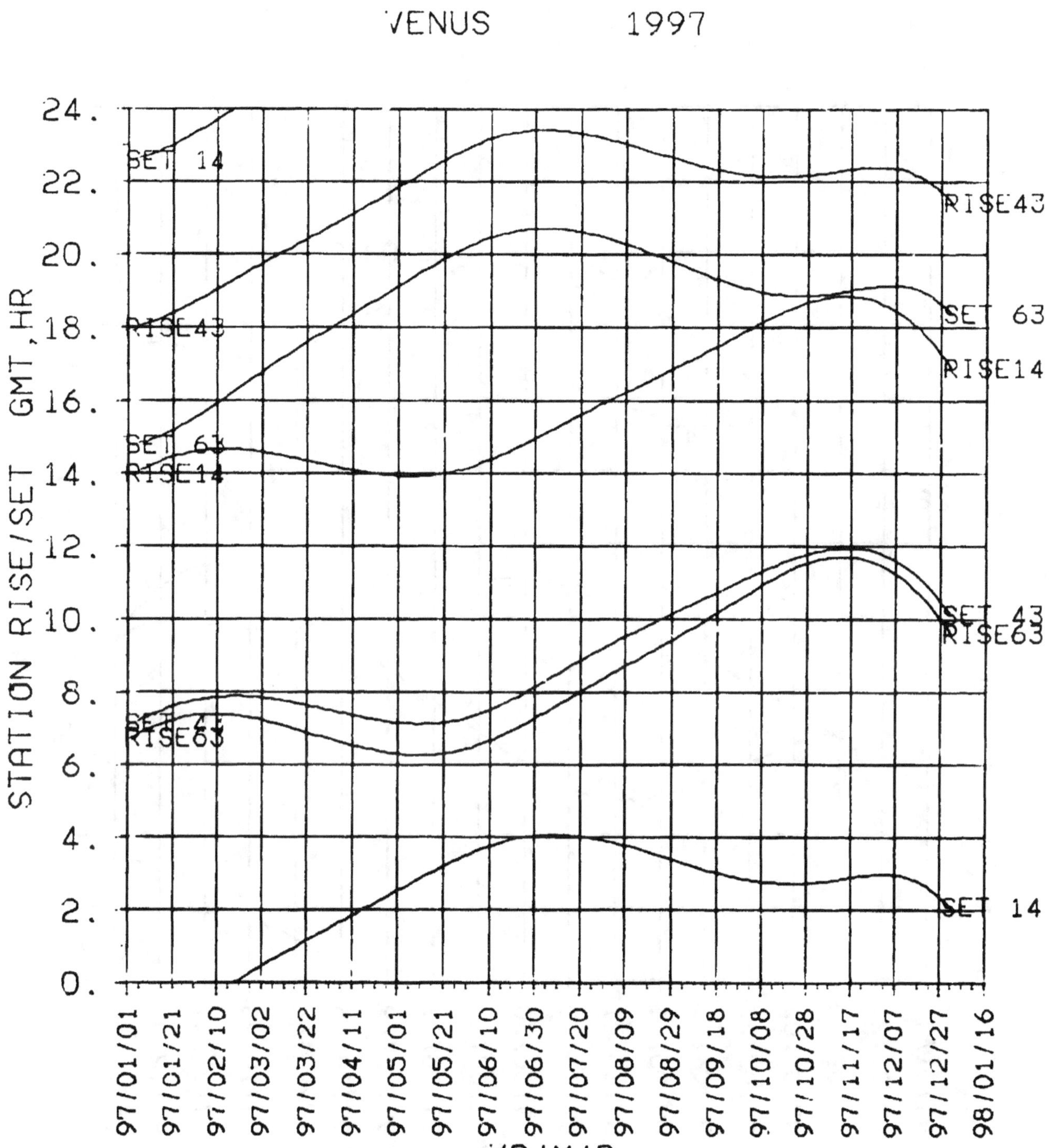

**STA R/S
NON-DSN
1997**

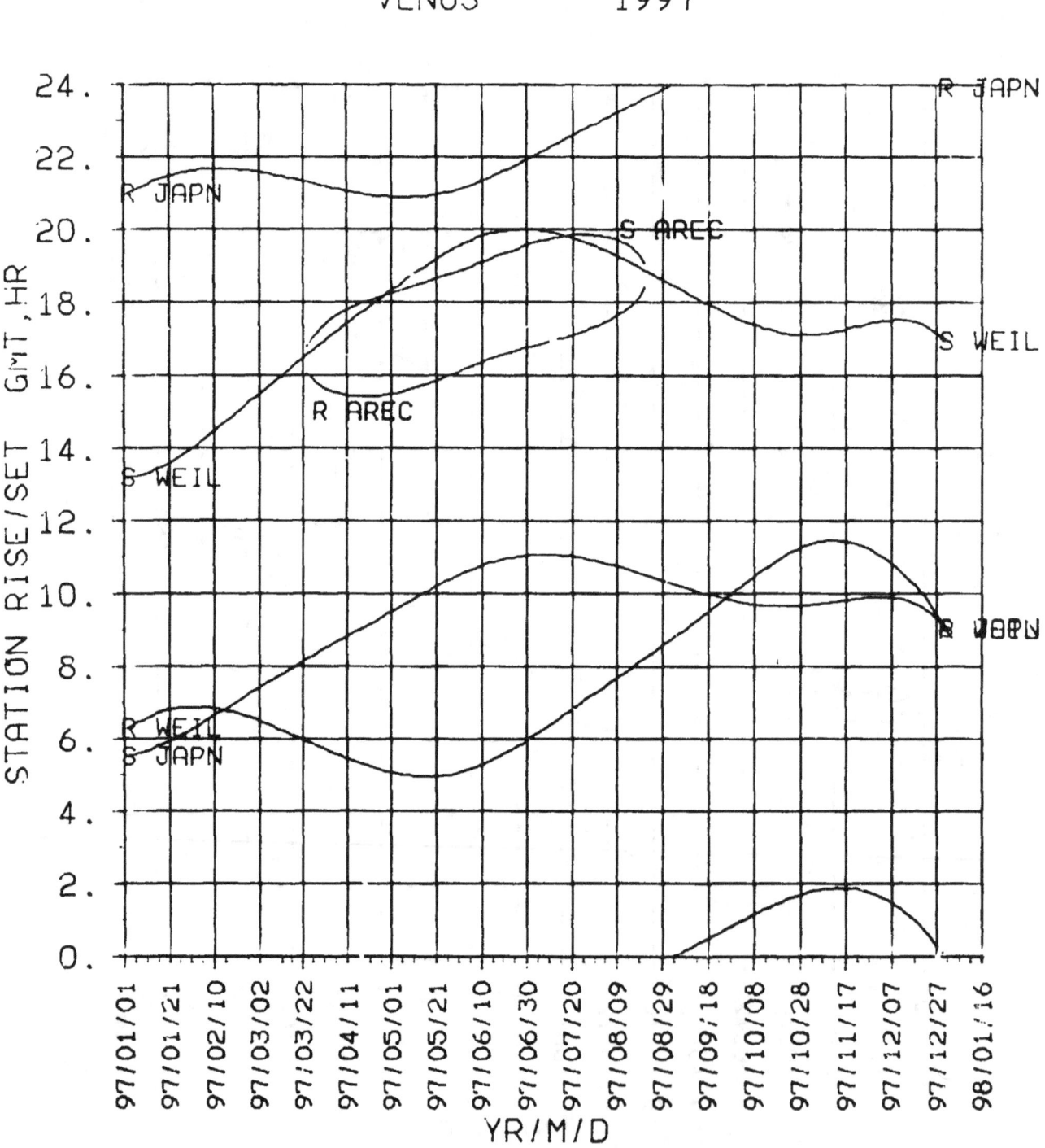

Venus

1998

DECLIN RT.ASC 1998

VENUS 1998

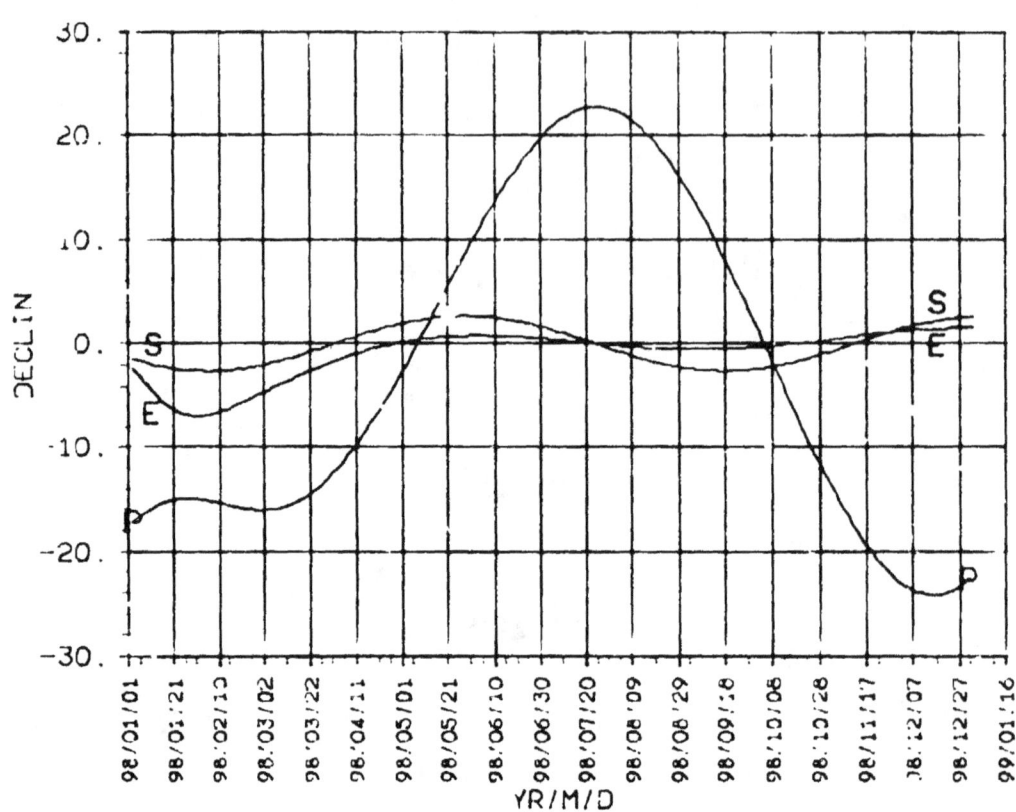

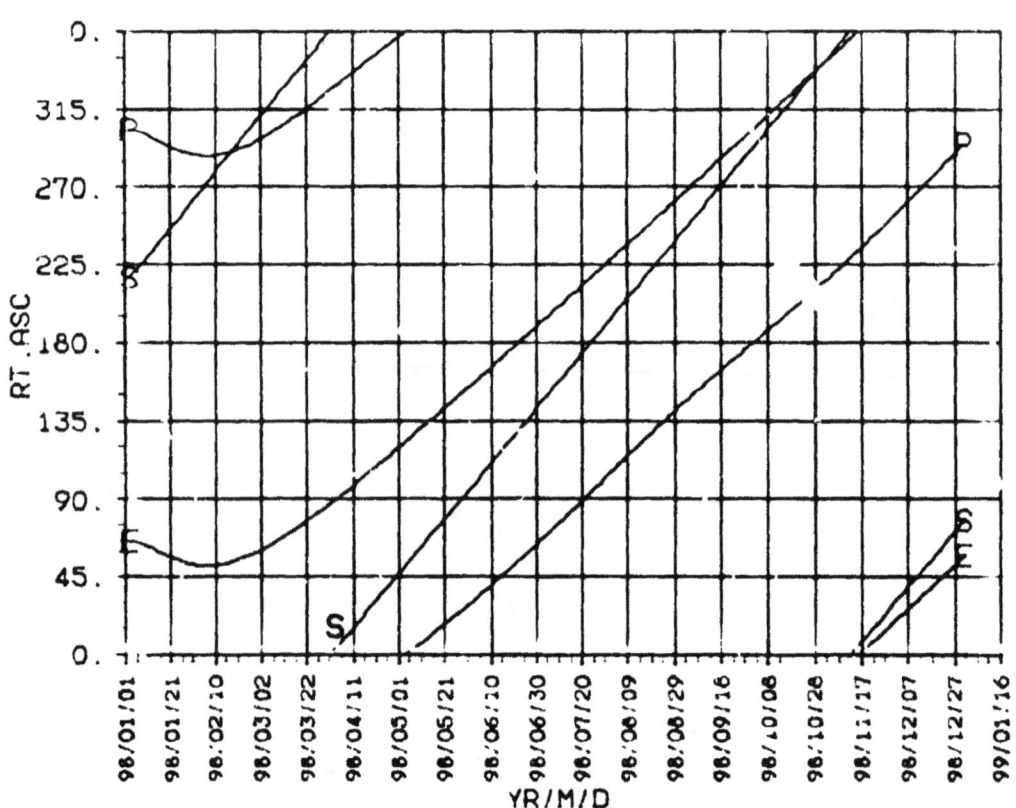

VENUS 1998

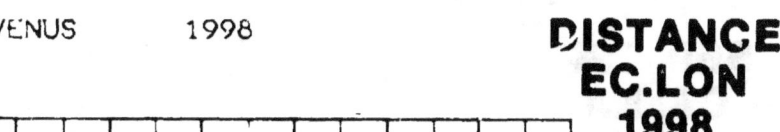
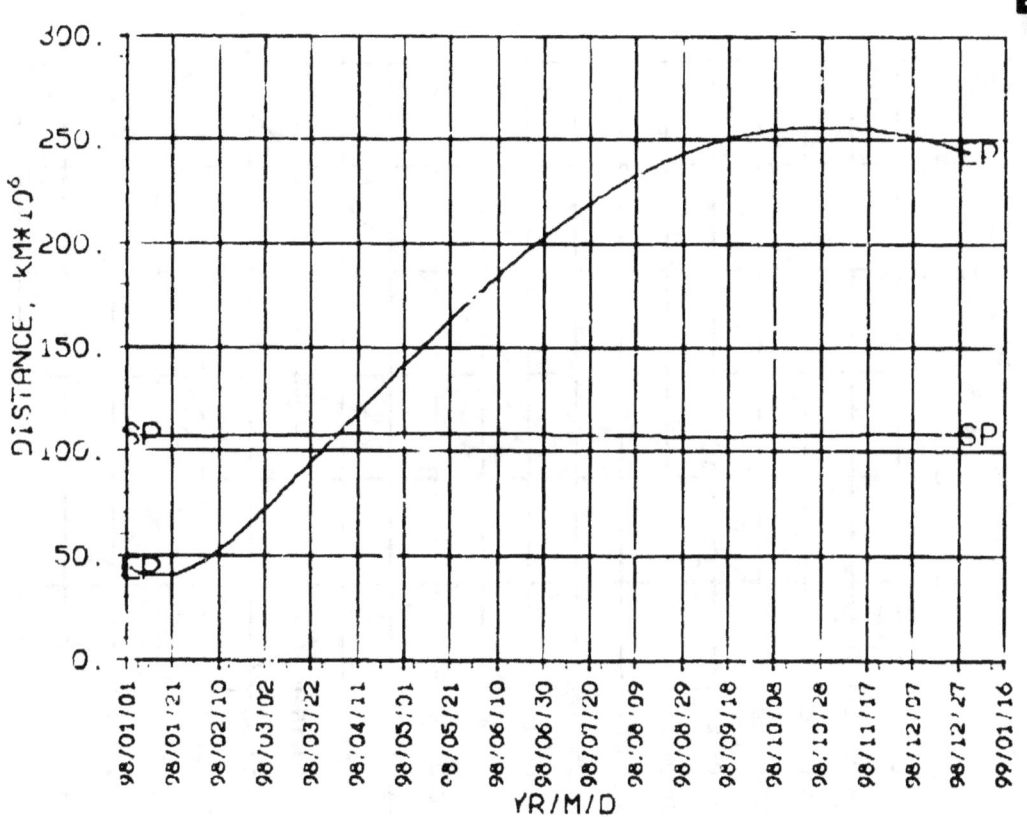
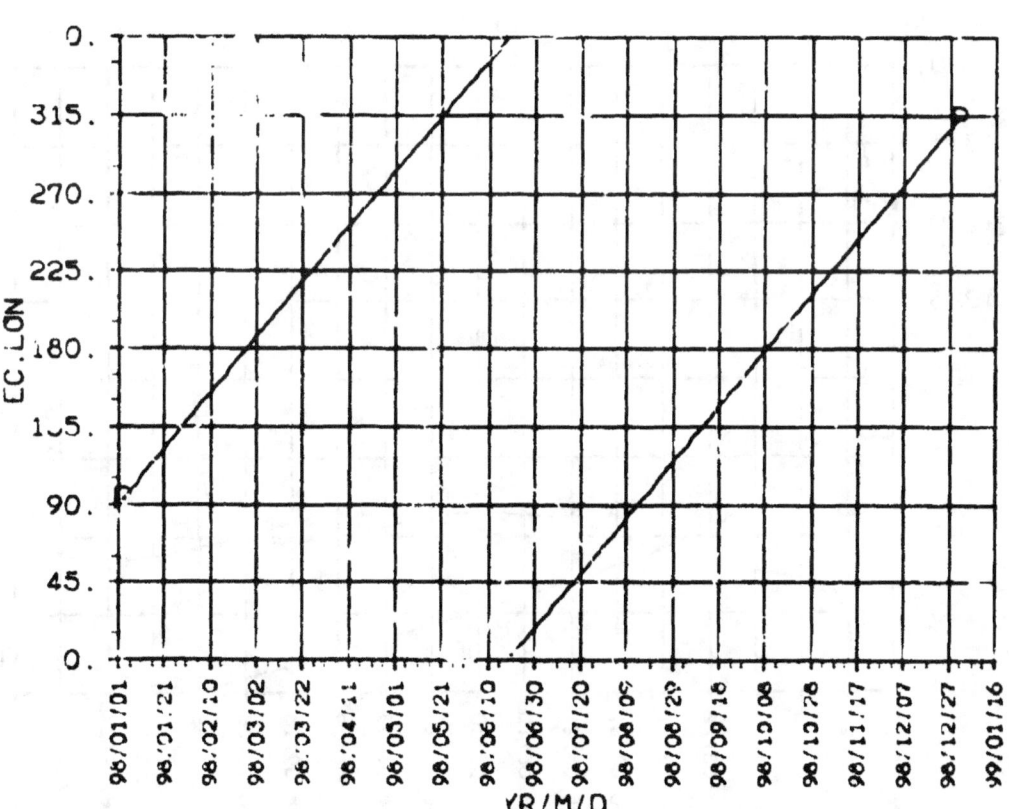

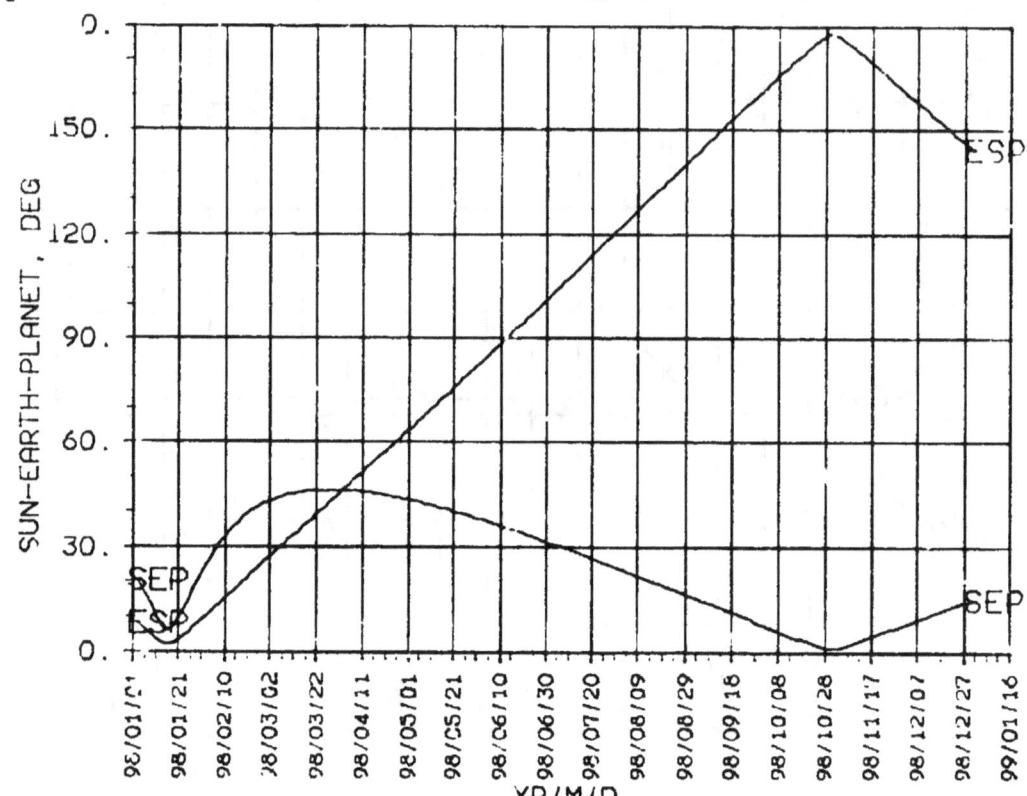
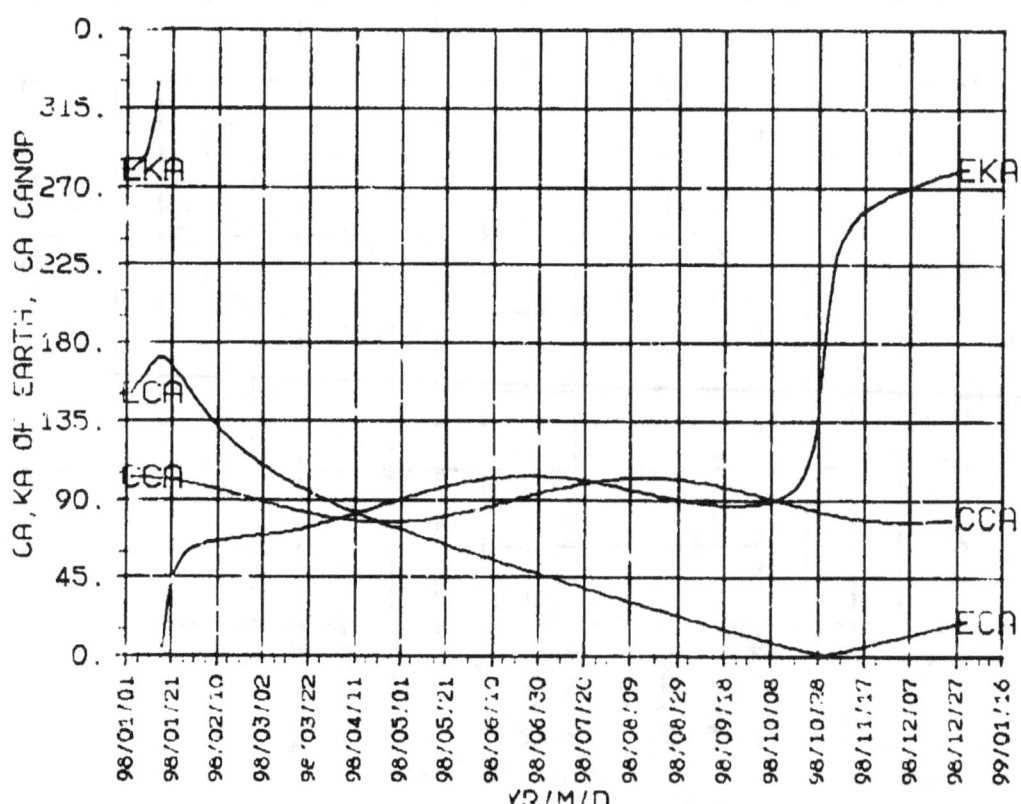

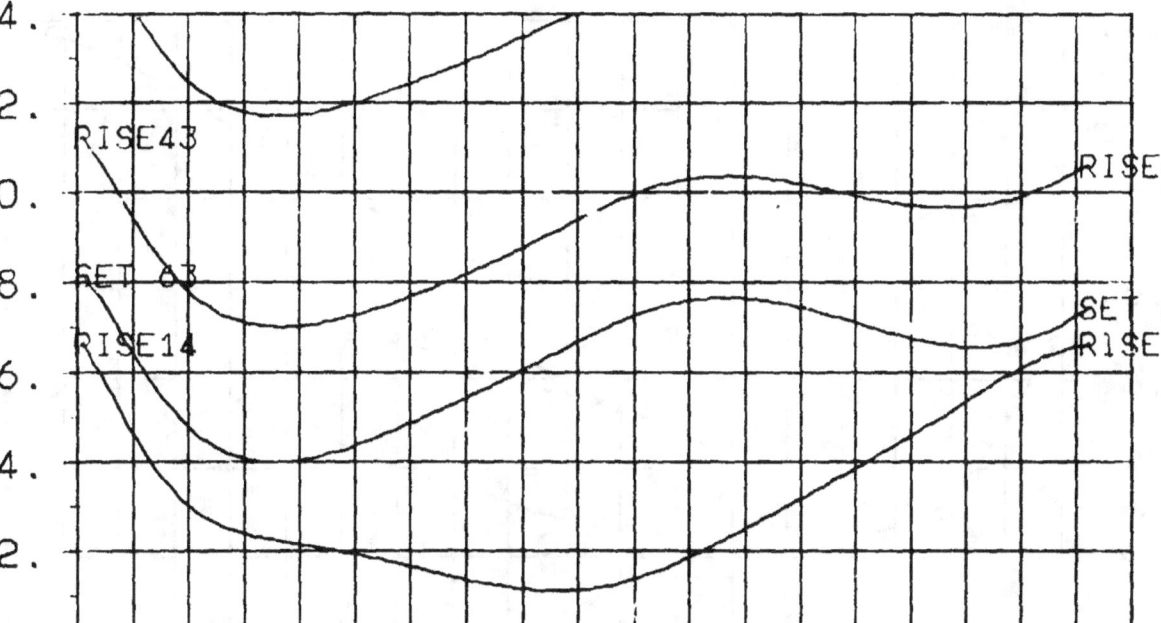

**STA R/S
NON-DSN
1998**

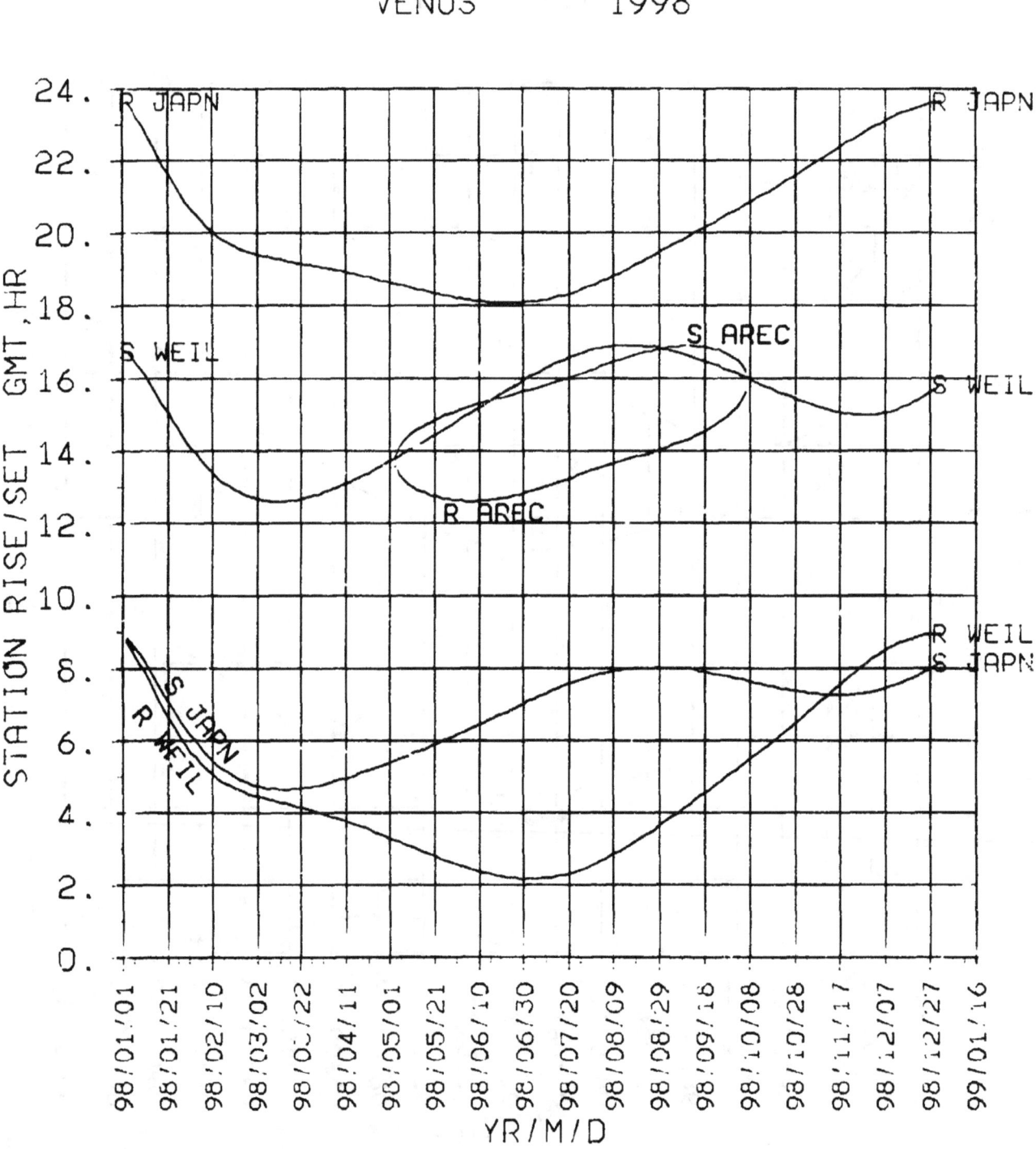

Venus

1999

DECLIN RT.ASC 1999

VENUS 1999

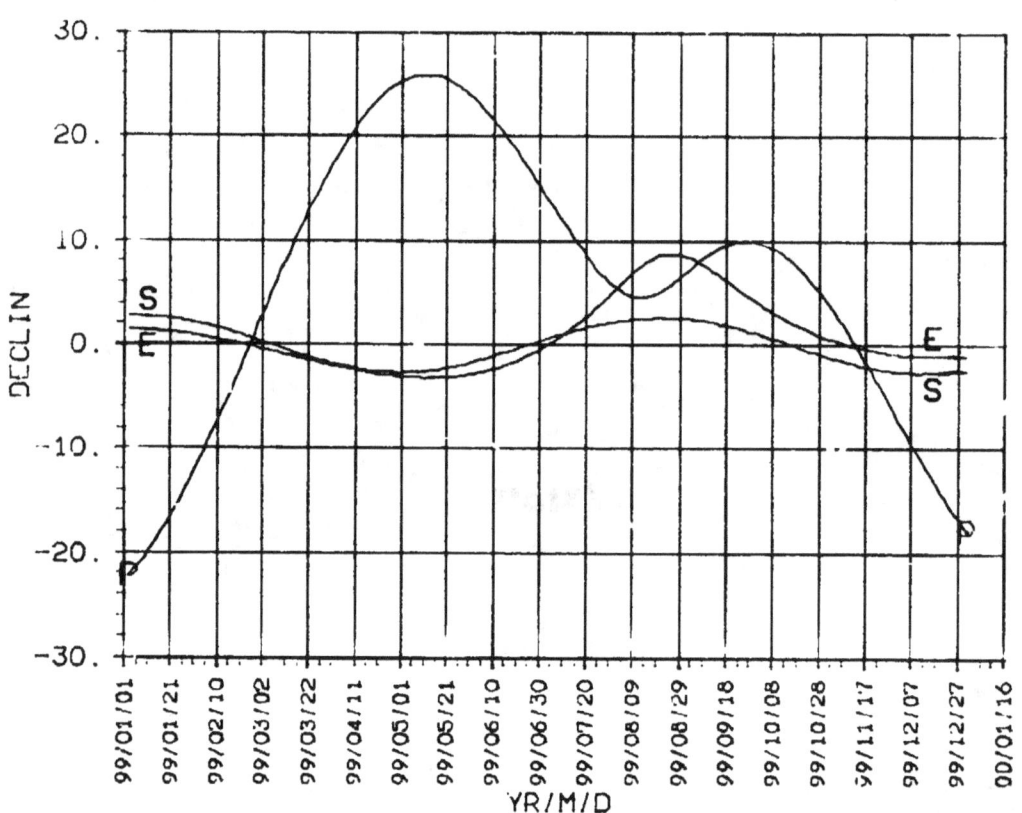

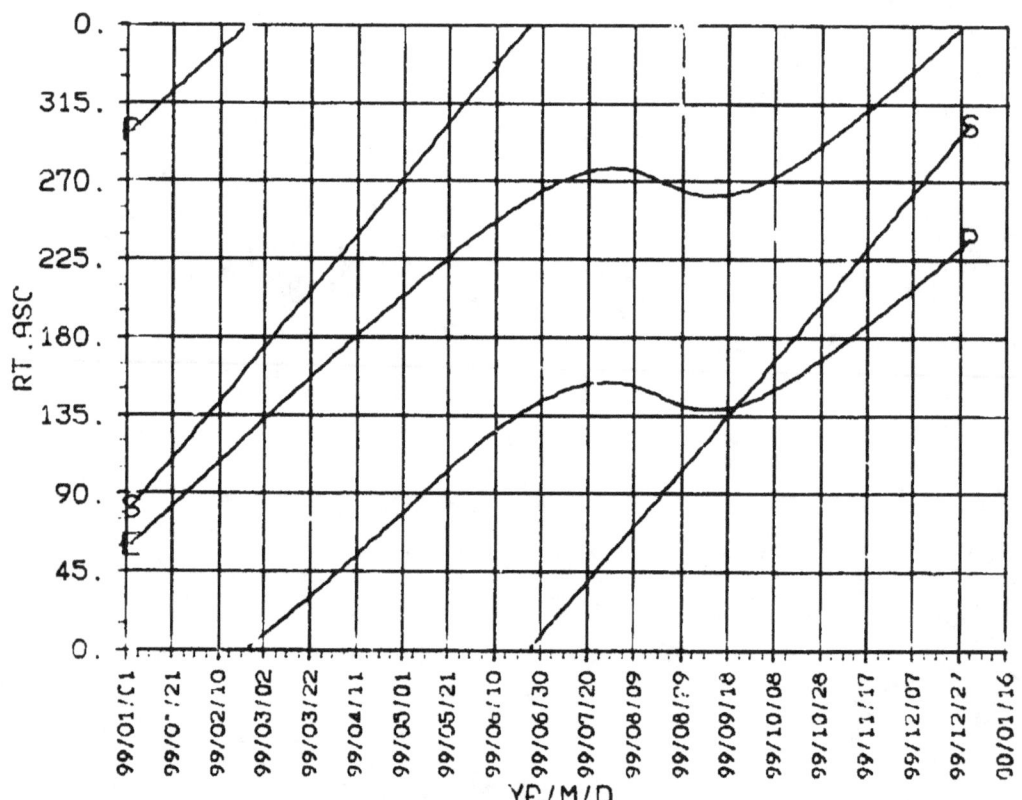

VENUS 1999

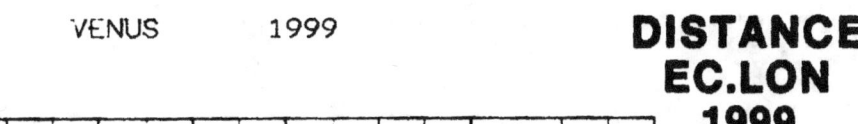

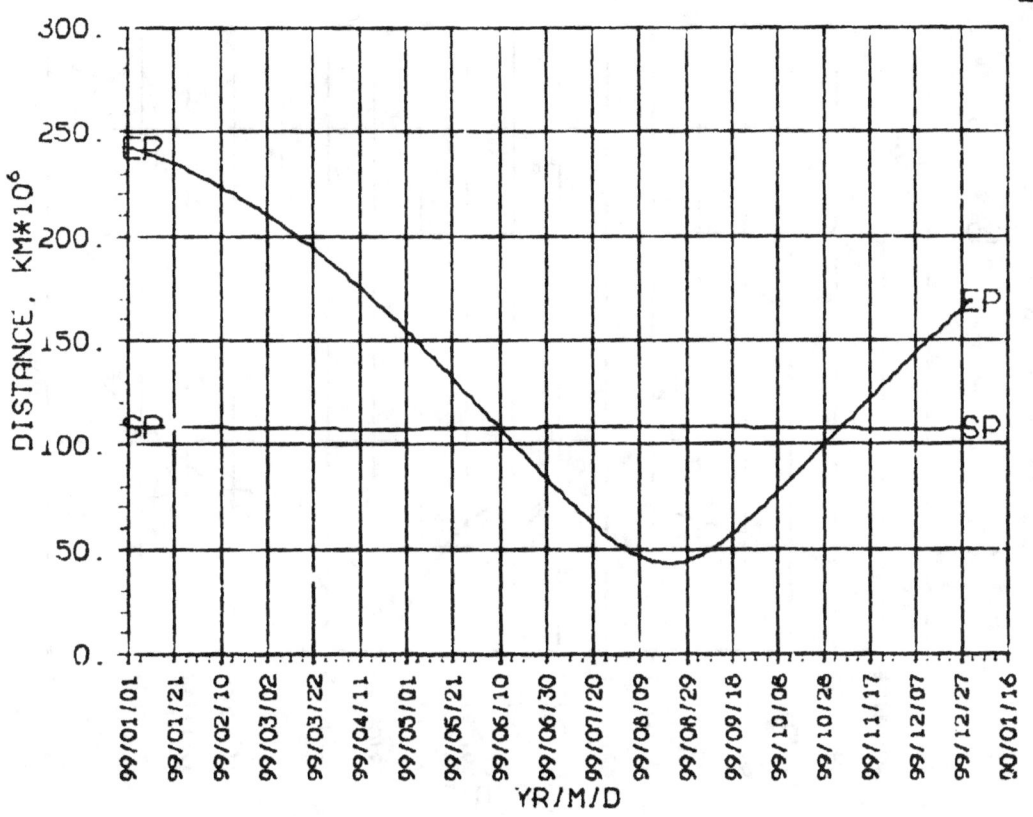

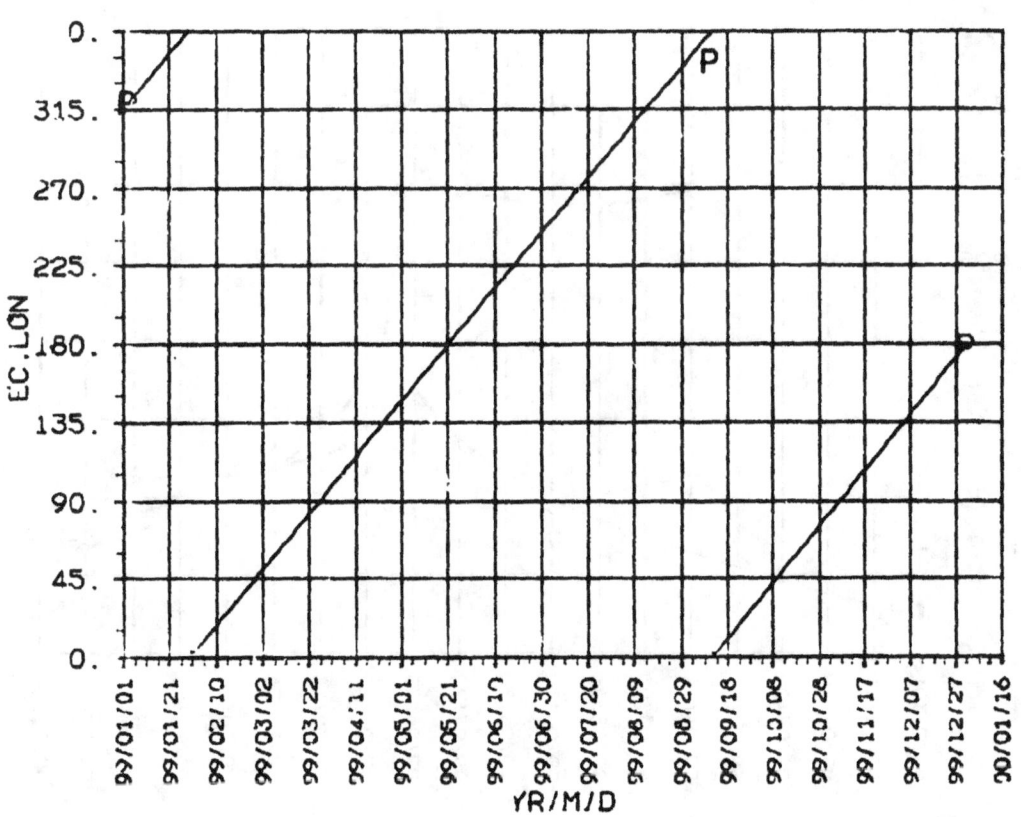

SEP, ESP CA, KA 1999

VENUS 1999

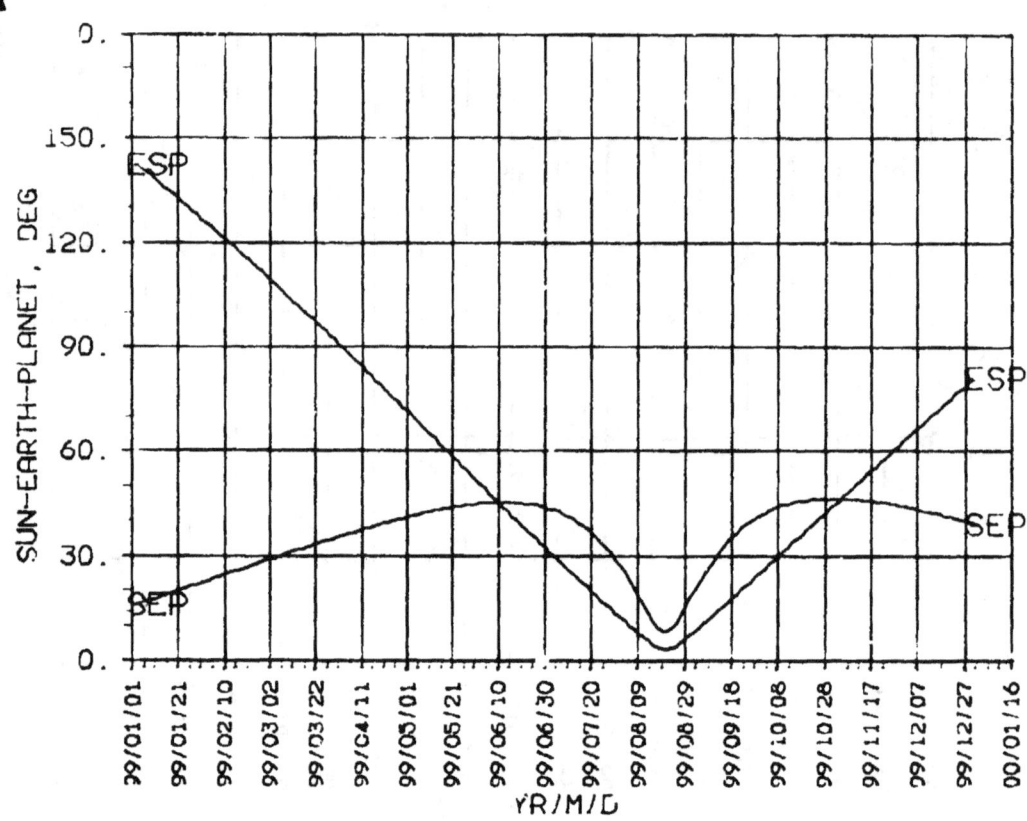

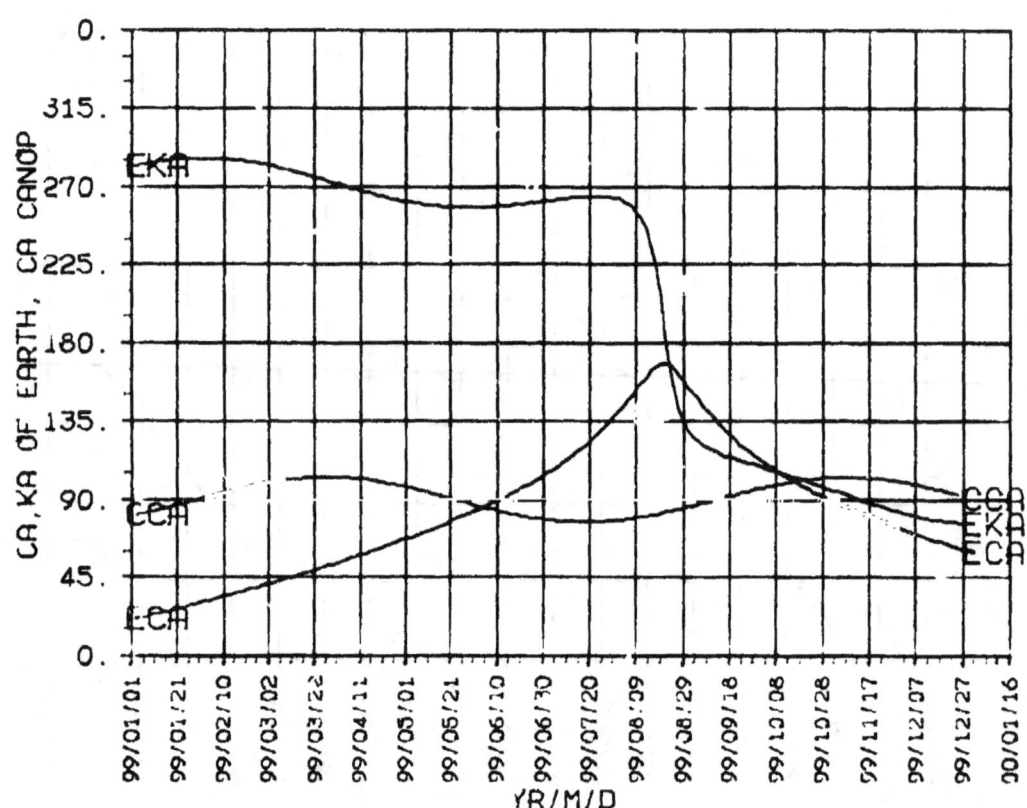

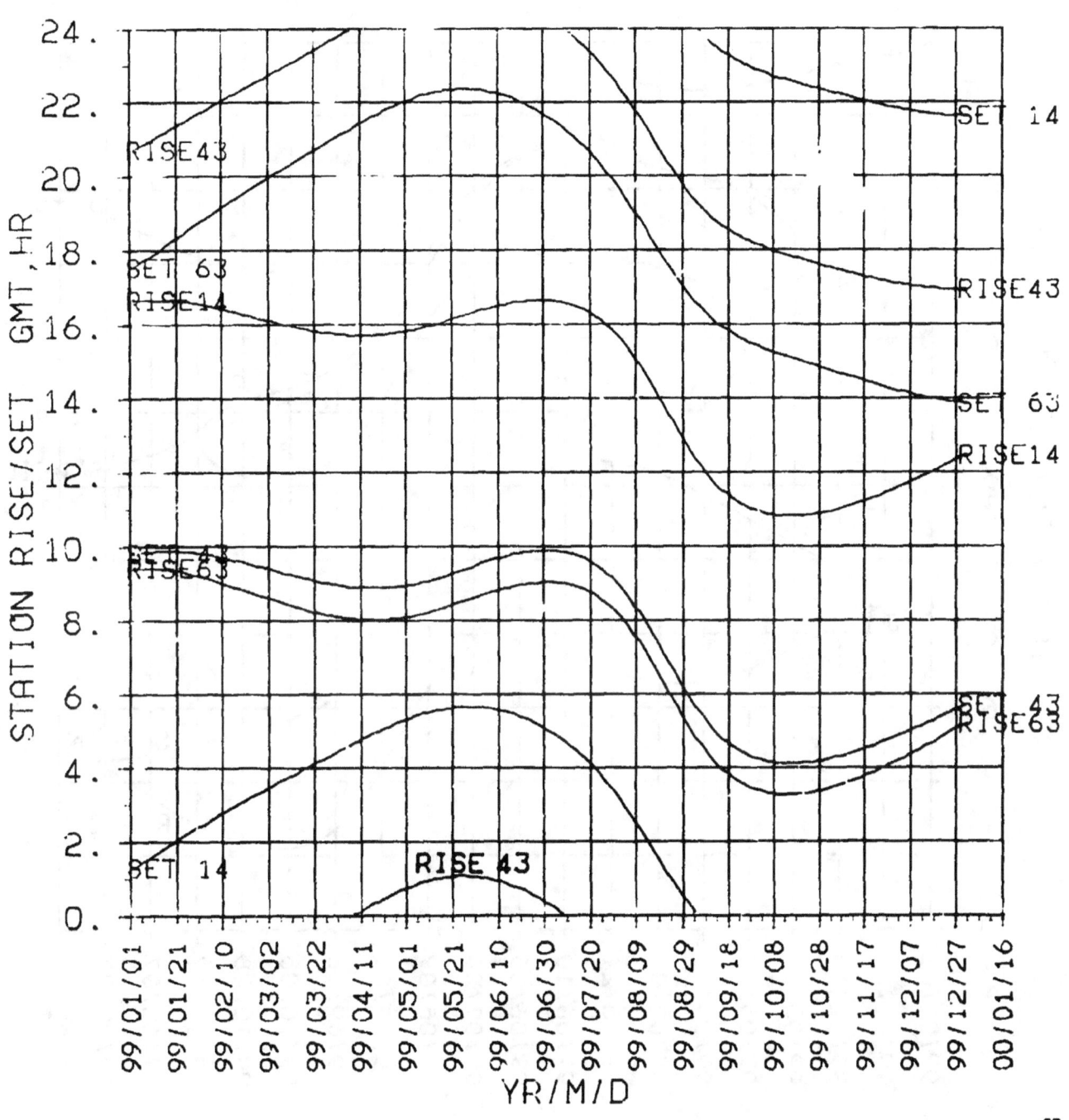

STA R/S NON-DSN 1999

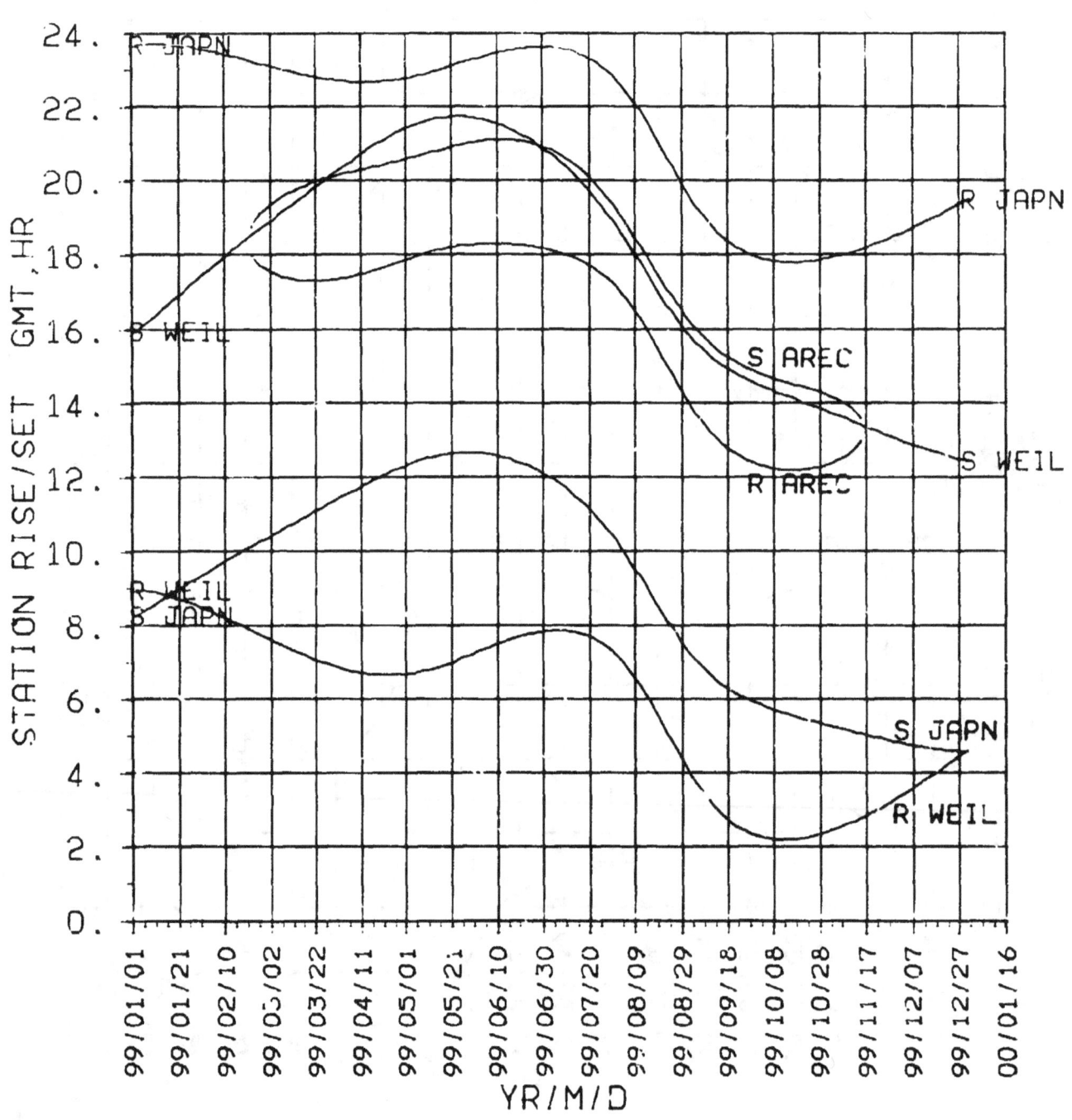

Venus

2000

DECLIN RT.ASC 2000

VENUS 2000

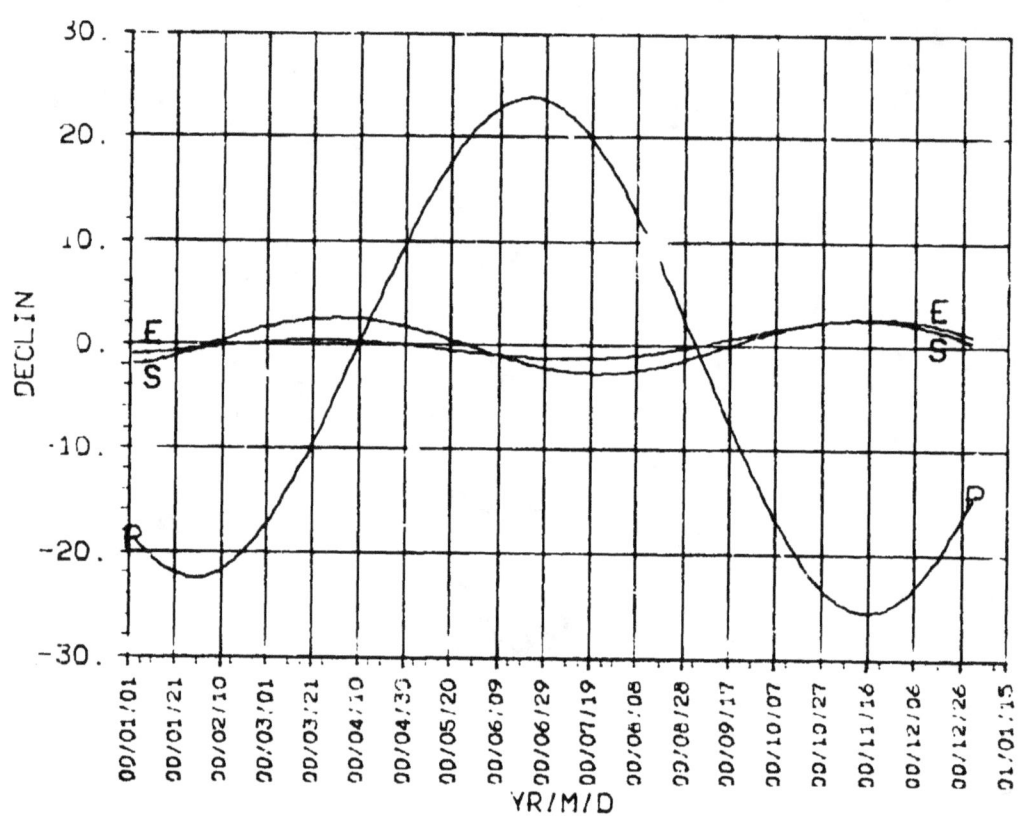

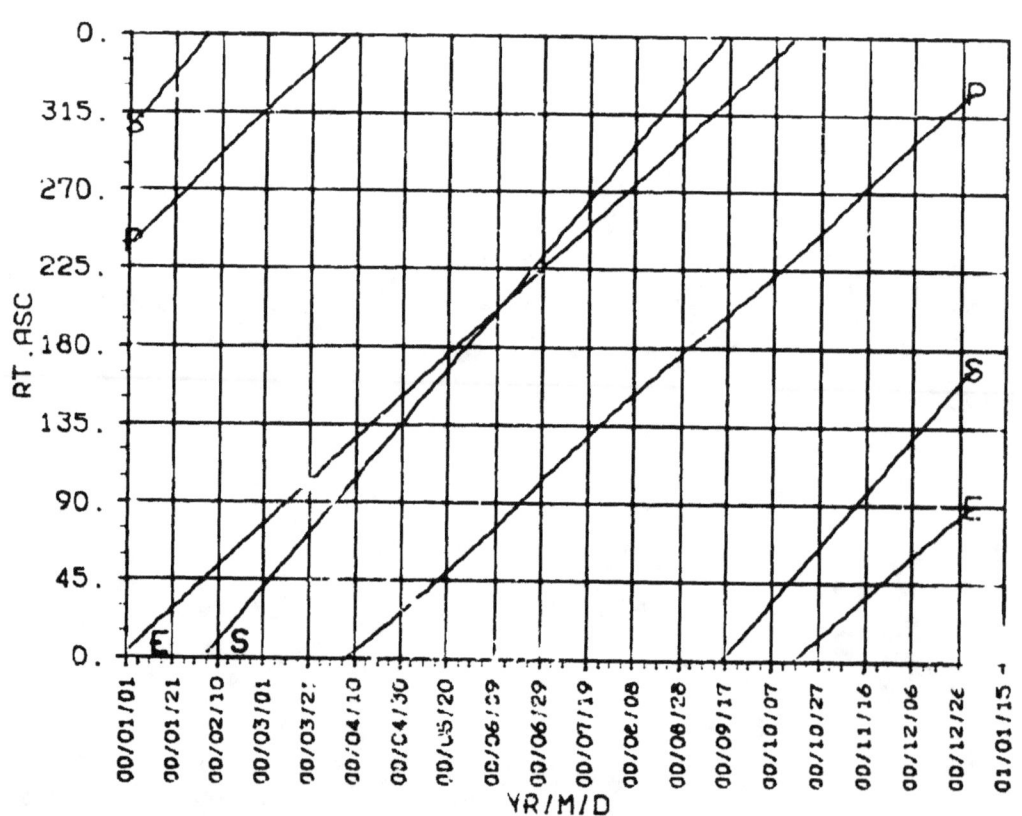

VENUS 2000

DISTANCE EC.LON 2000

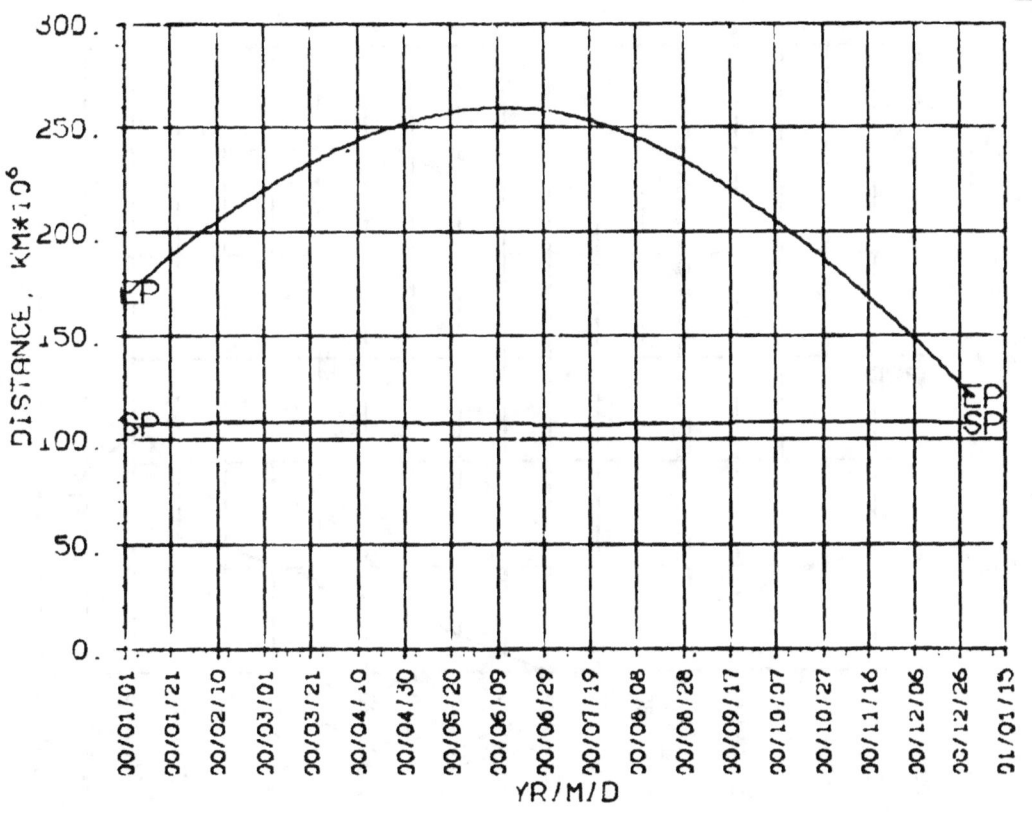

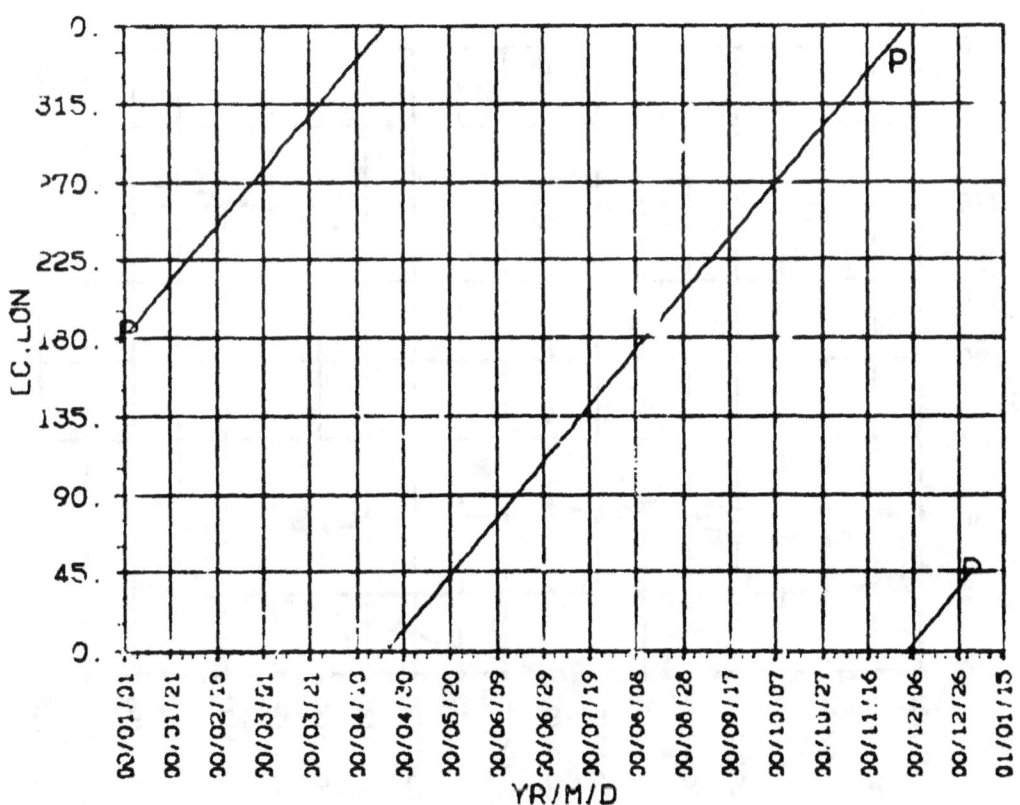

SEP, ESP CA, KA 2000

VENUS 2000

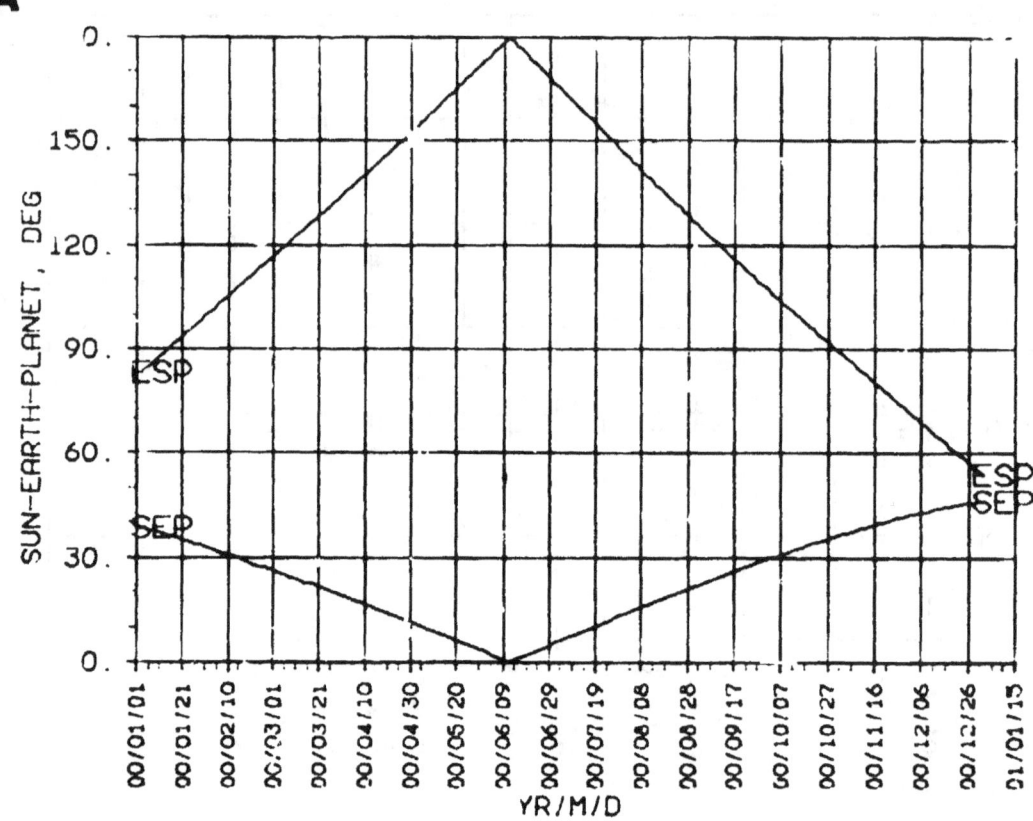

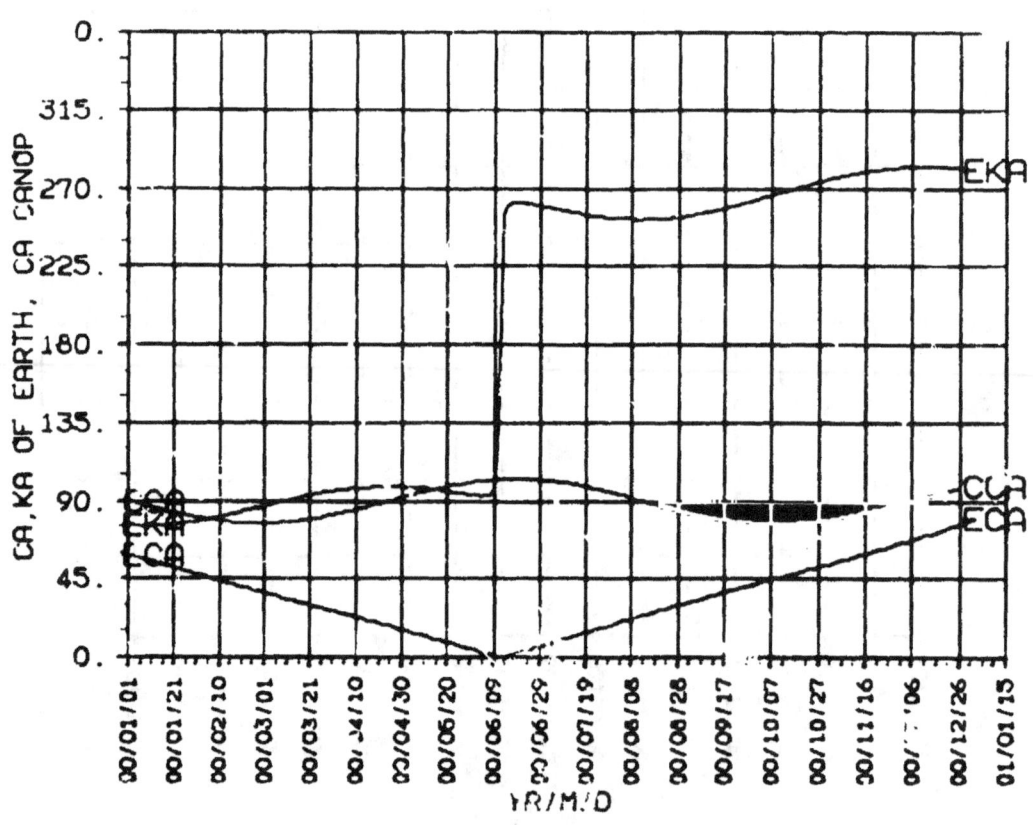

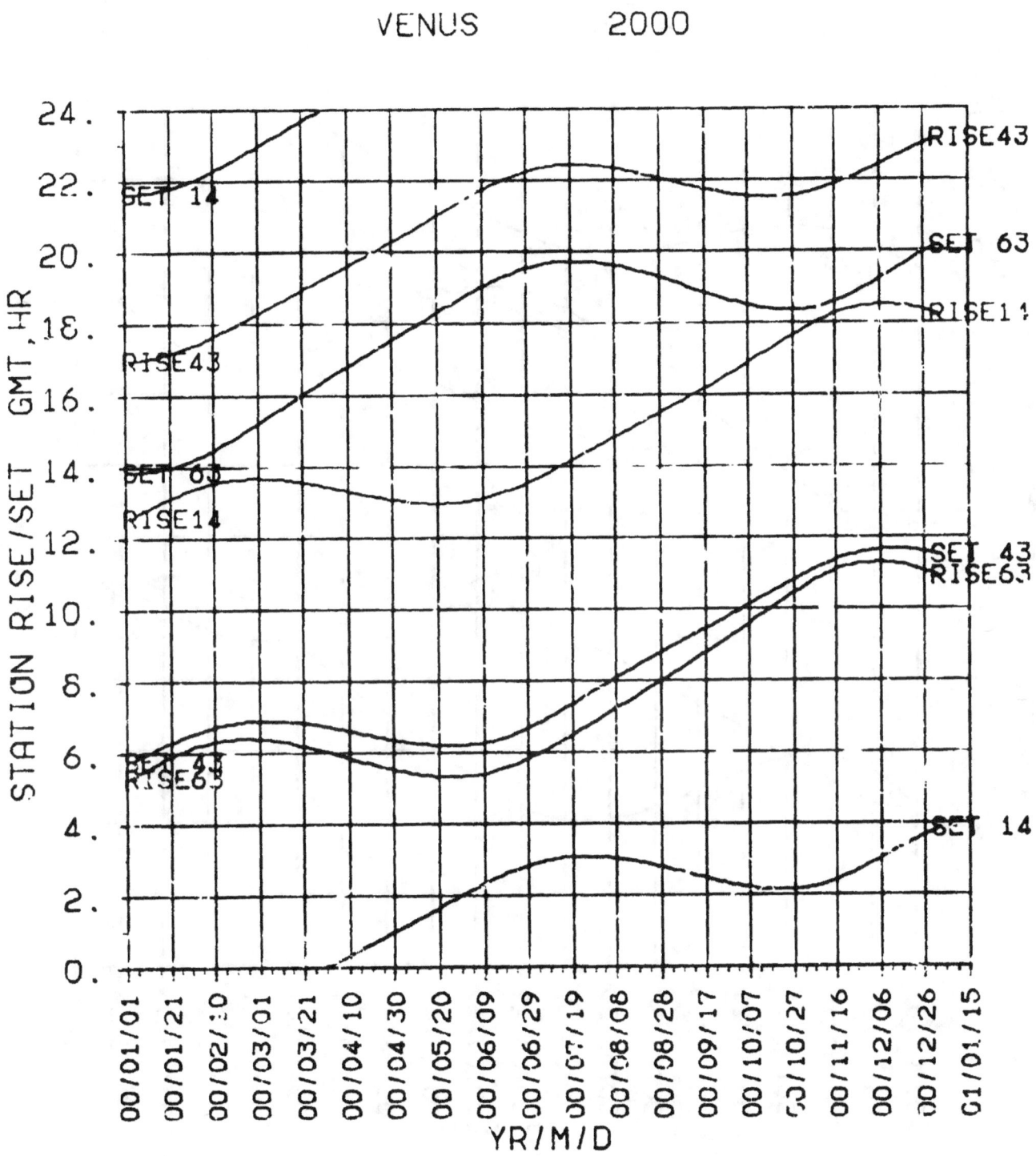

**STA R/S
NON-DSN
2000**

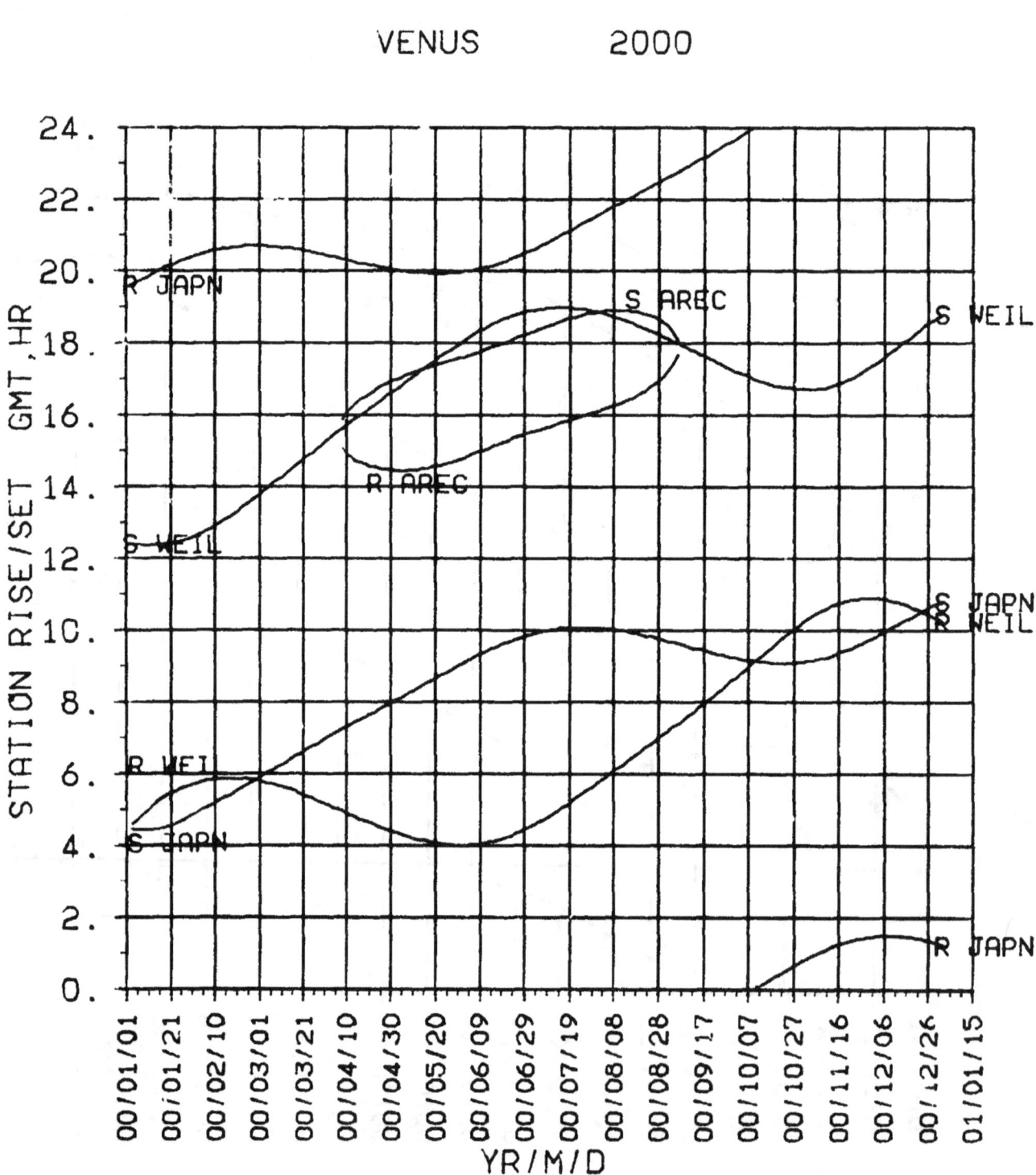

Venus

2001

DECLIN RT.ASC 2001

VENUS 2001

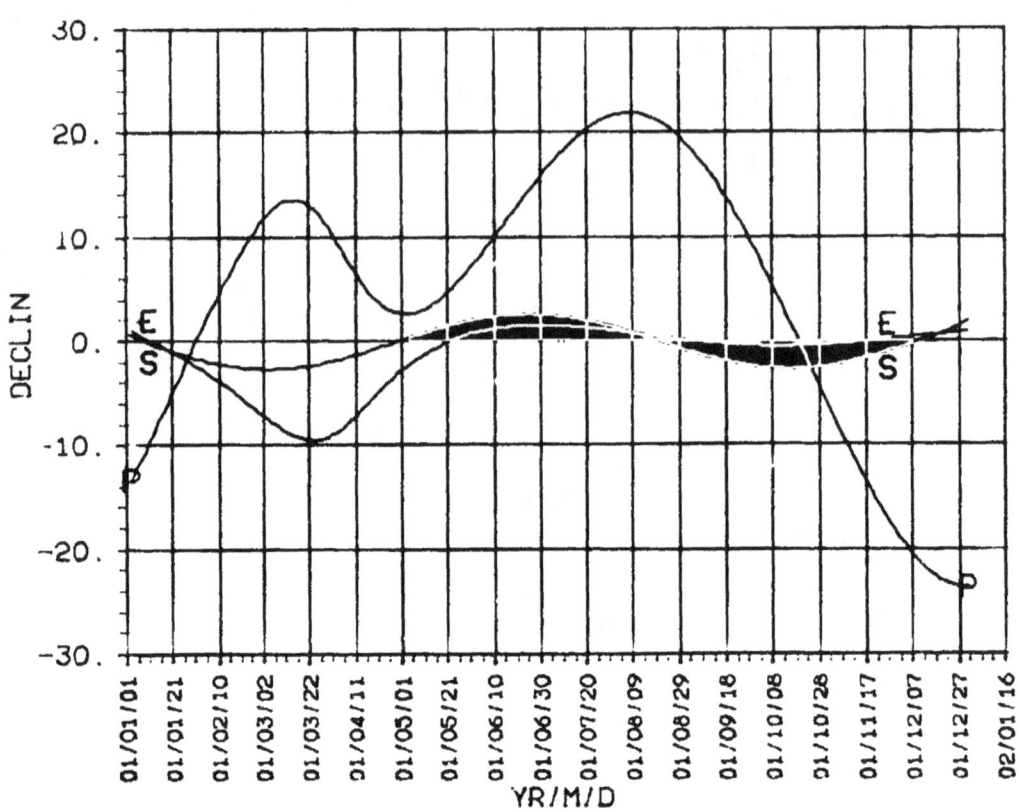

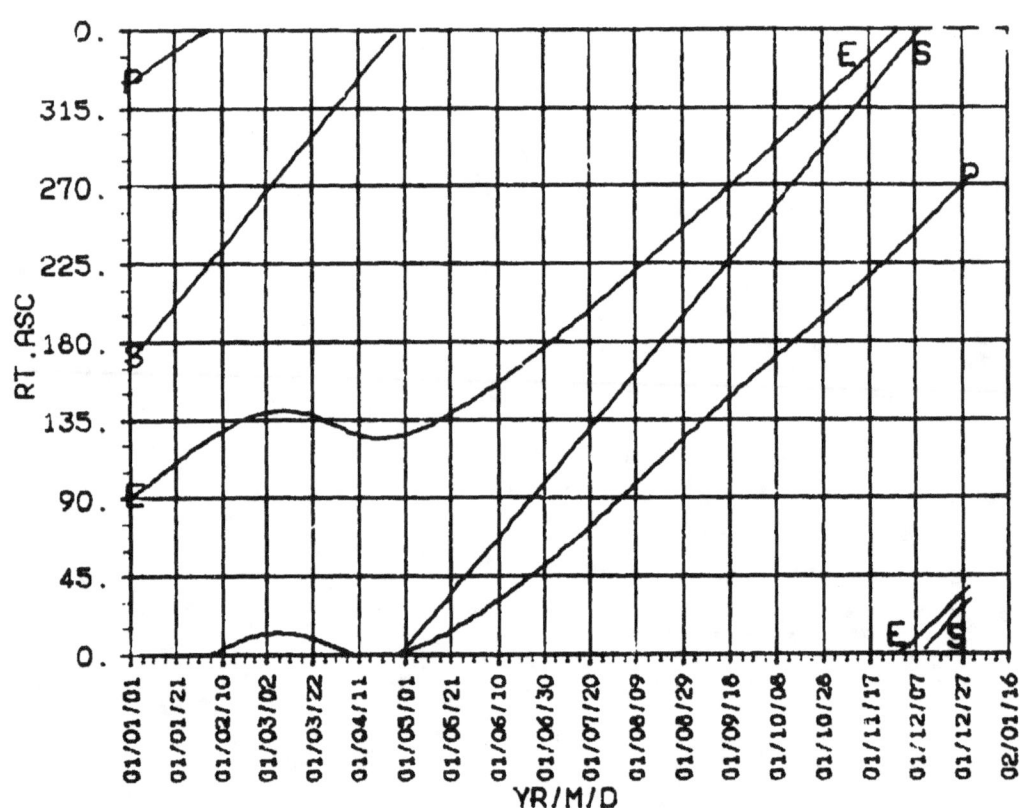

VENUS 2001

DISTANCE EC.LON 2001

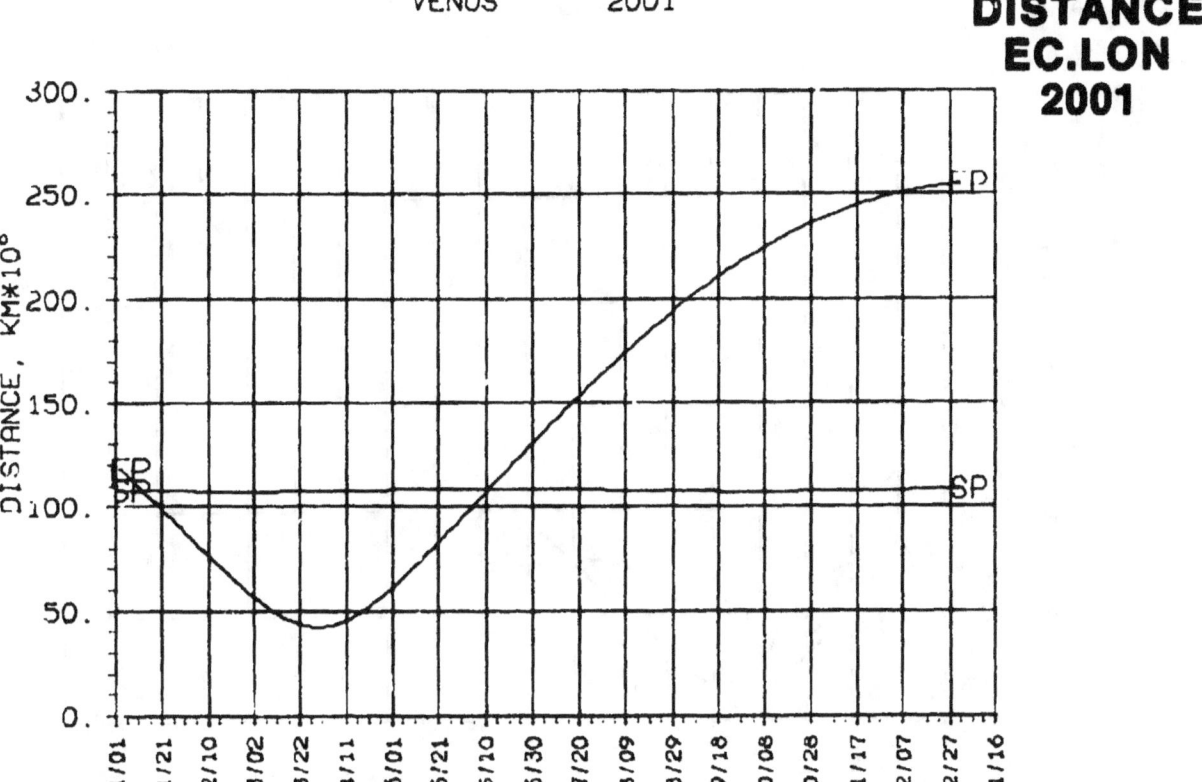

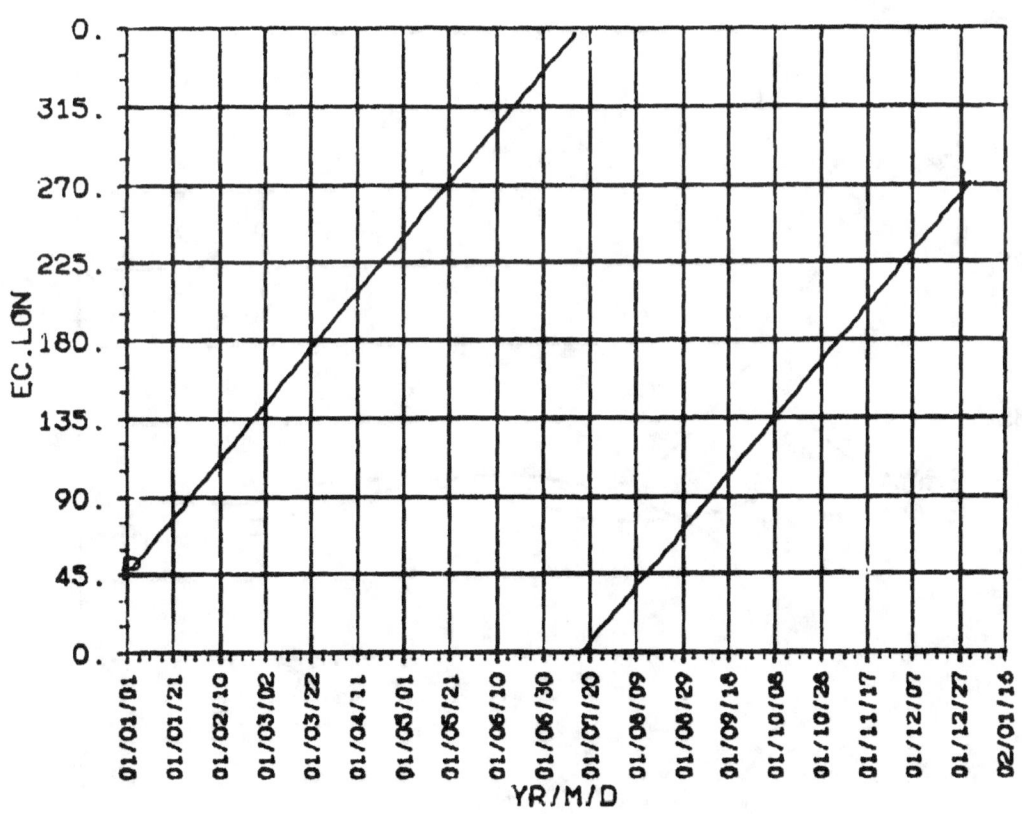

**SEP, ESP
CA, KA
2001**

VENUS 2001

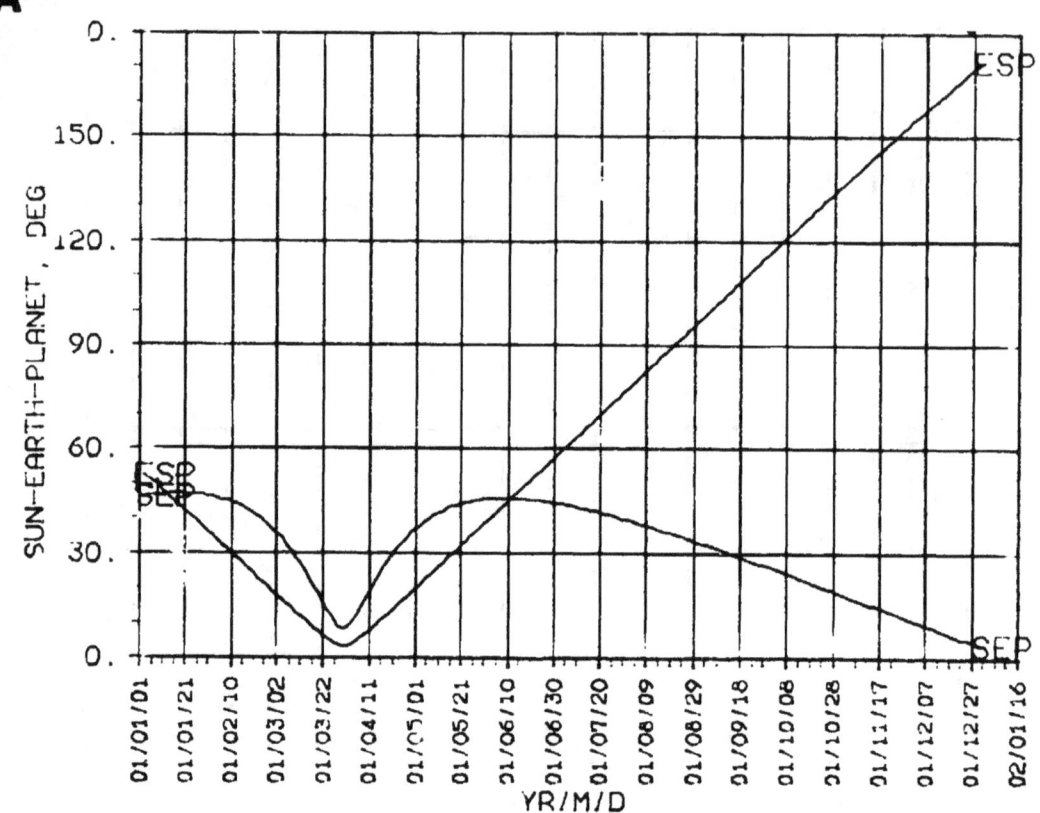

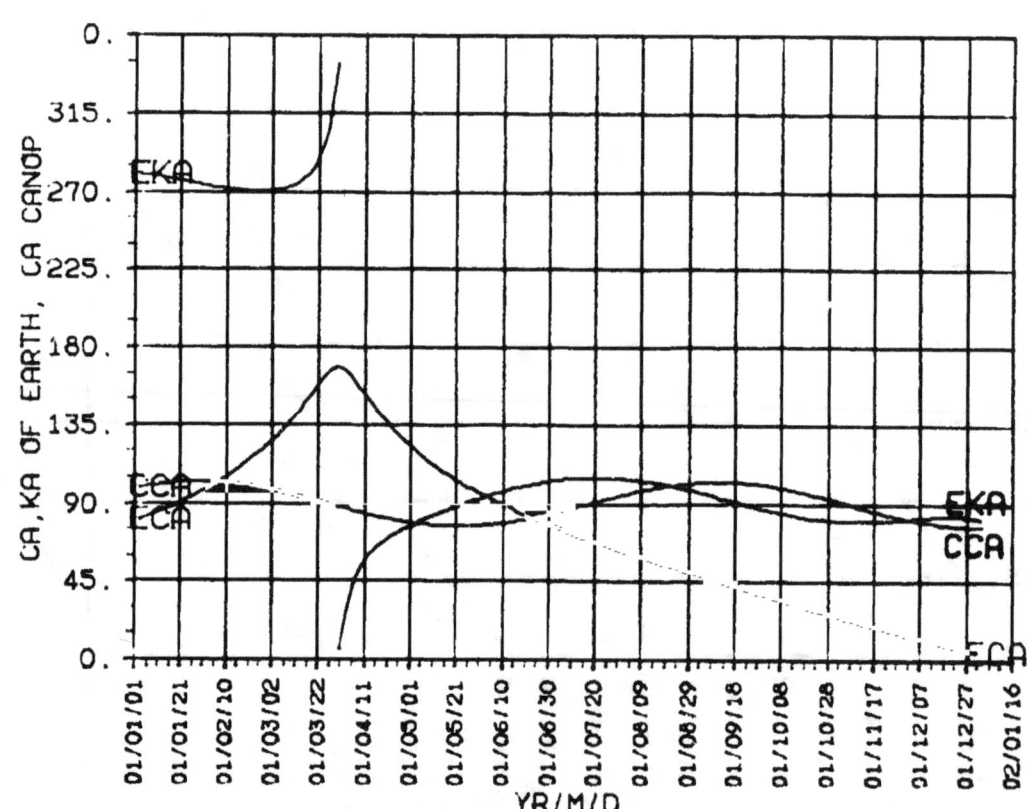

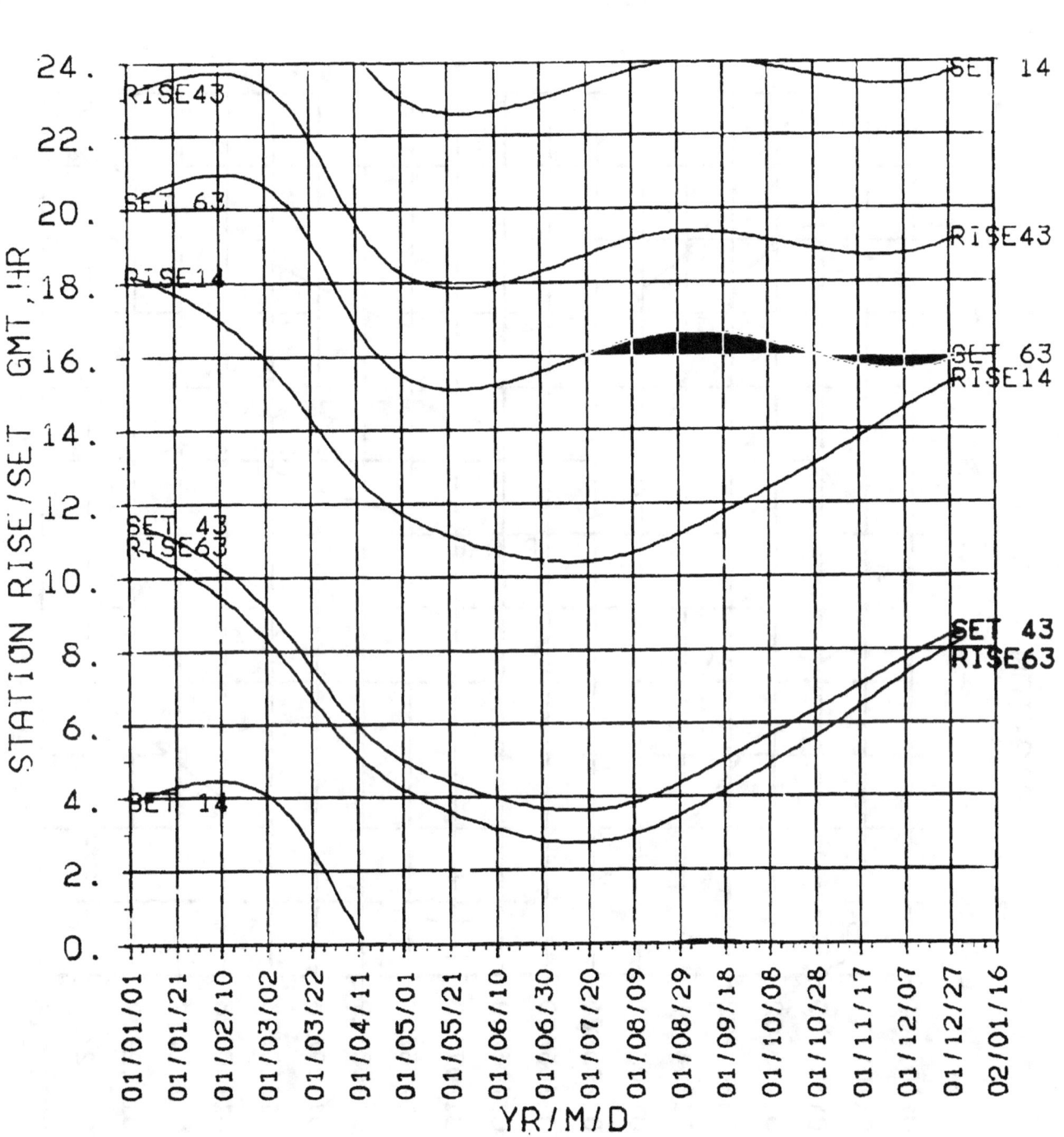

**STA R/S
NON-DSN
2001**

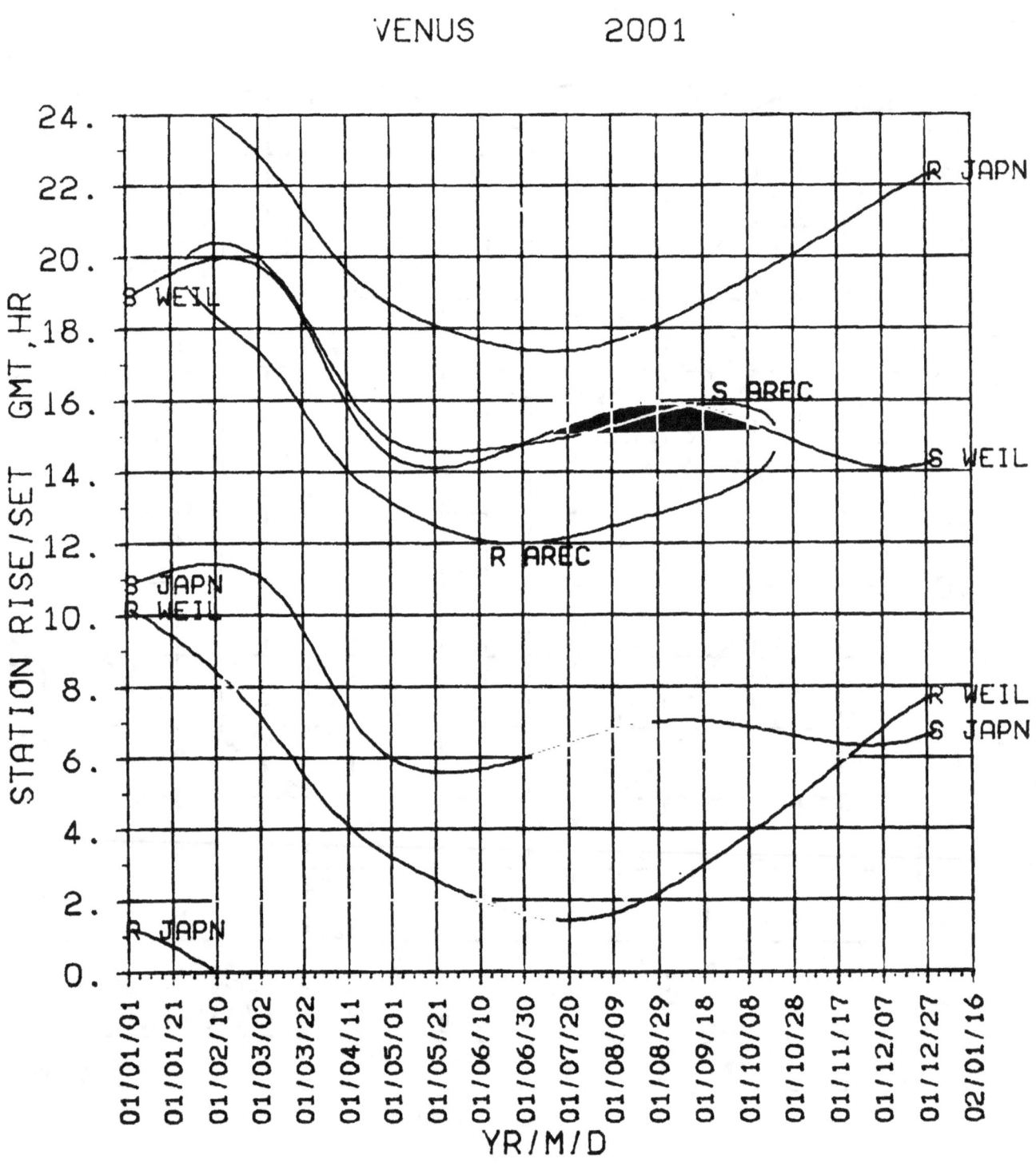

Venus

2002

DECLIN RT.ASC 2002

VENUS 2002

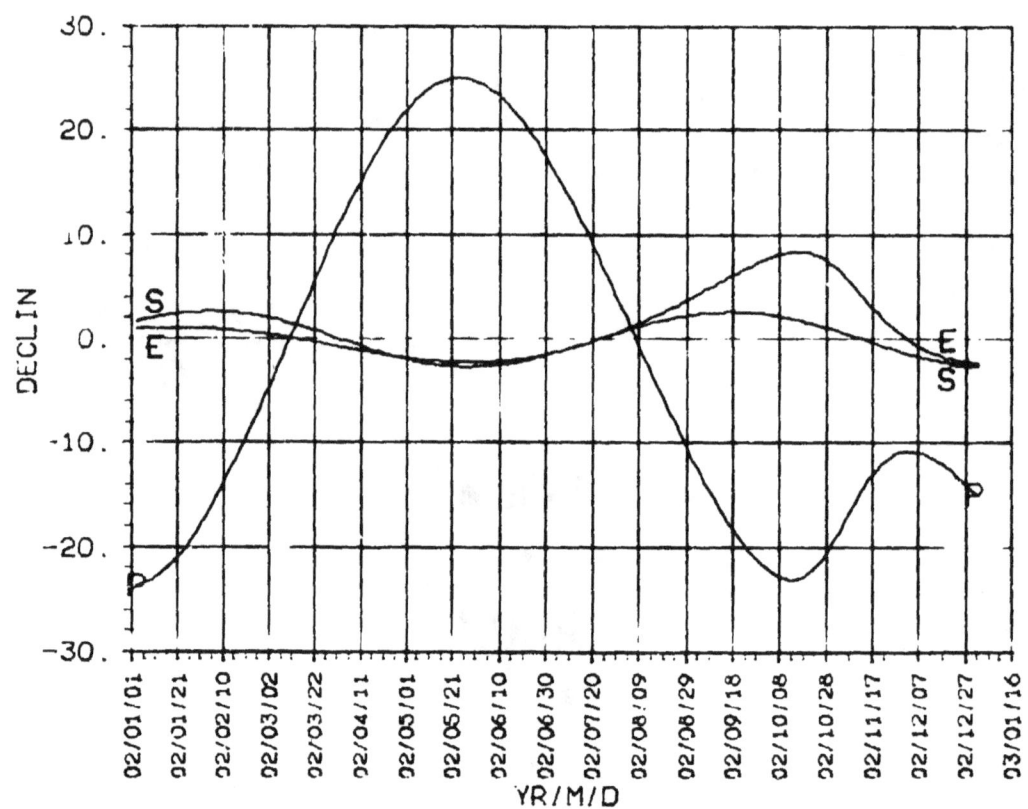

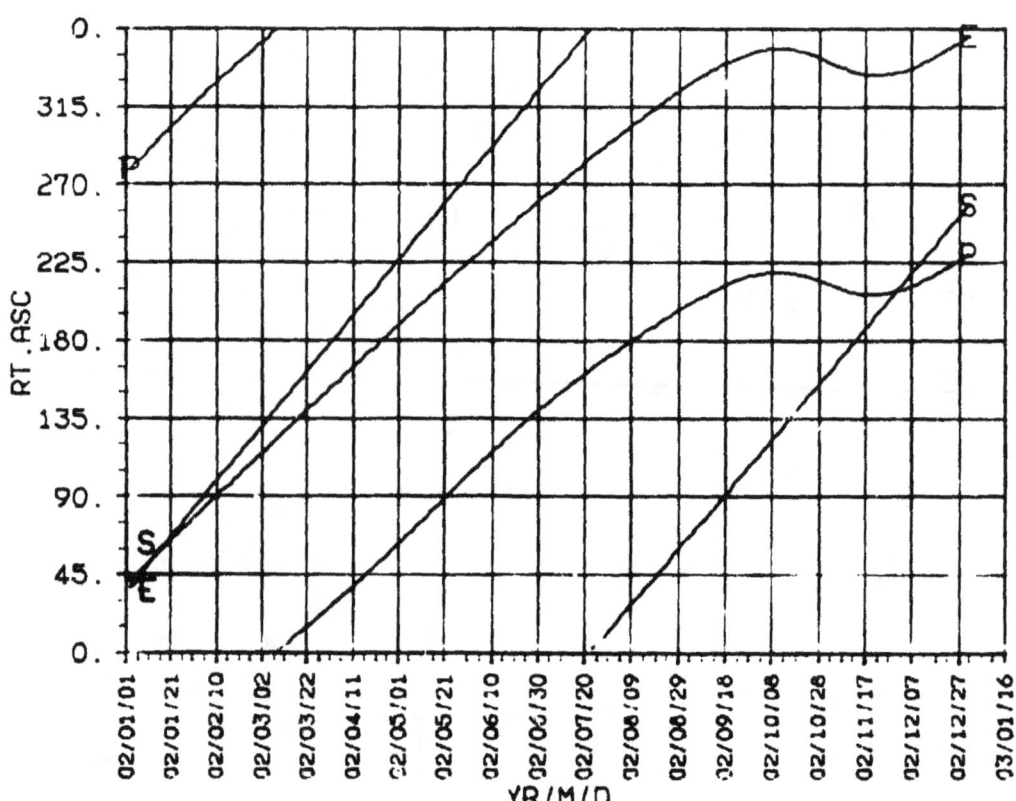

VENUS 2002

**DISTANCE
EC.LON
2002**

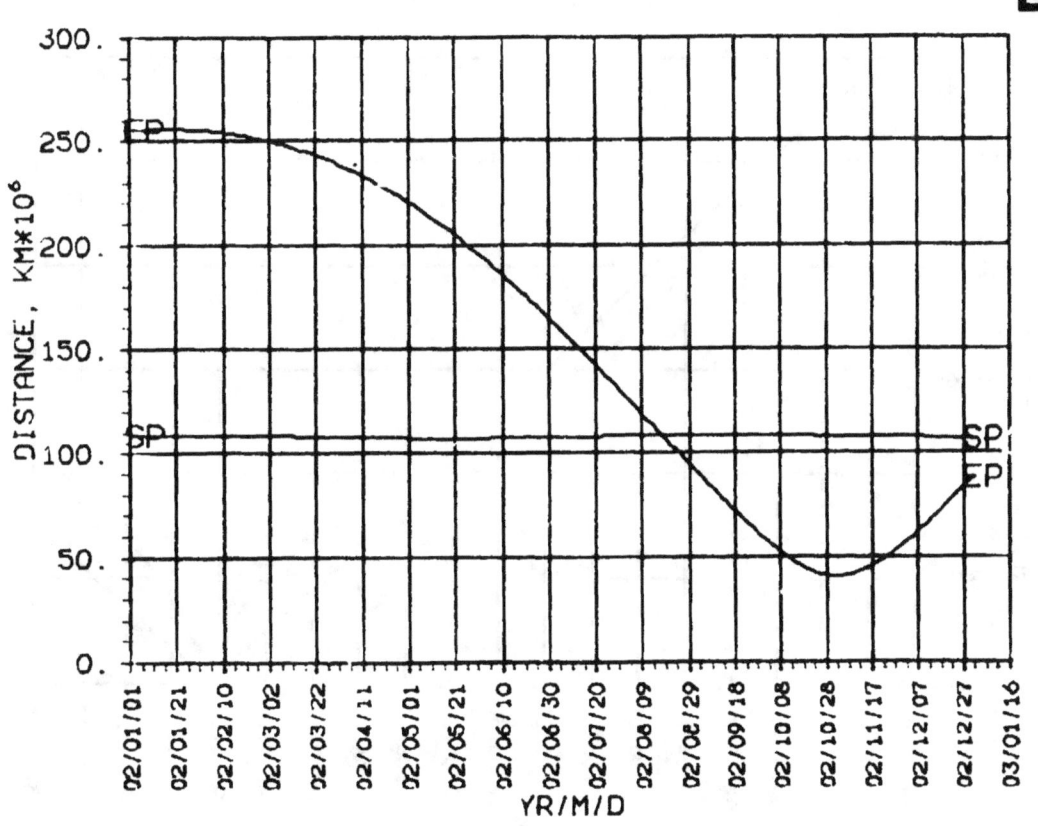

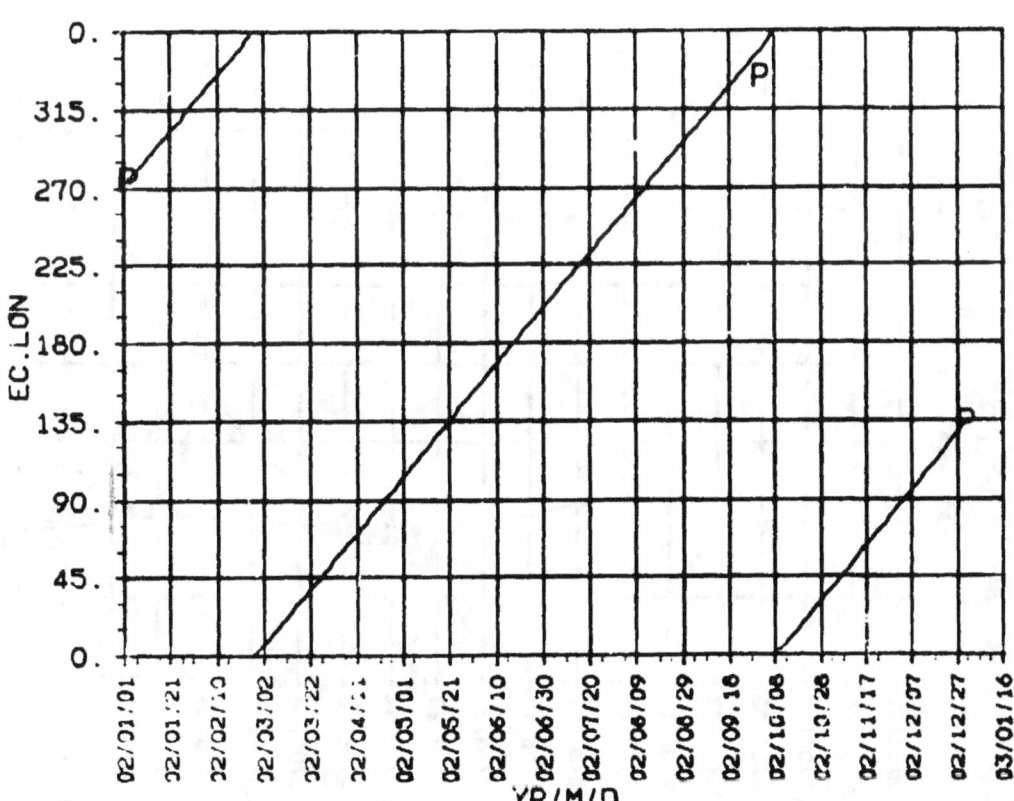

SEP, ESP CA, KA 2002

VENUS 2002

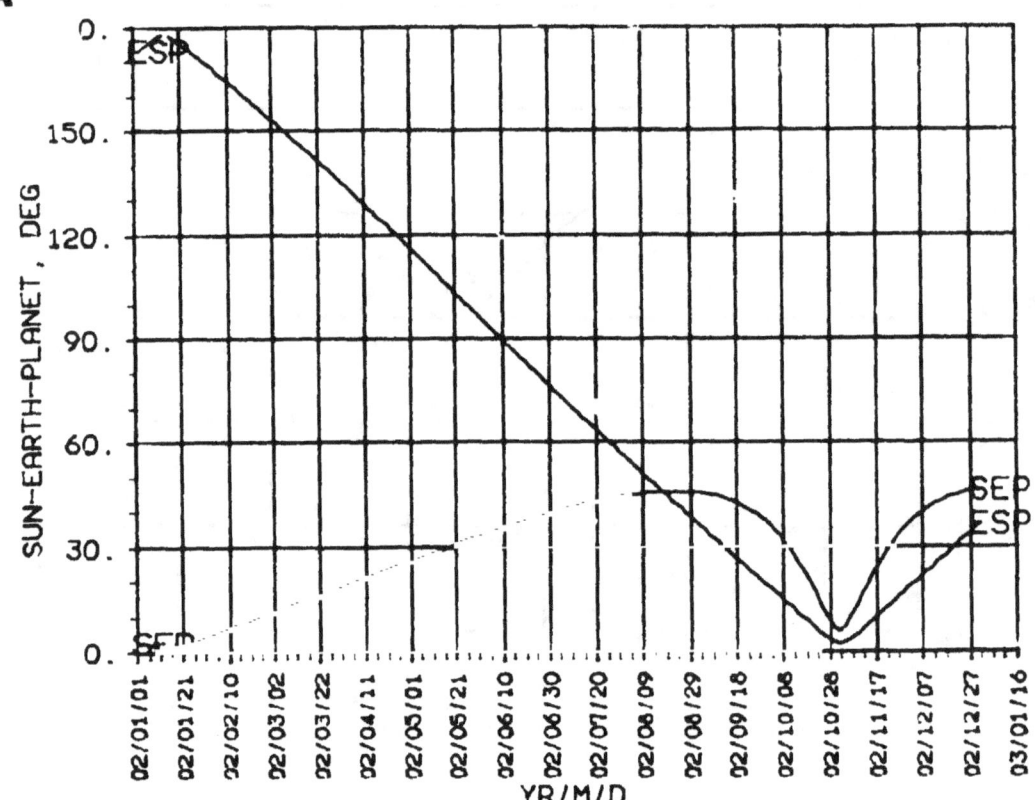

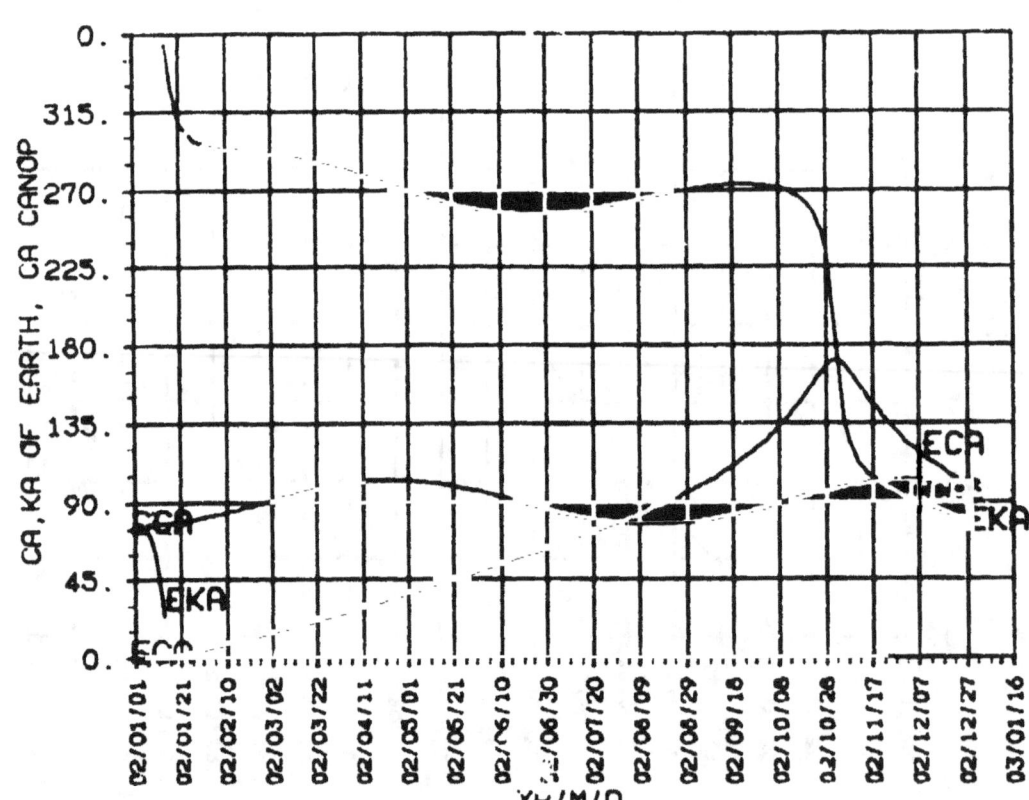

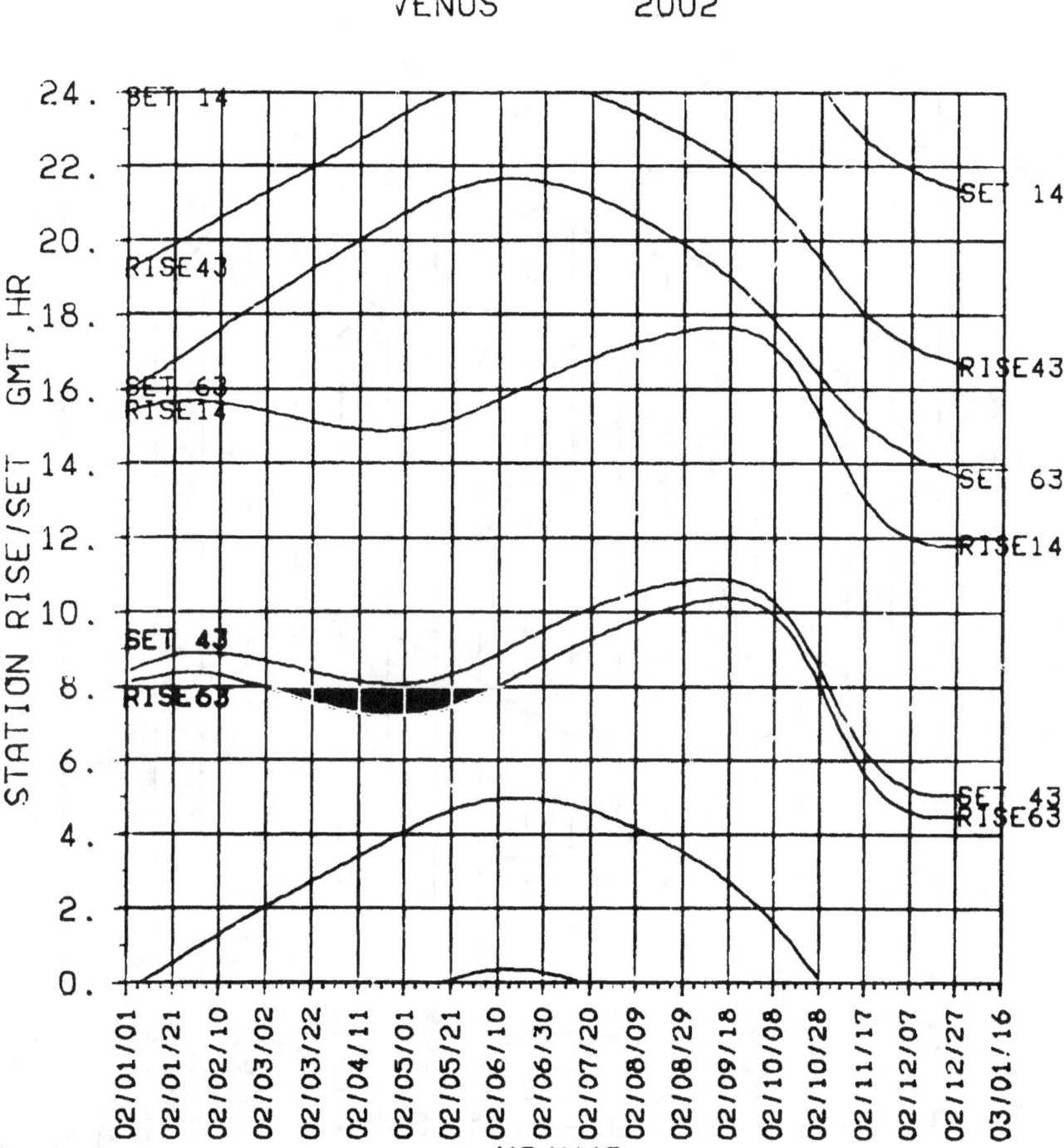

**STA R/S
NON-DSN
2002**

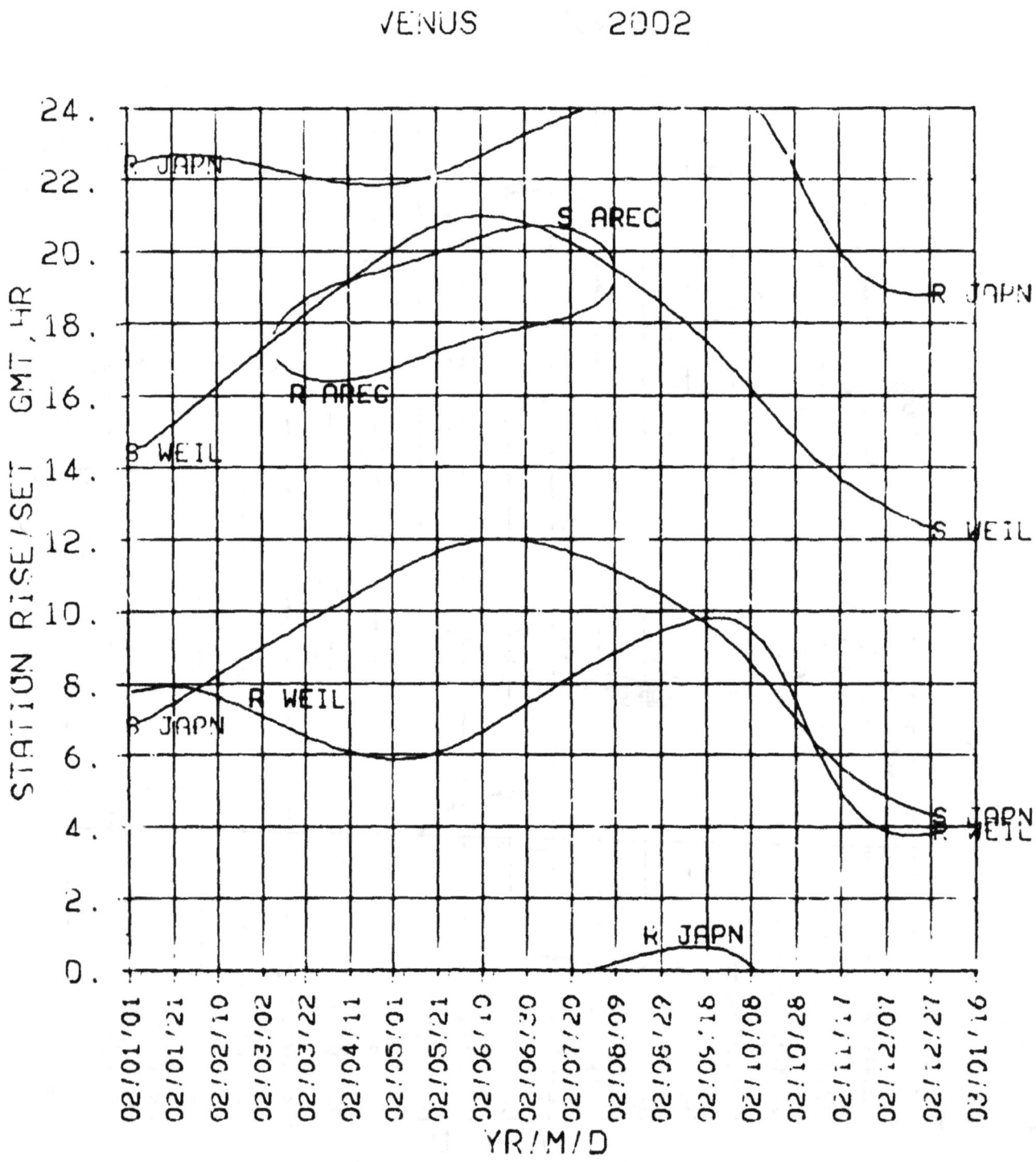

Venus

2003

DECLIN RT.ASC 2003

VENUS 2003

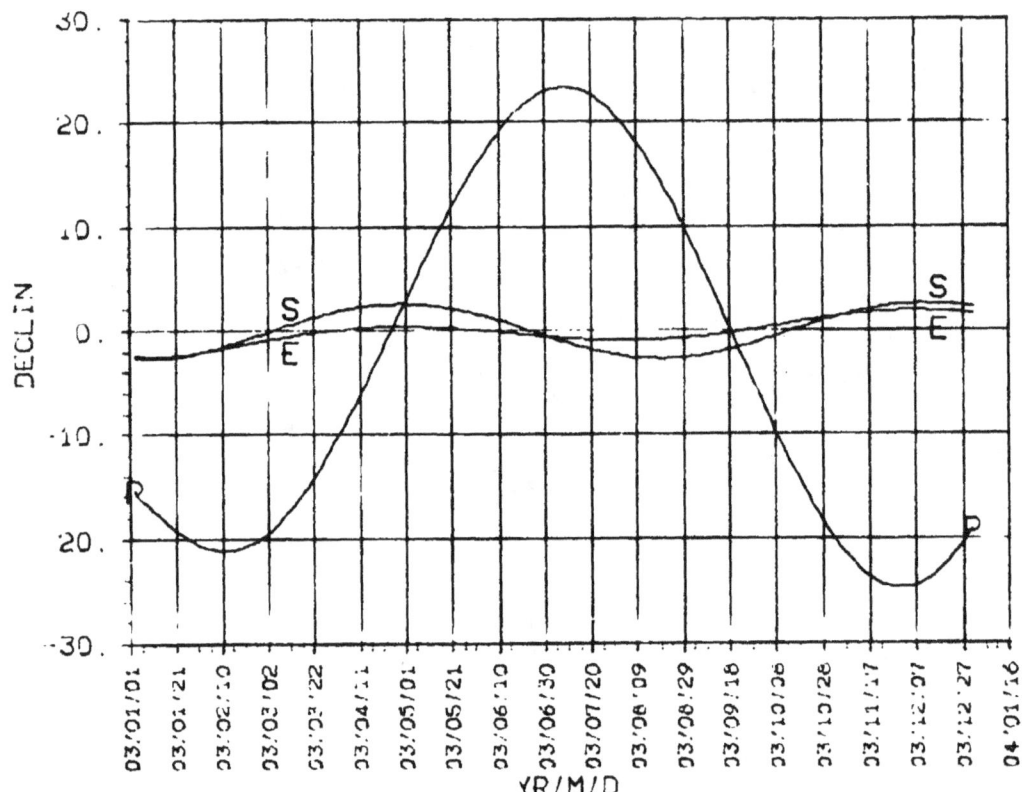

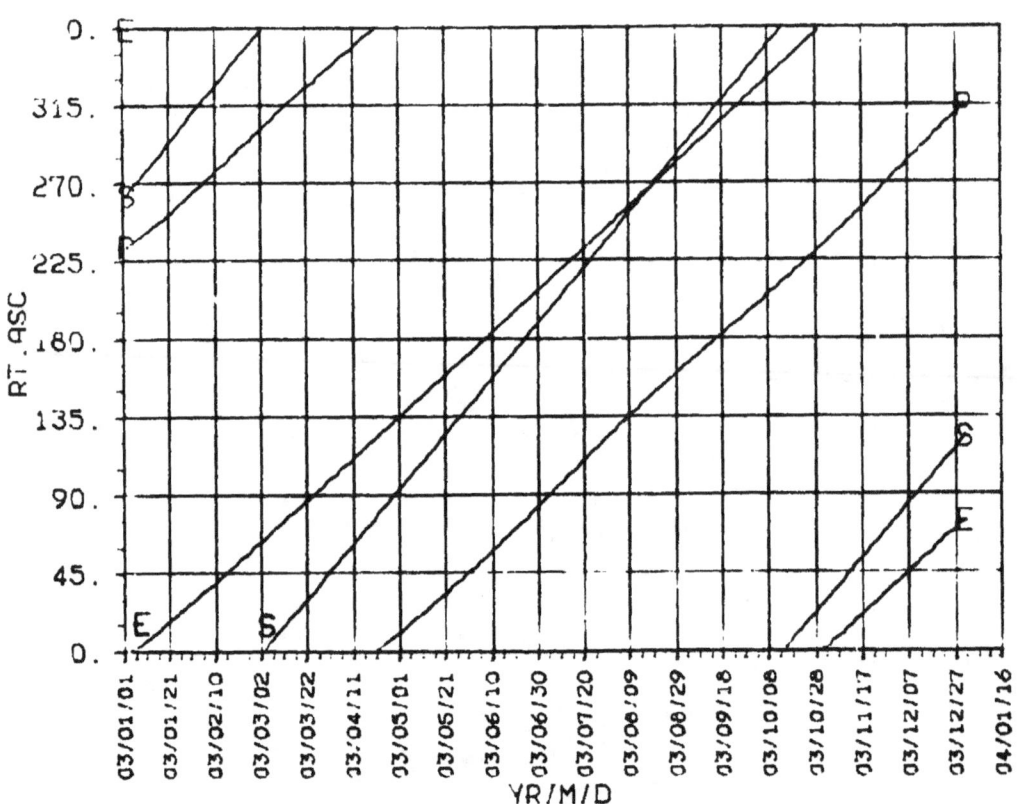

VENUS 2003

DISTANCE EC.LON 2003

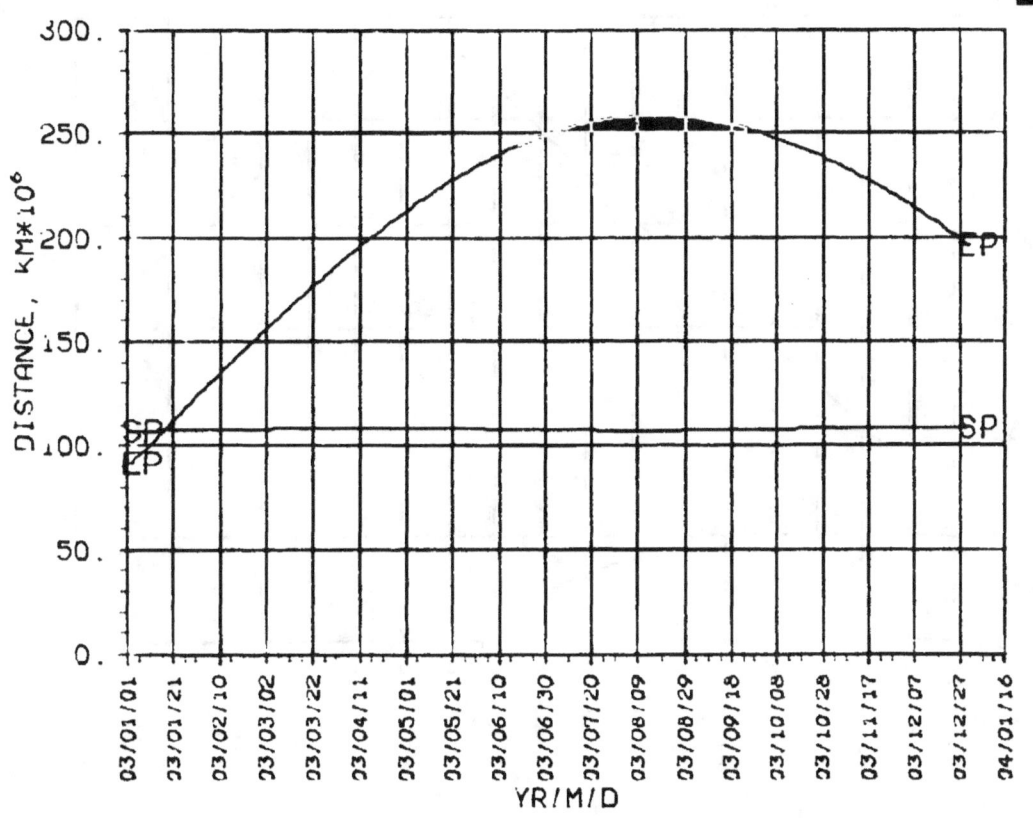

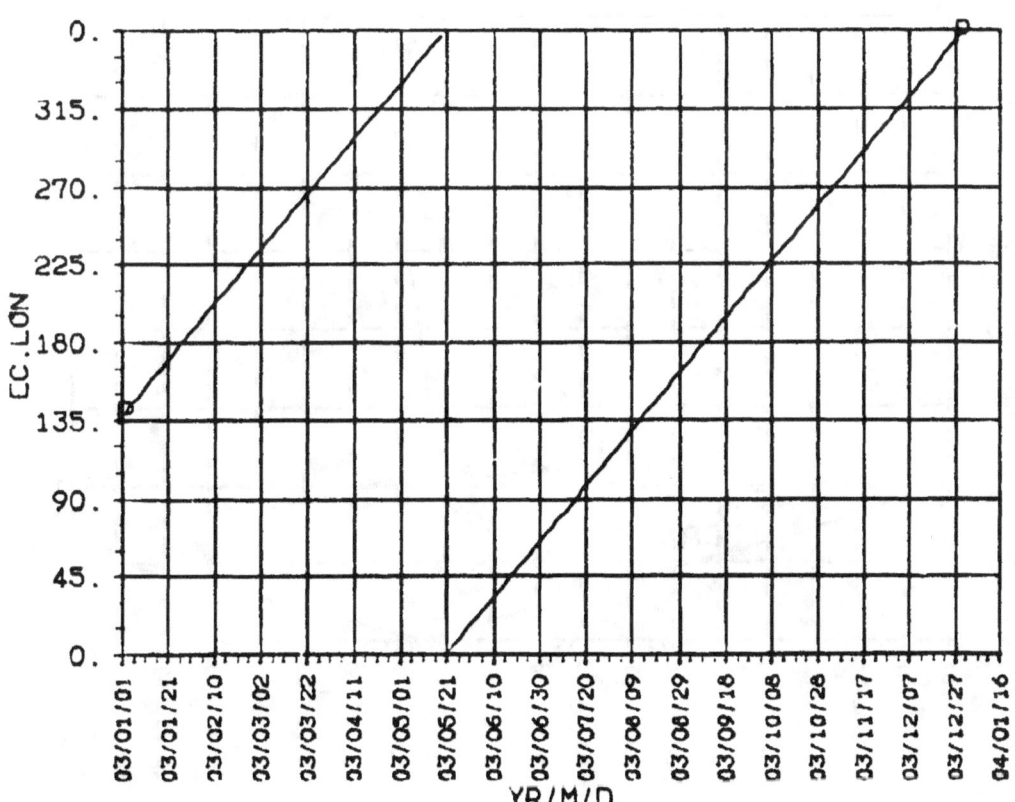

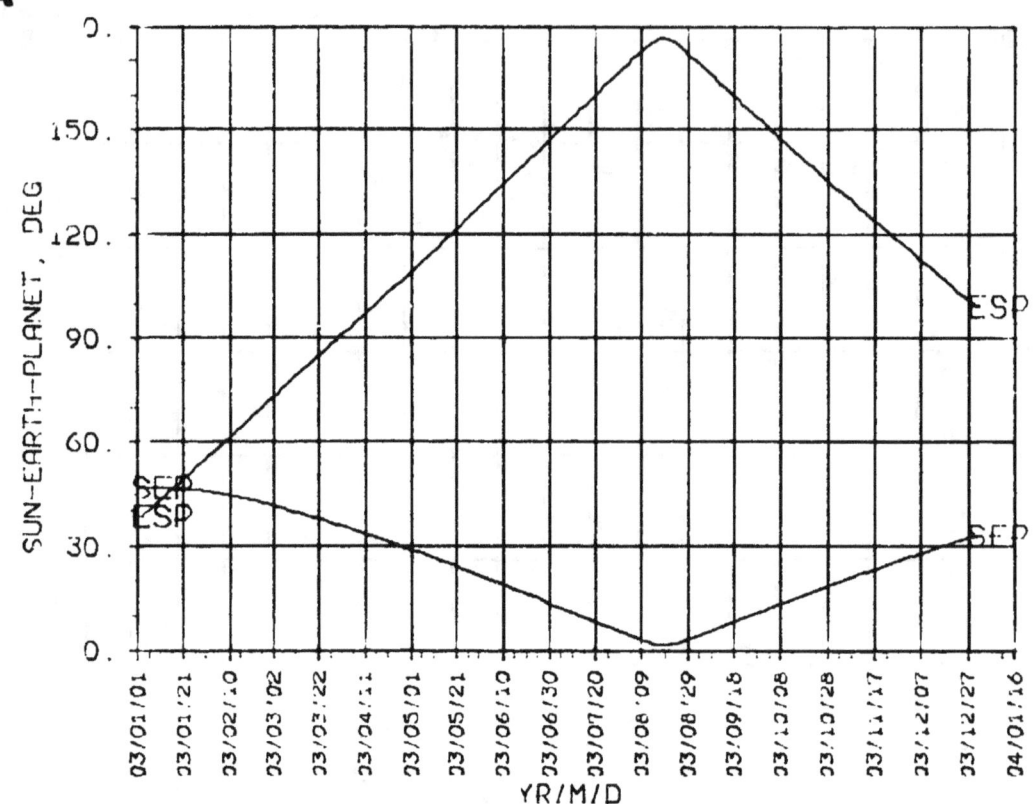
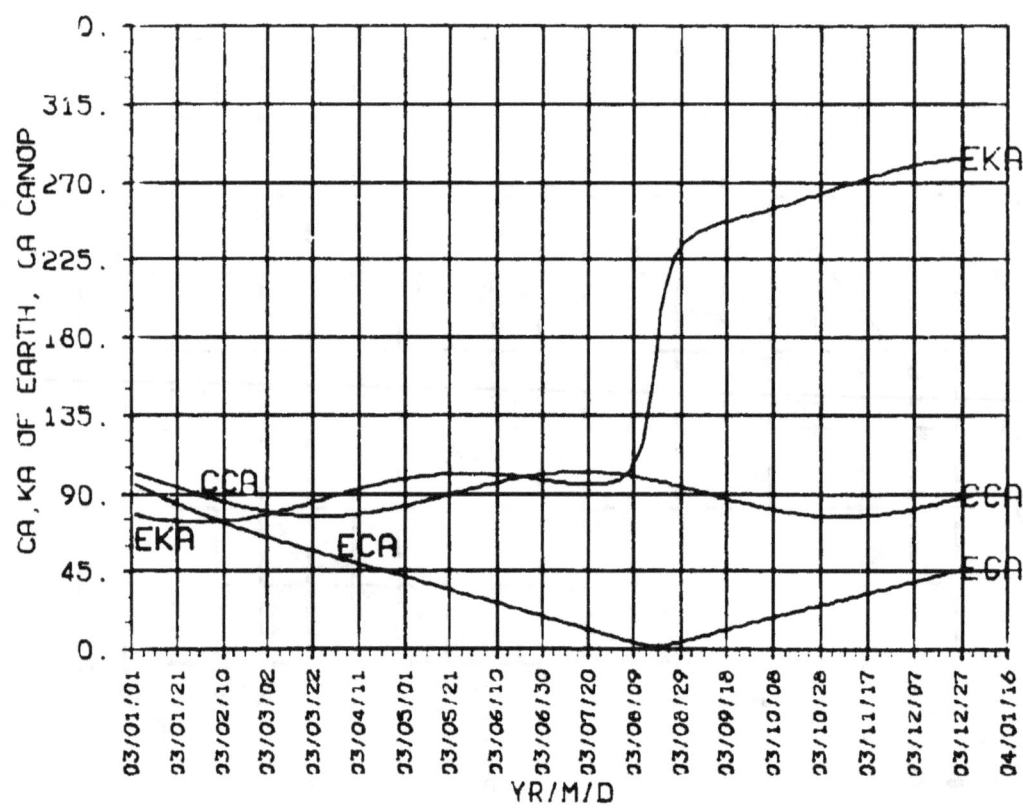

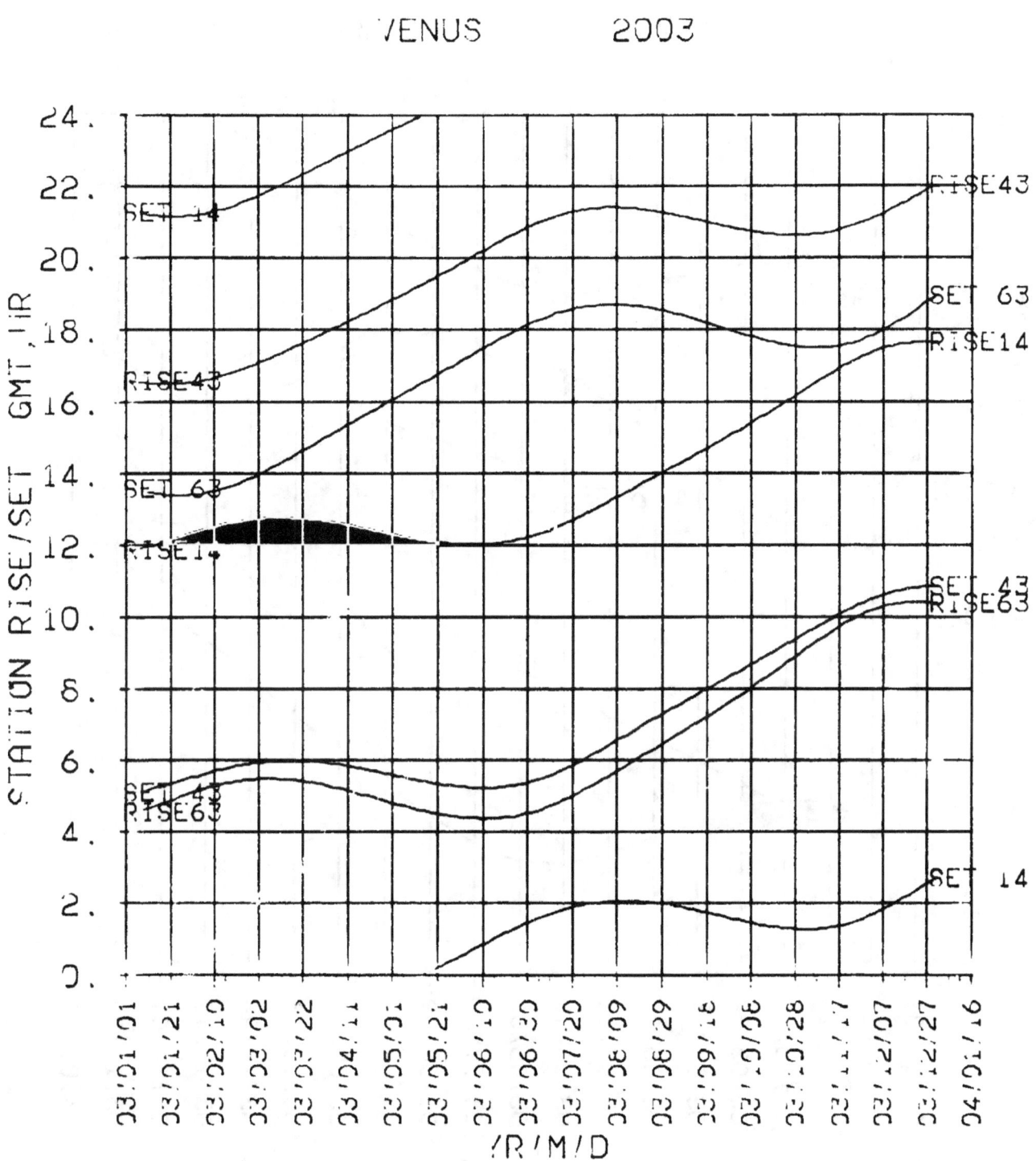

STA R/S NON-DSN 2003

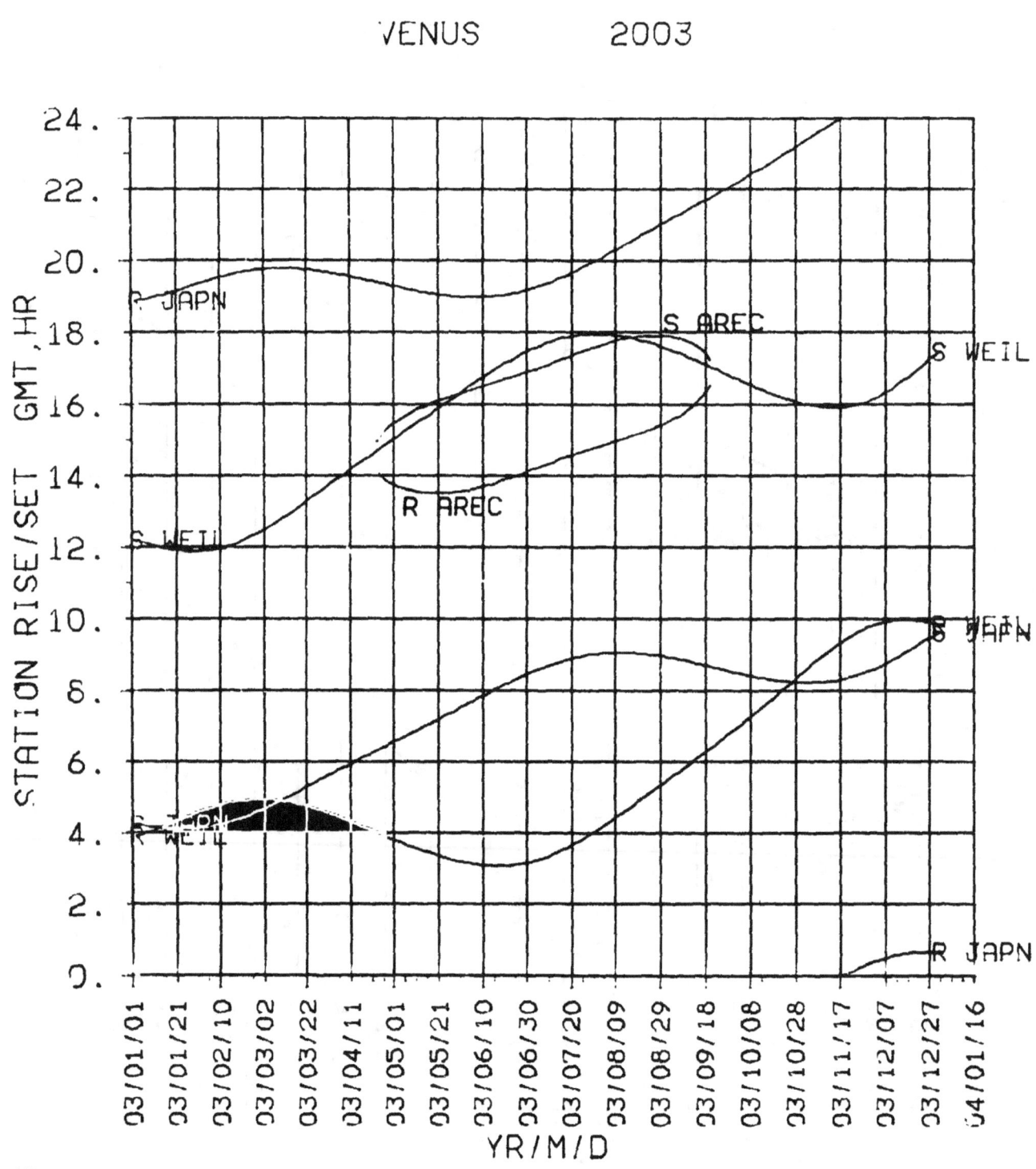

Venus

2004

DECLIN RT.ASC 2004

VENUS 2004

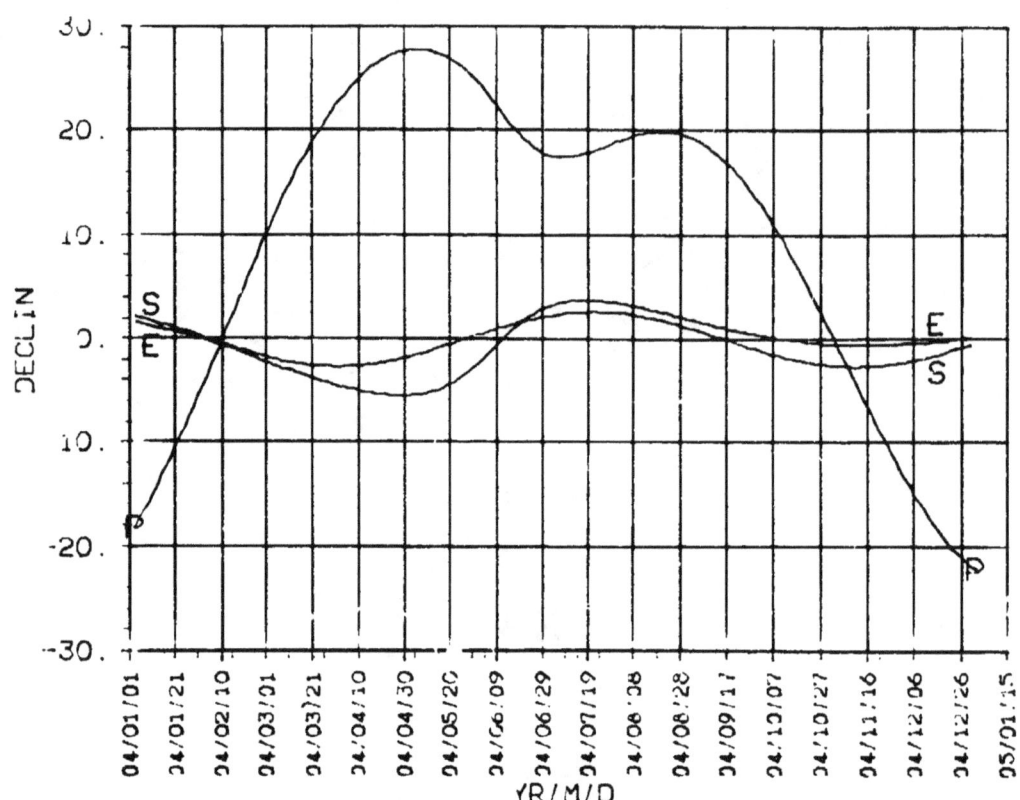

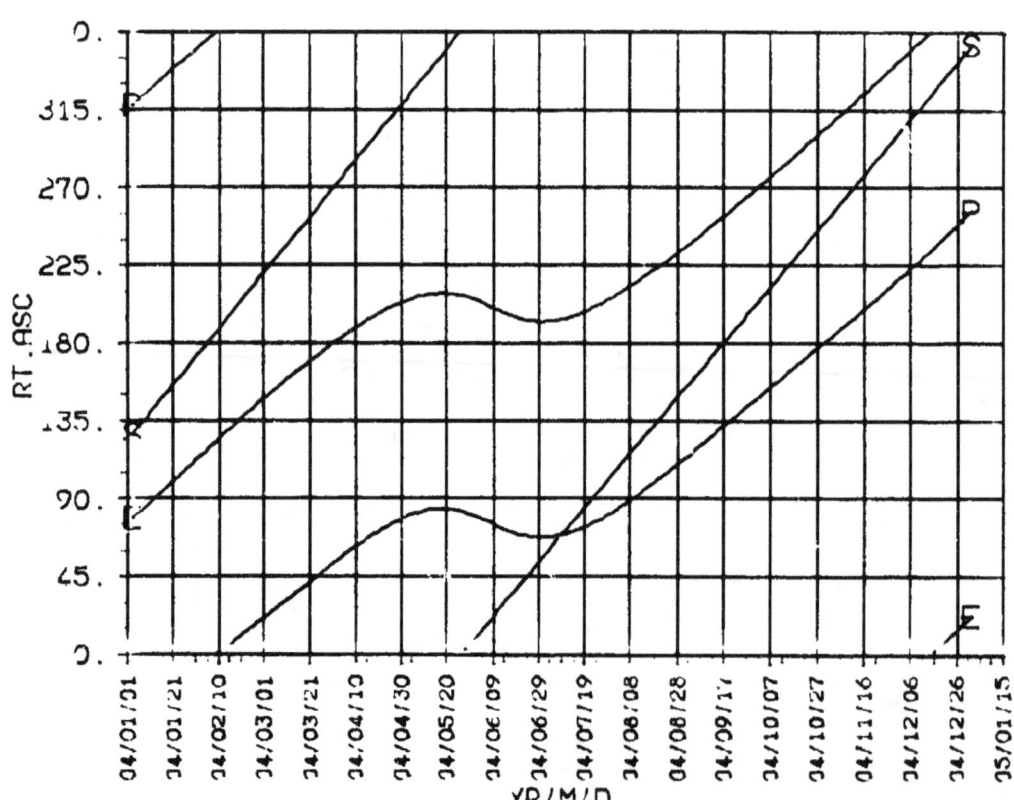

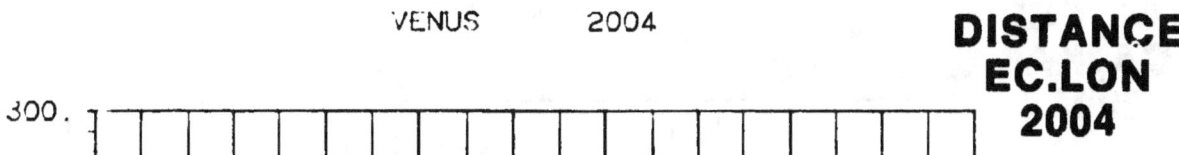
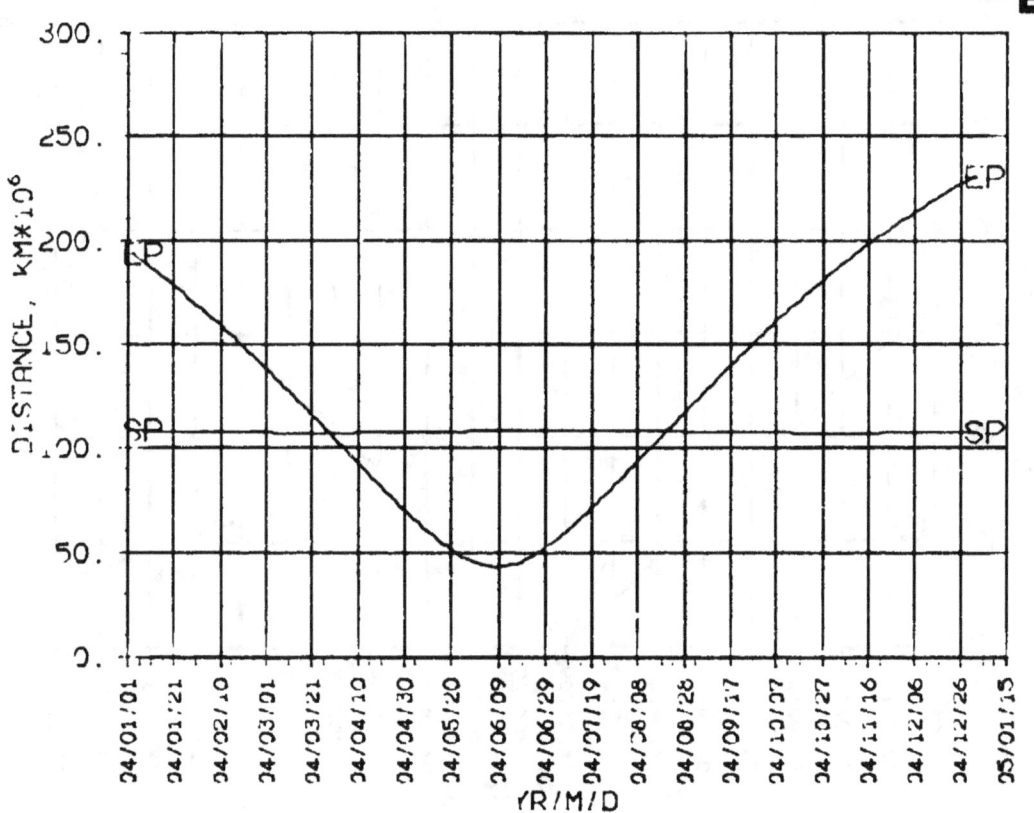
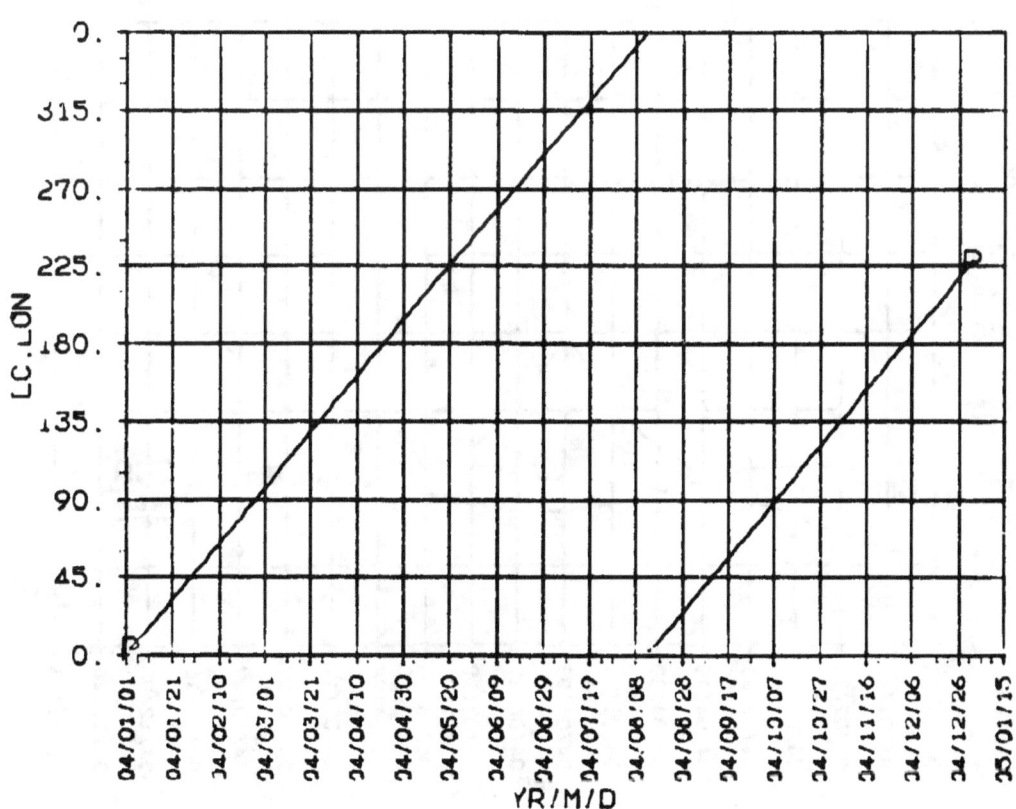

**SEP, ESP
CA, KA
2004**

VENUS   2004

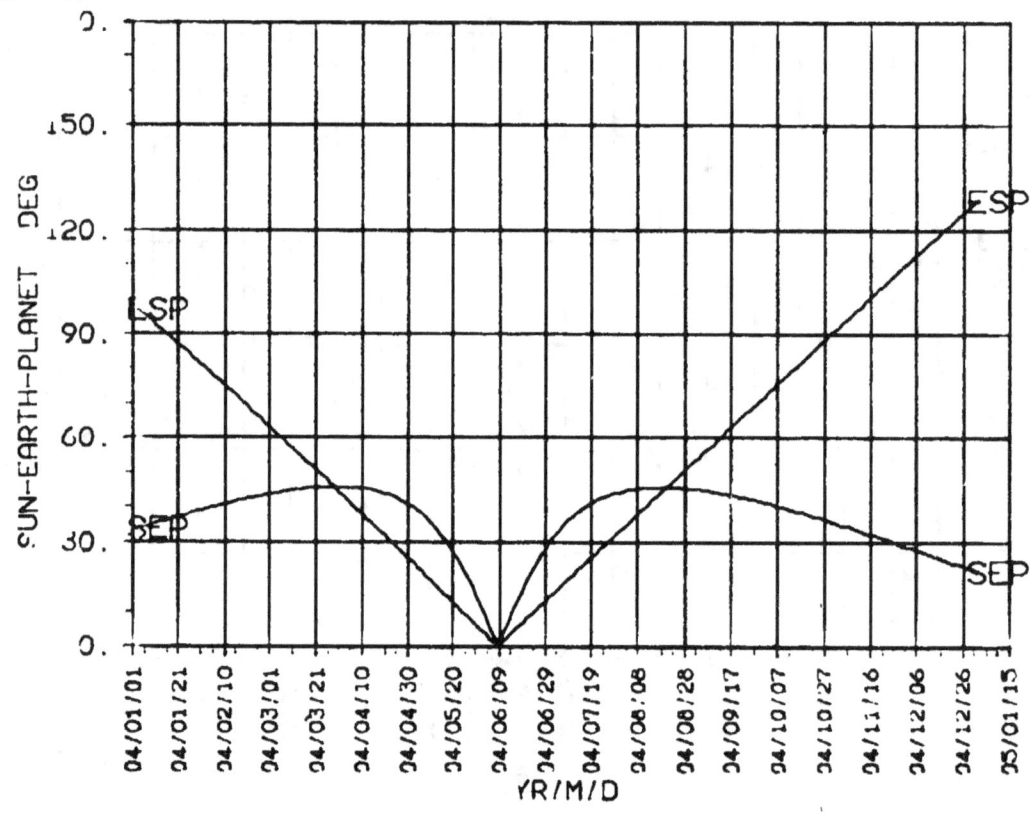

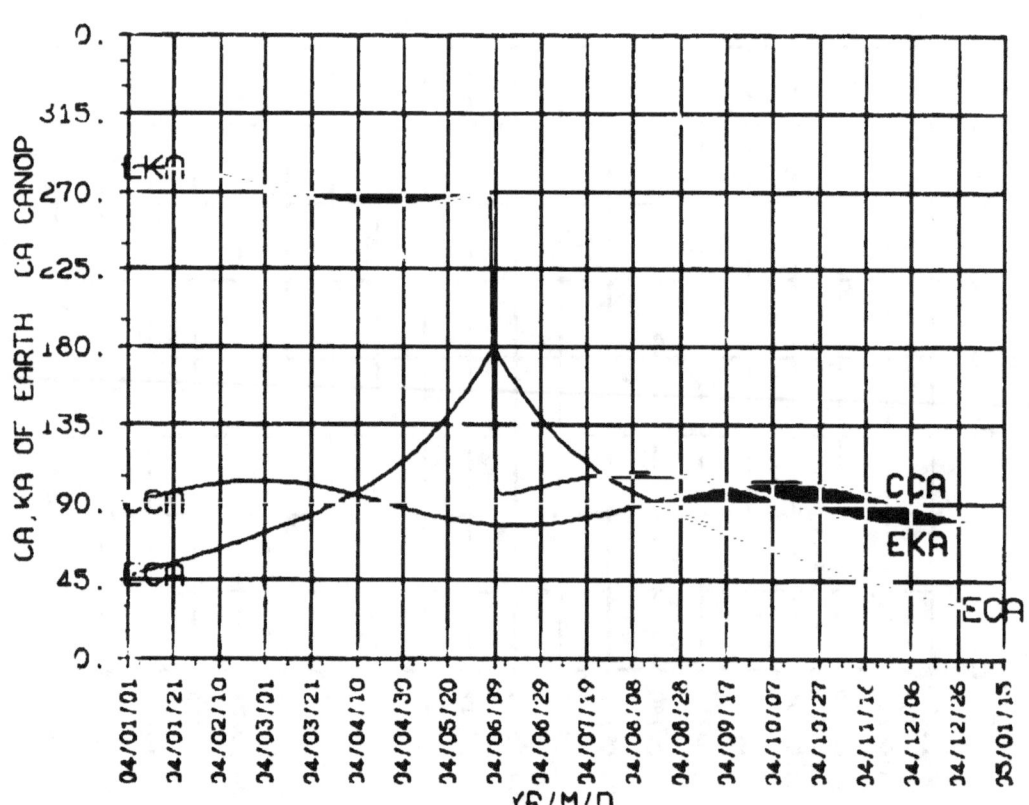

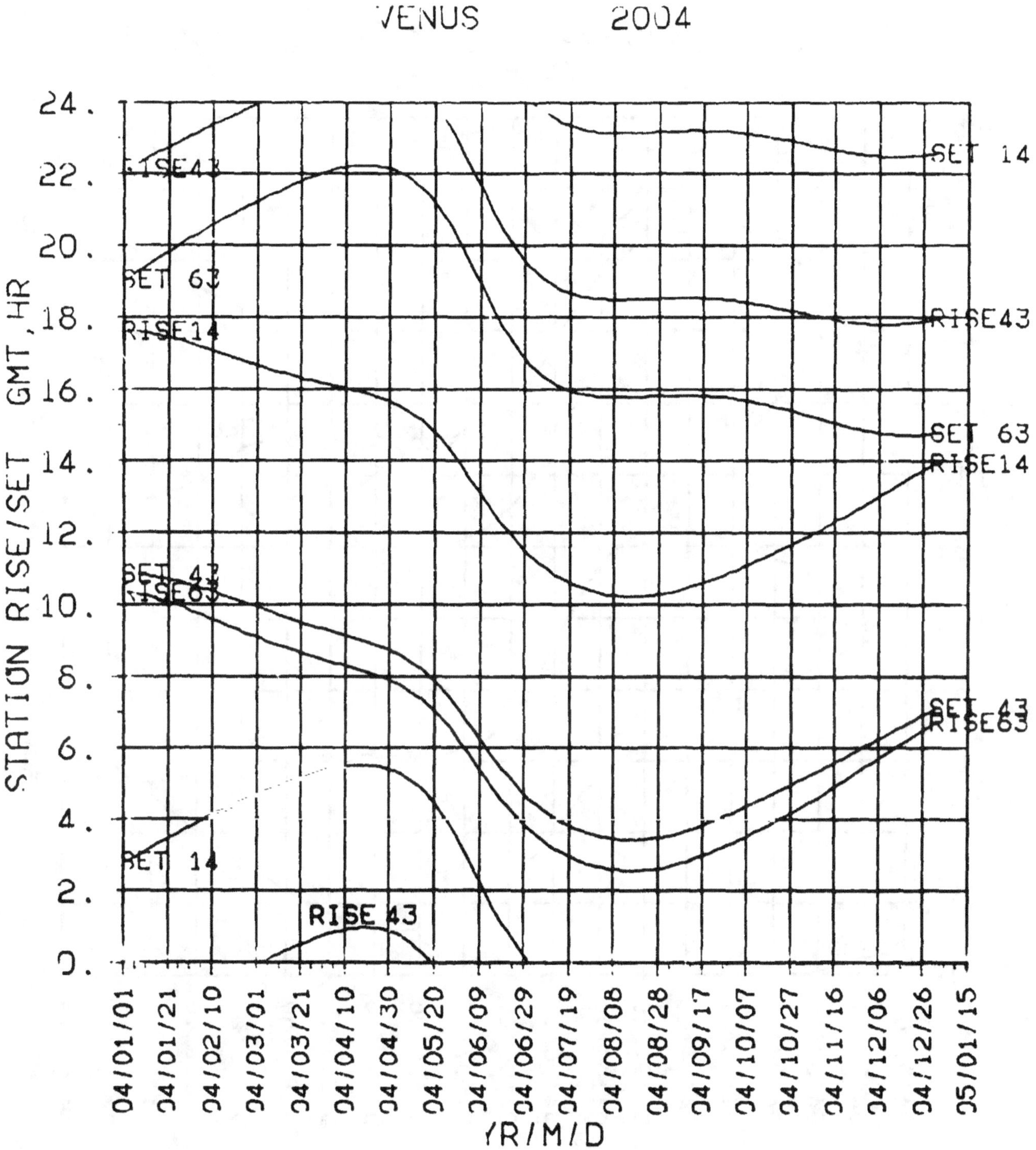
VENUS 2004
STATION RISE/SET GMT,HR
YR/M/D
24.
22.
20.
18.
16.
14.
12.
10.
8.
6.
4.
2.
0.
04/01/01
04/01/21
04/02/10
04/03/01
04/03/21
04/04/10
04/04/30
04/05/20
04/06/09
04/06/29
04/07/19
04/08/08
04/08/28
04/09/17
04/10/07
04/10/27
04/11/16
04/12/06
04/12/26
05/01/15
RISE43
SET 63
RISE14
SET 43
RISE63
SET 14
RISE 43
SET 14
RISE43
SET 63
RISE14
SET 43
RISE63

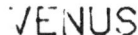

STA R/S NON-DSN 2004

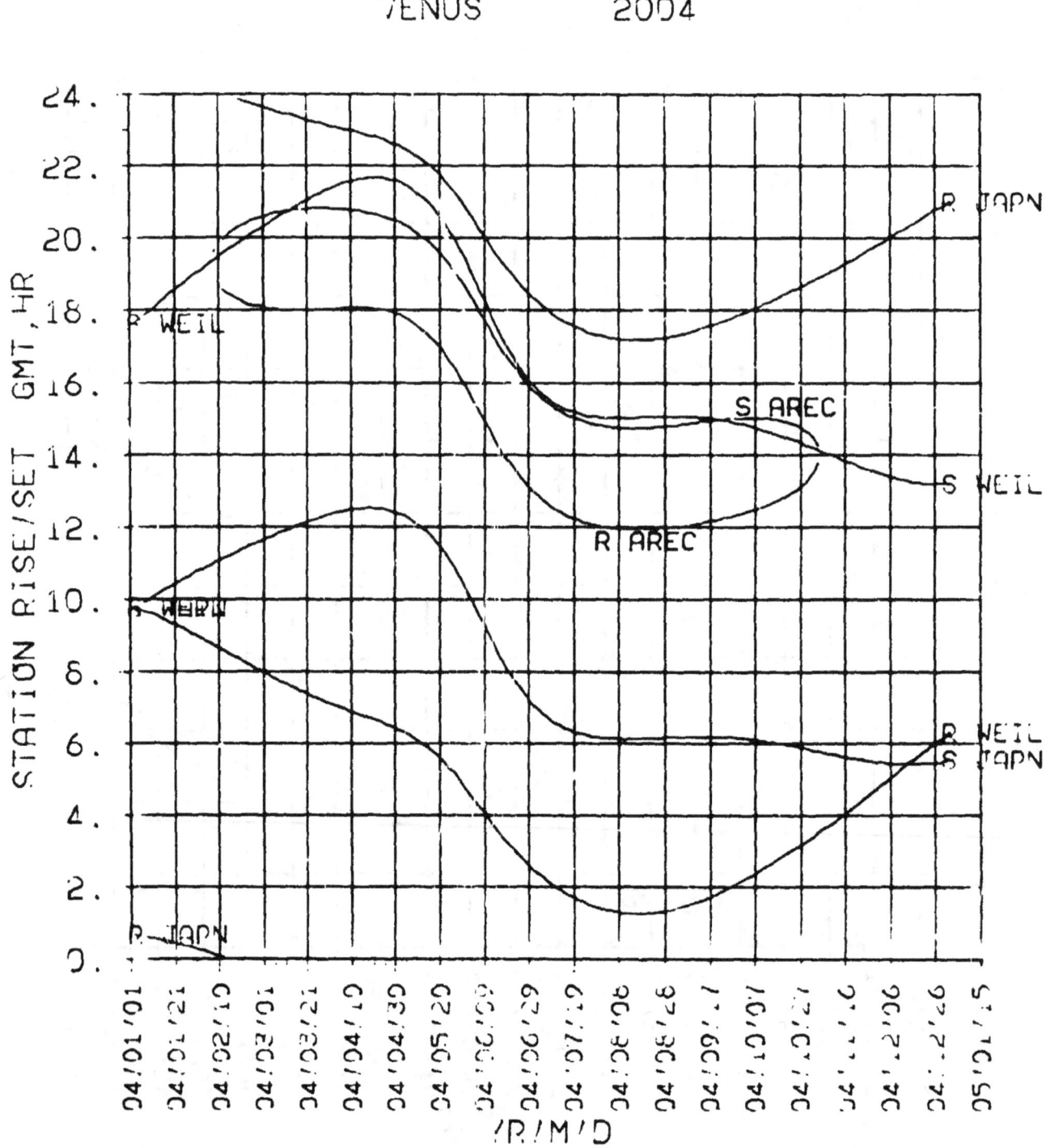

Venus

2005

DECLIN RT.ASC 2005

VENUS 2005

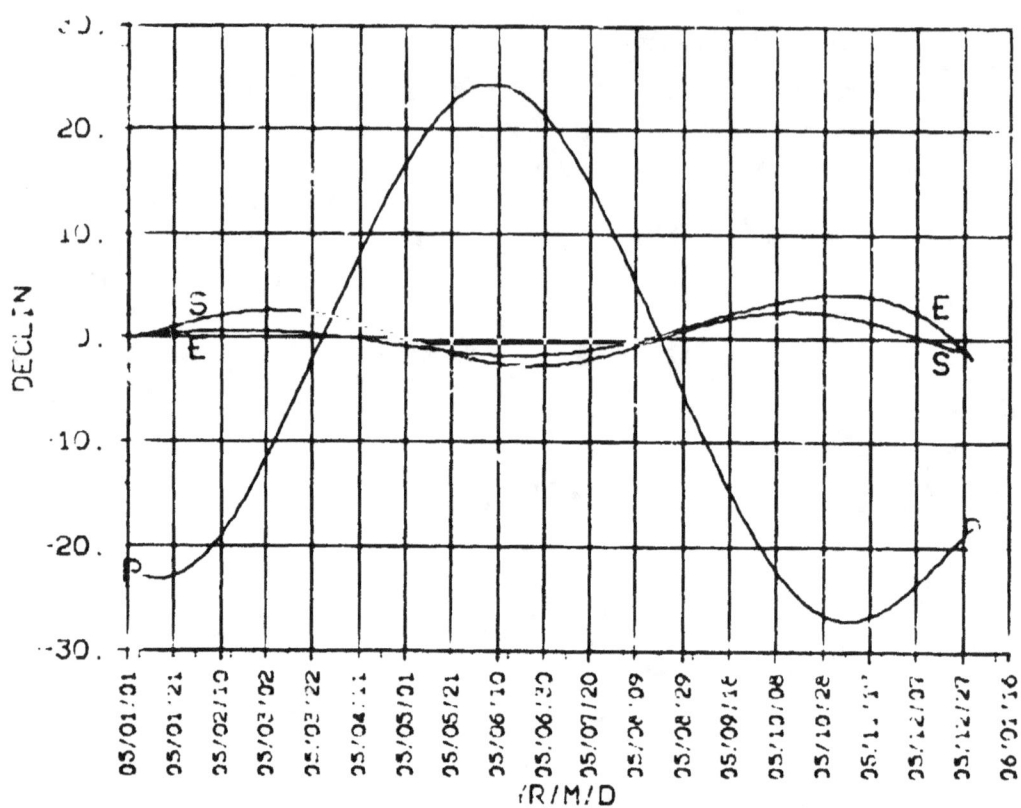

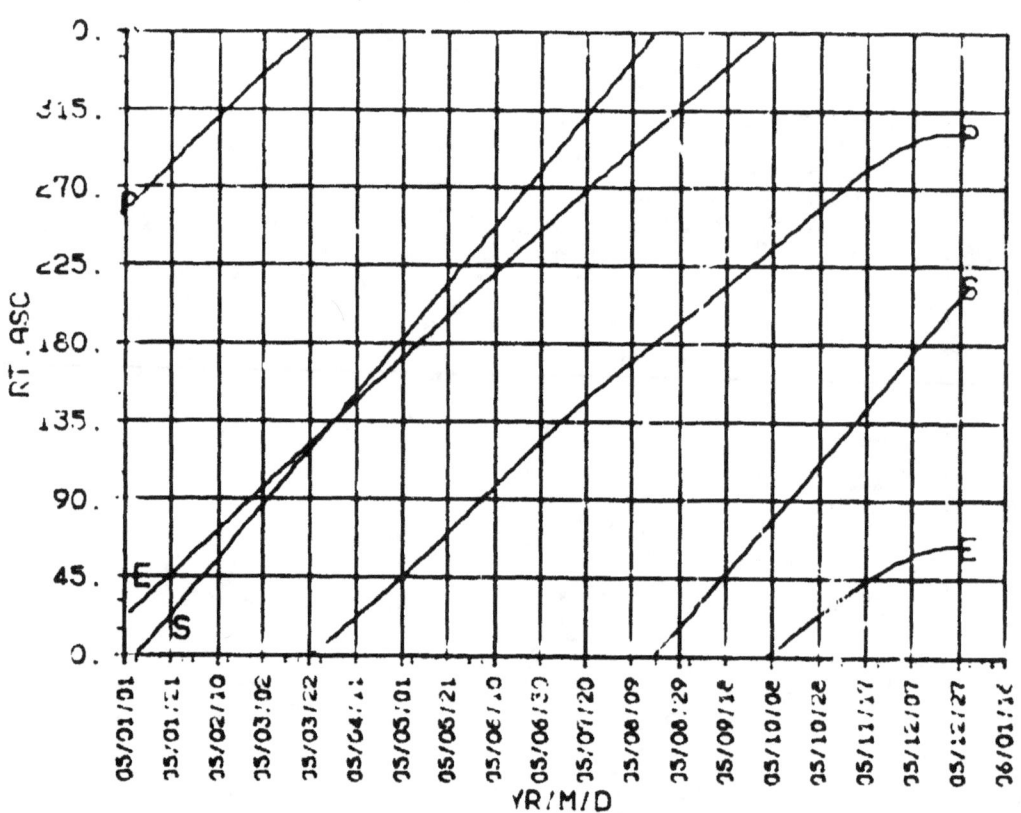

DISTANCE EC.LON 2005

VENUS 2005

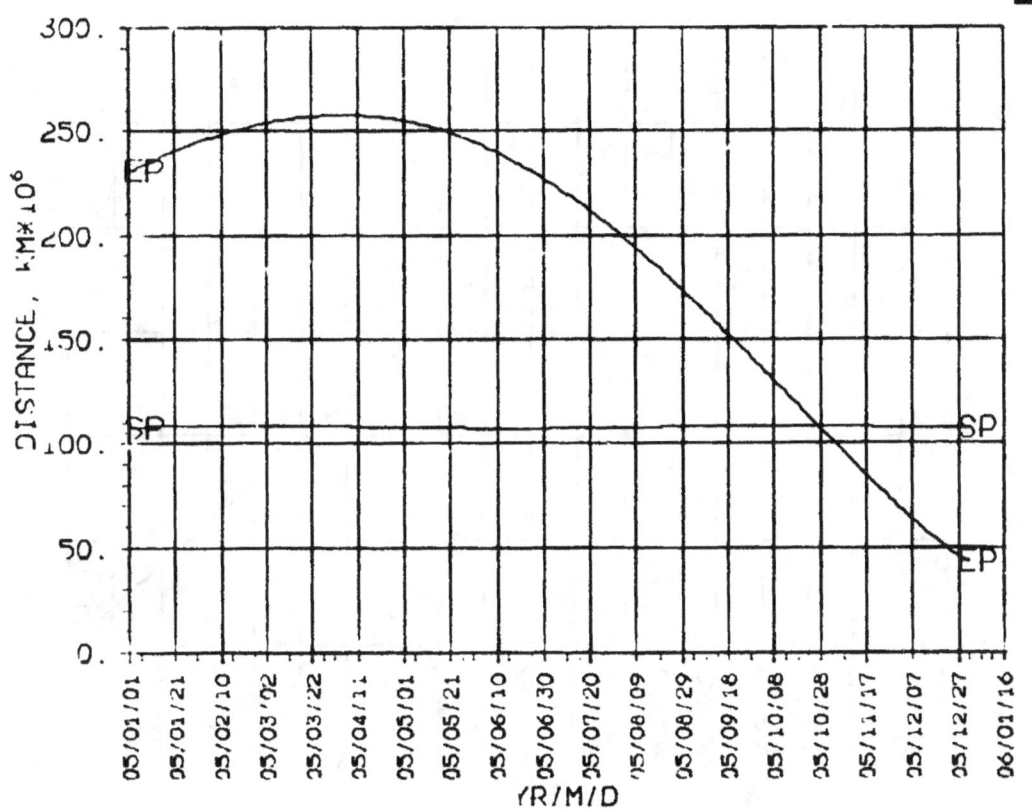

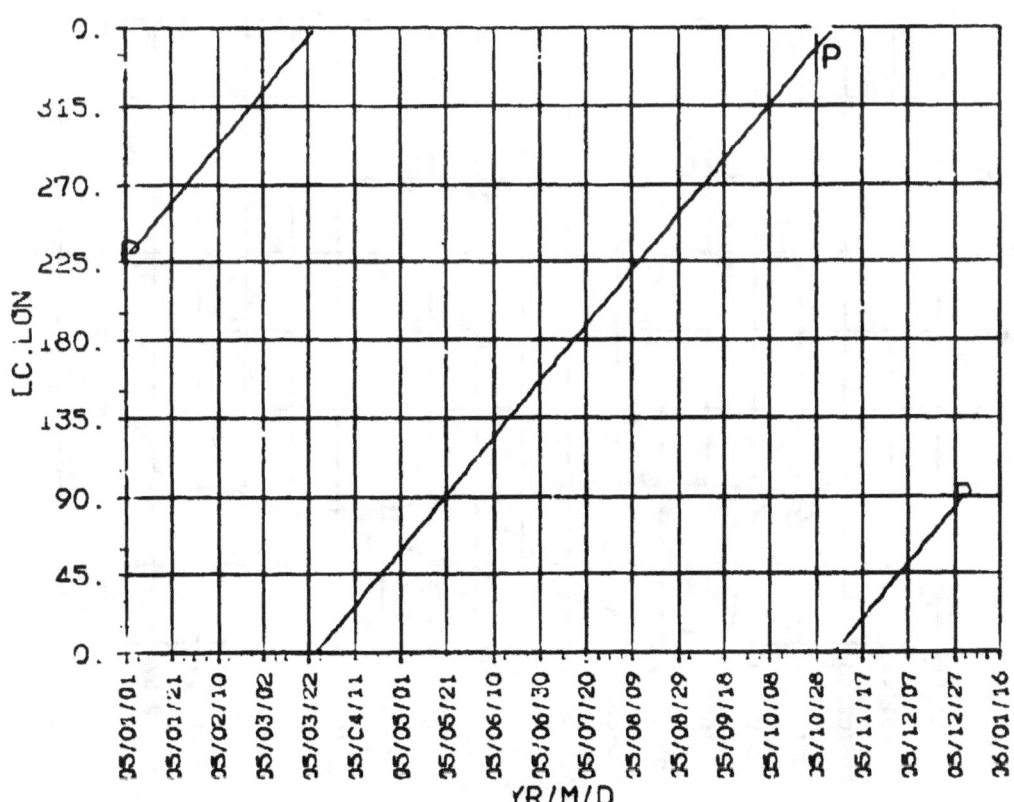

SEP, ESP CA, KA 2005

VENUS 2005

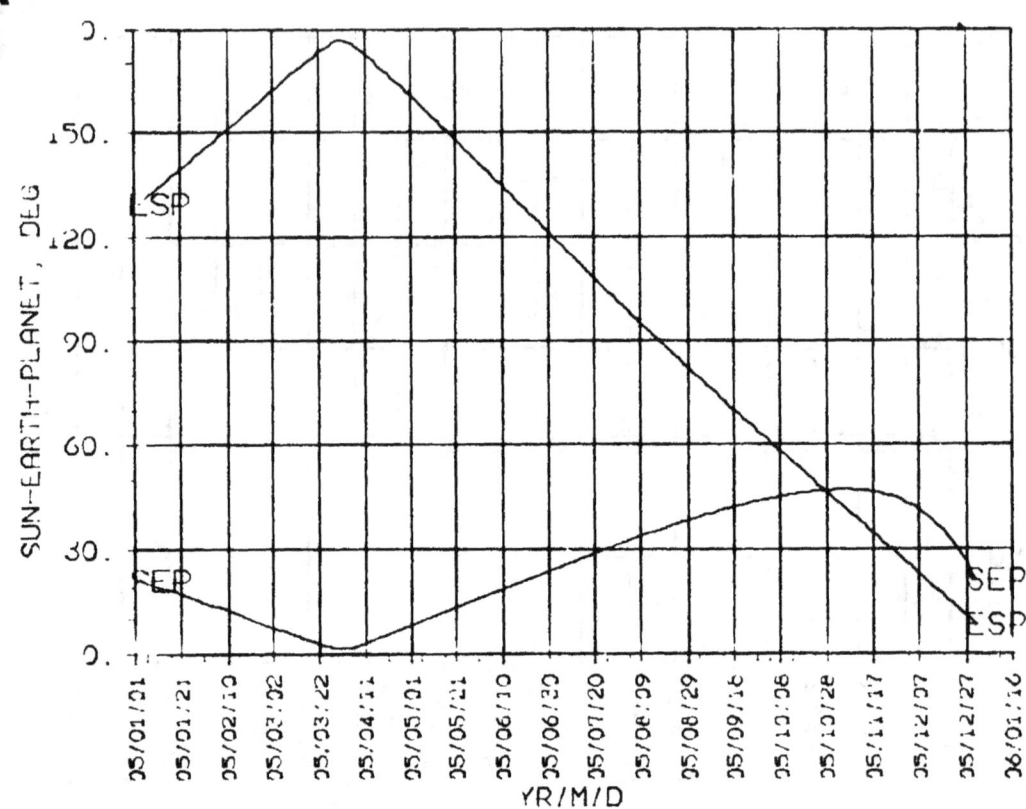

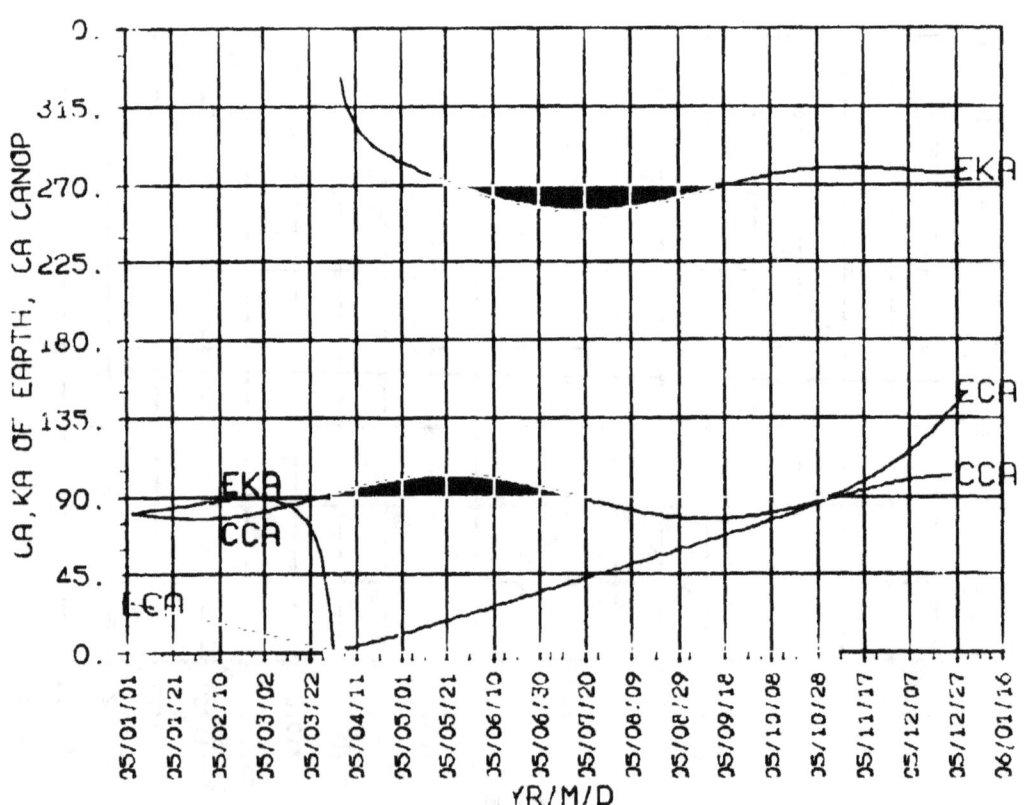

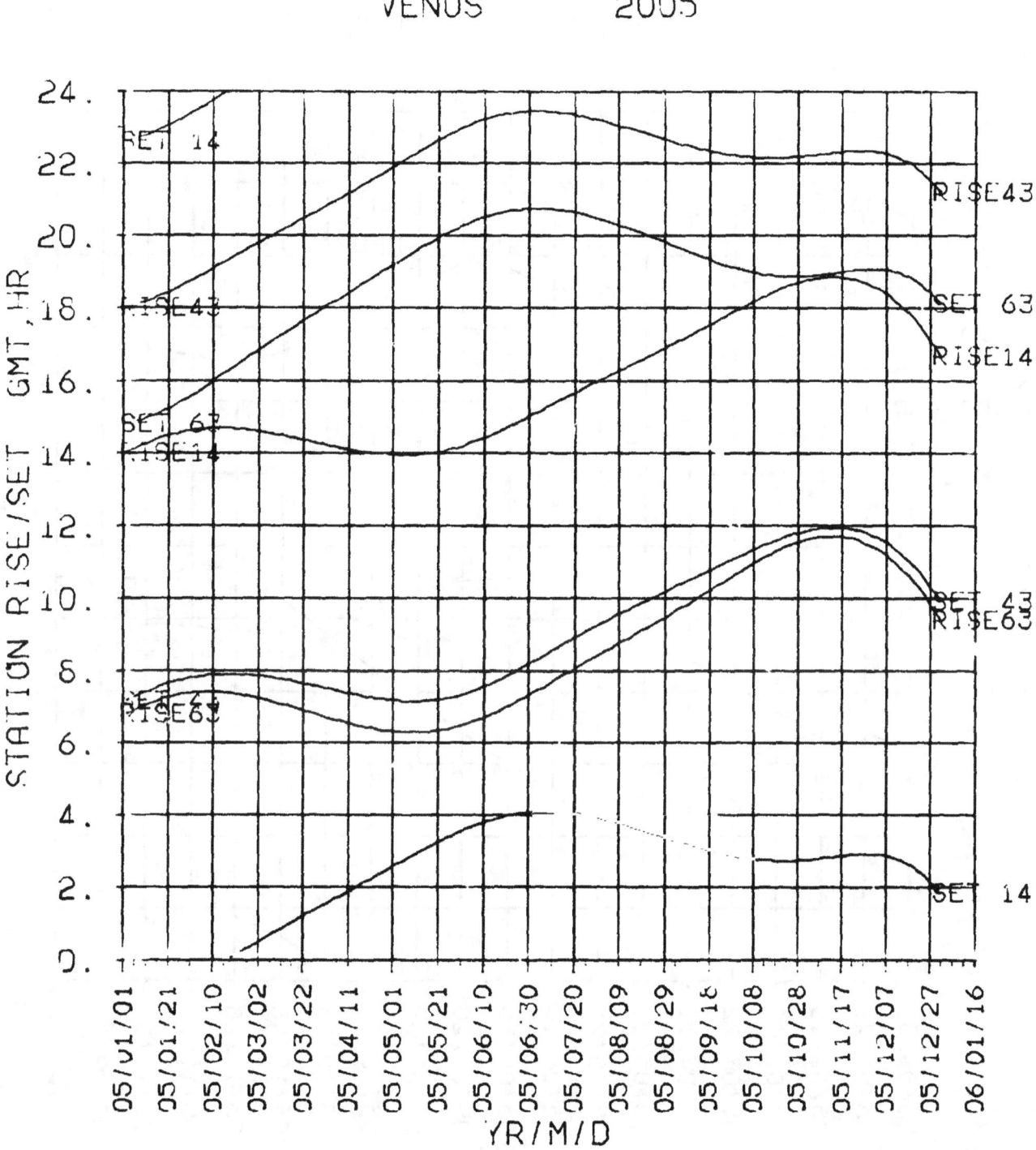

**STA R/S
NON-DSN
2005**

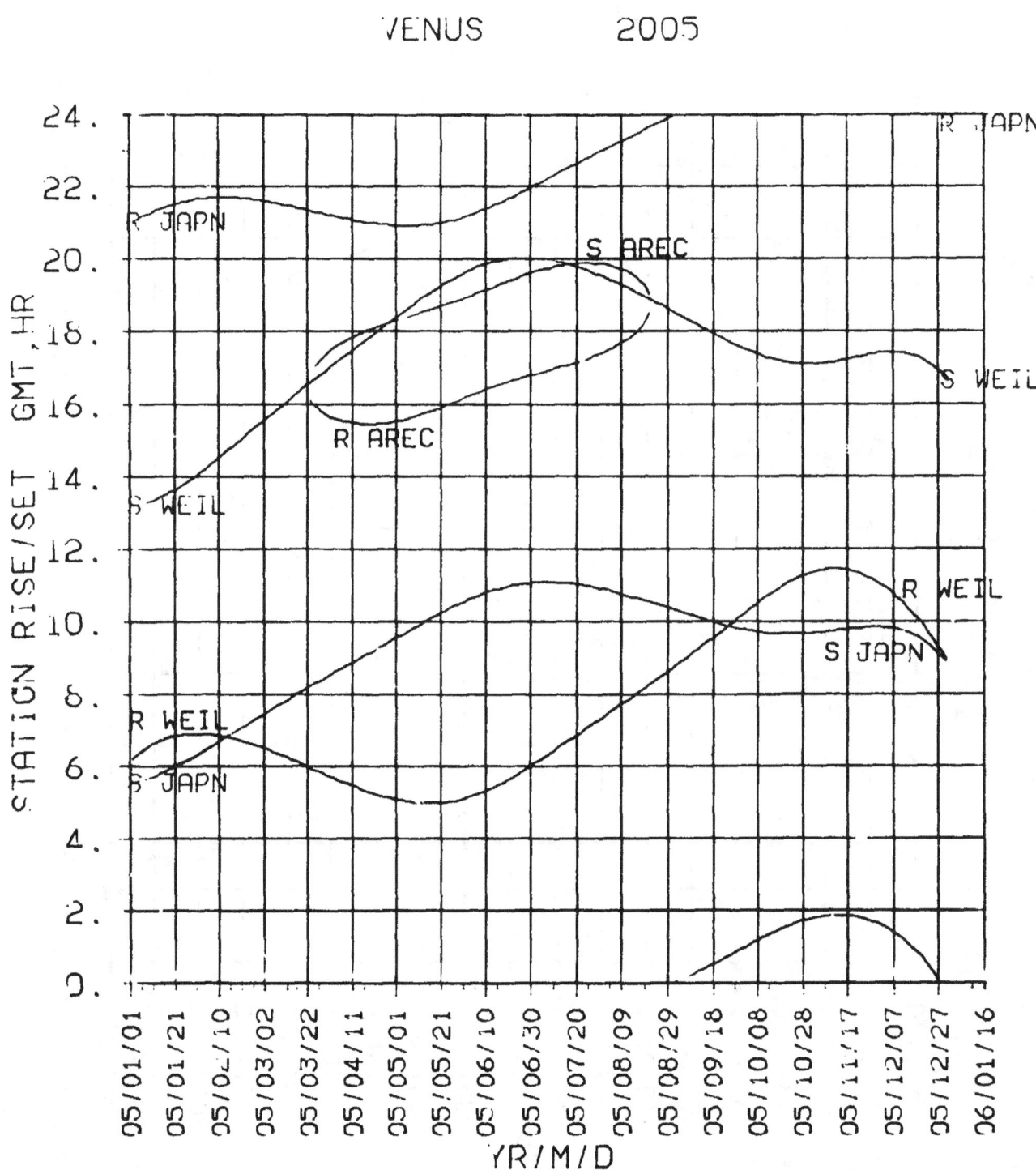

Venus

2006

DECLIN RT.ASC 2006

VENUS 2006

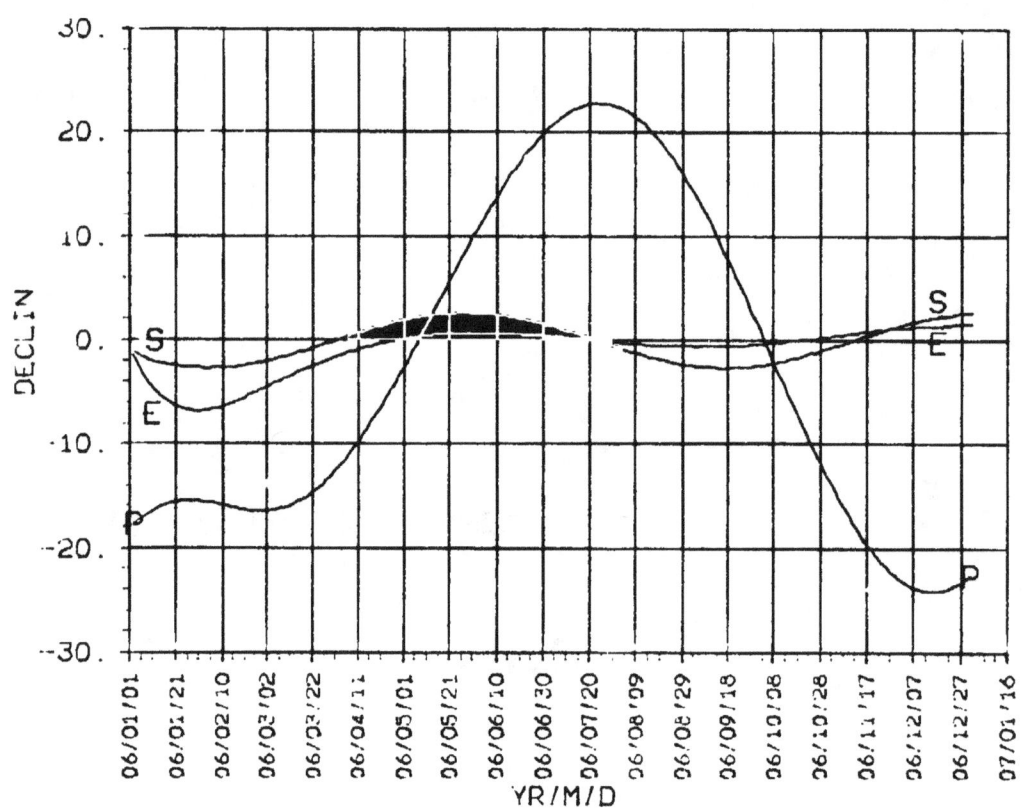

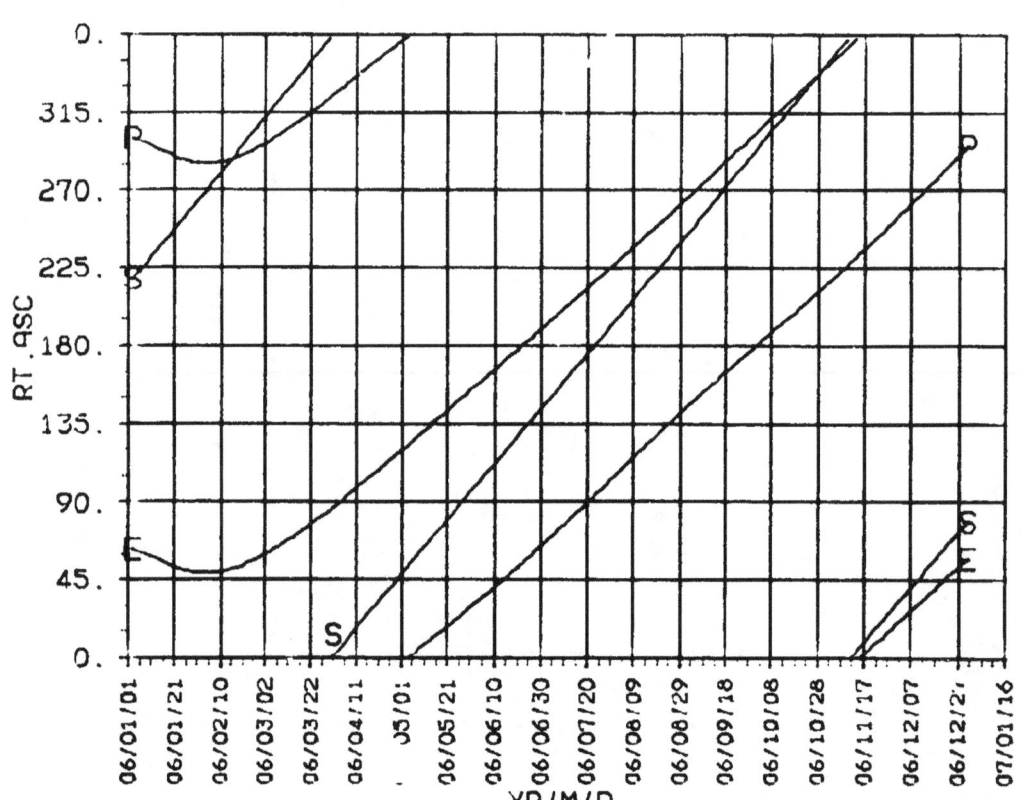

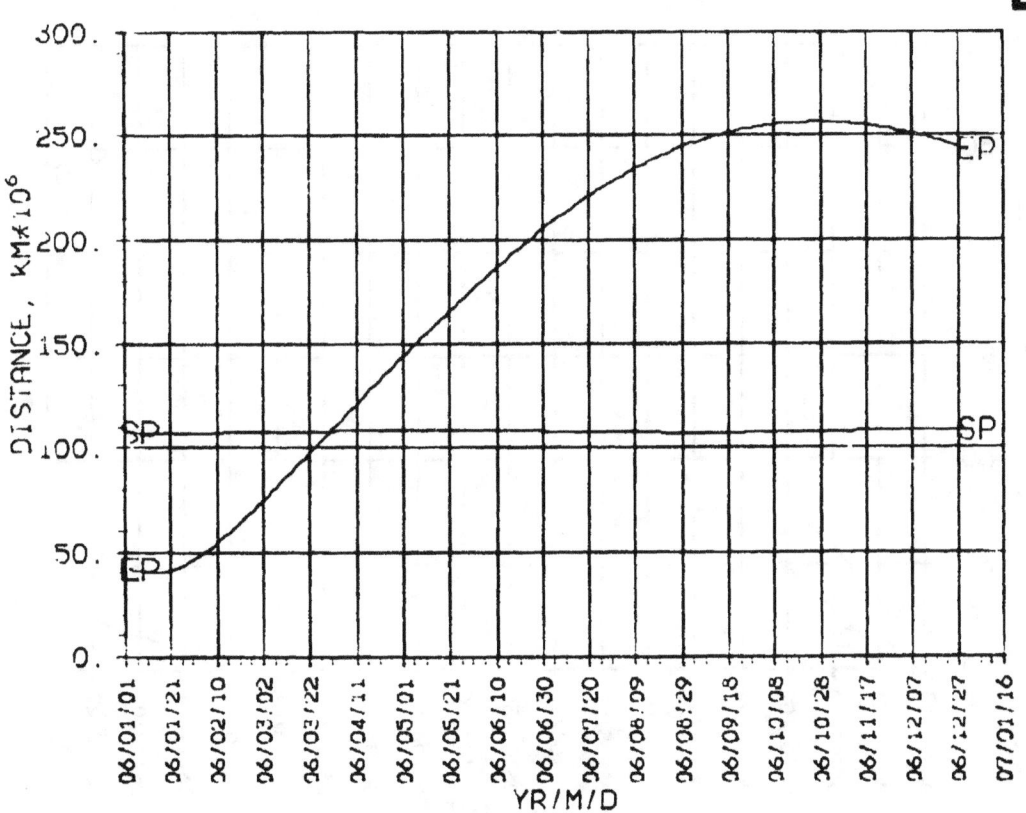
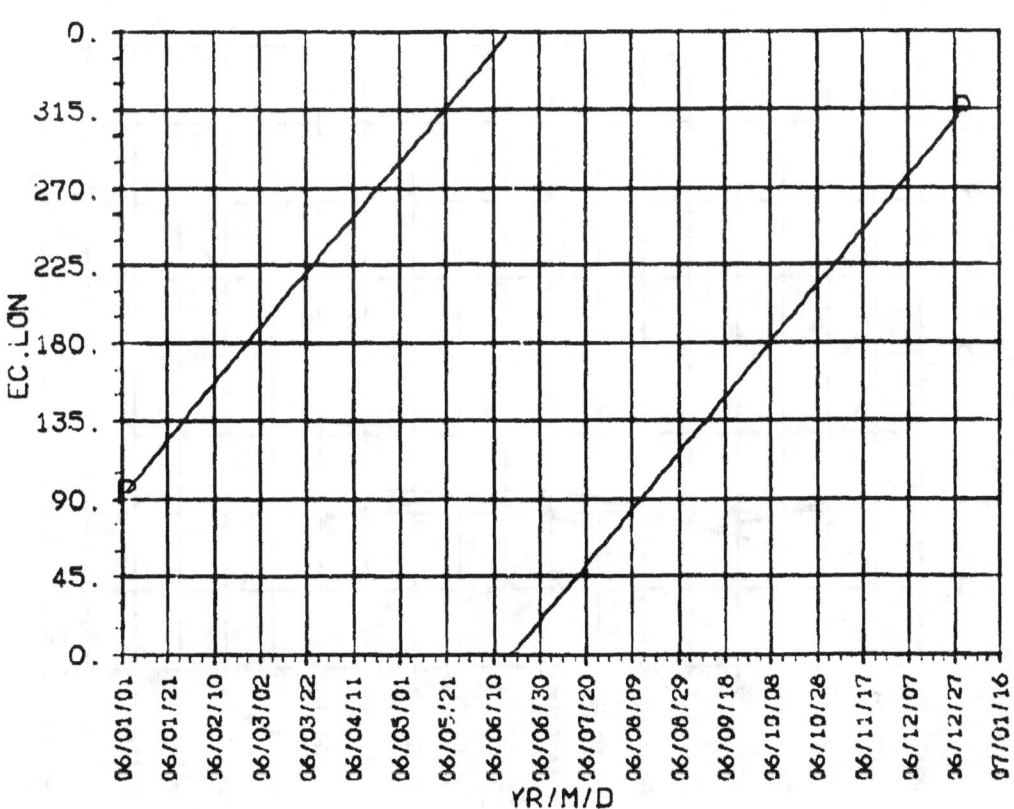

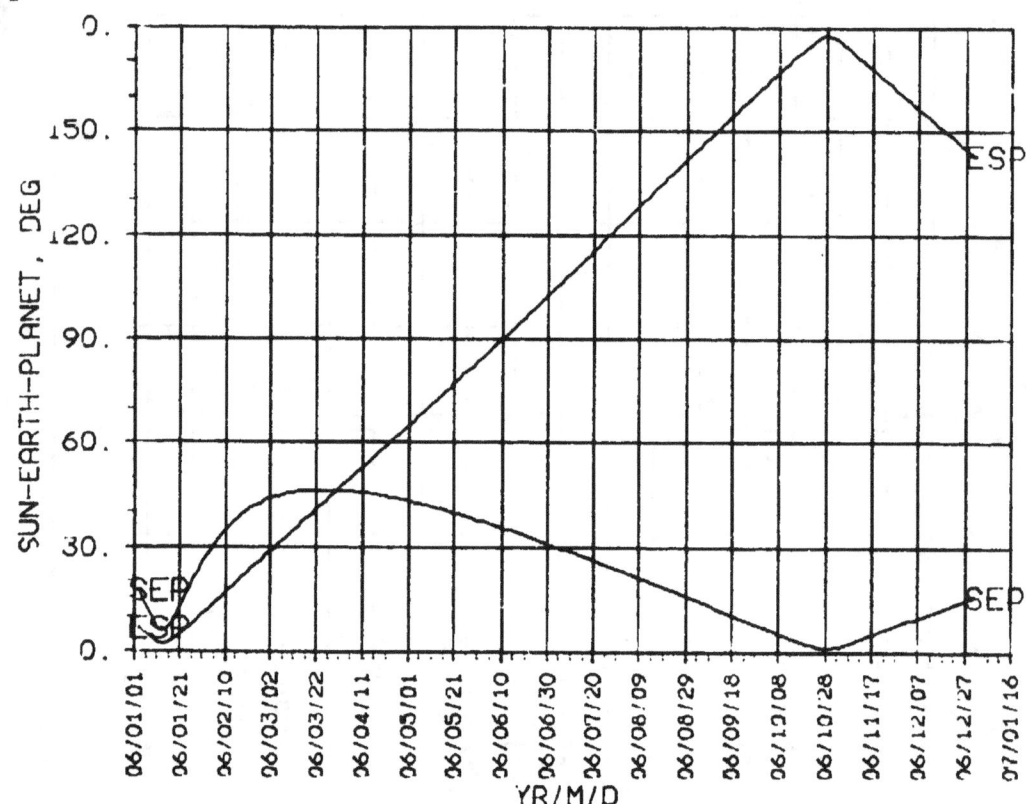
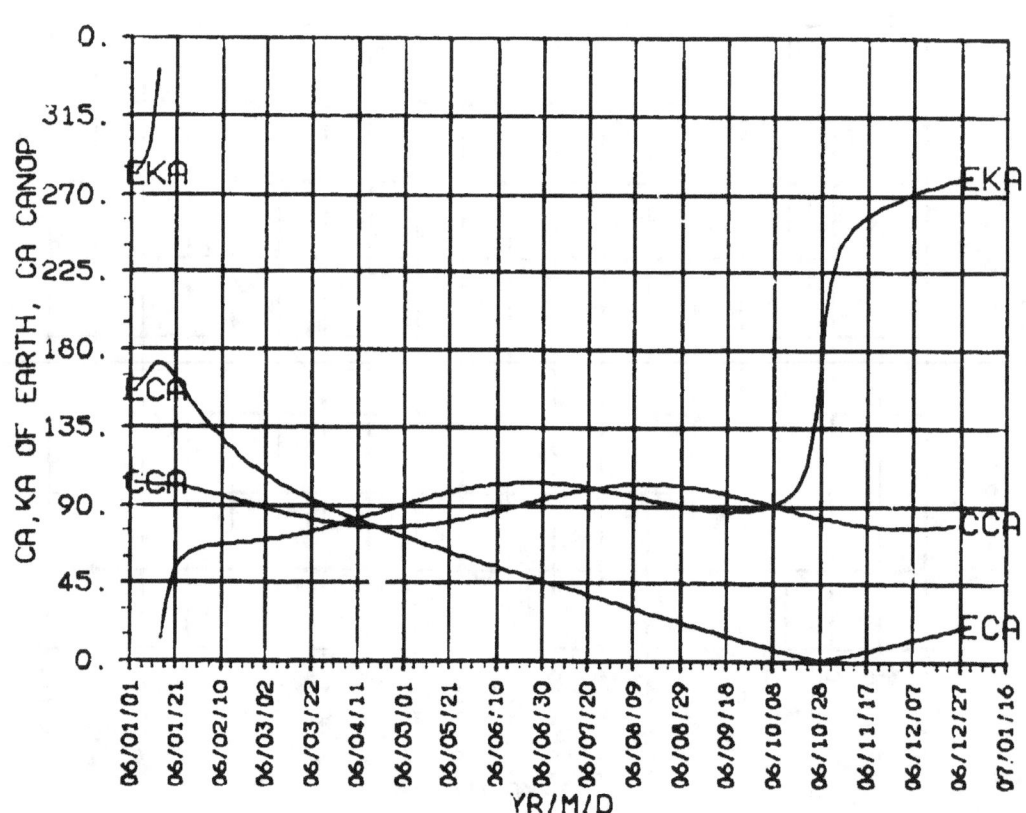

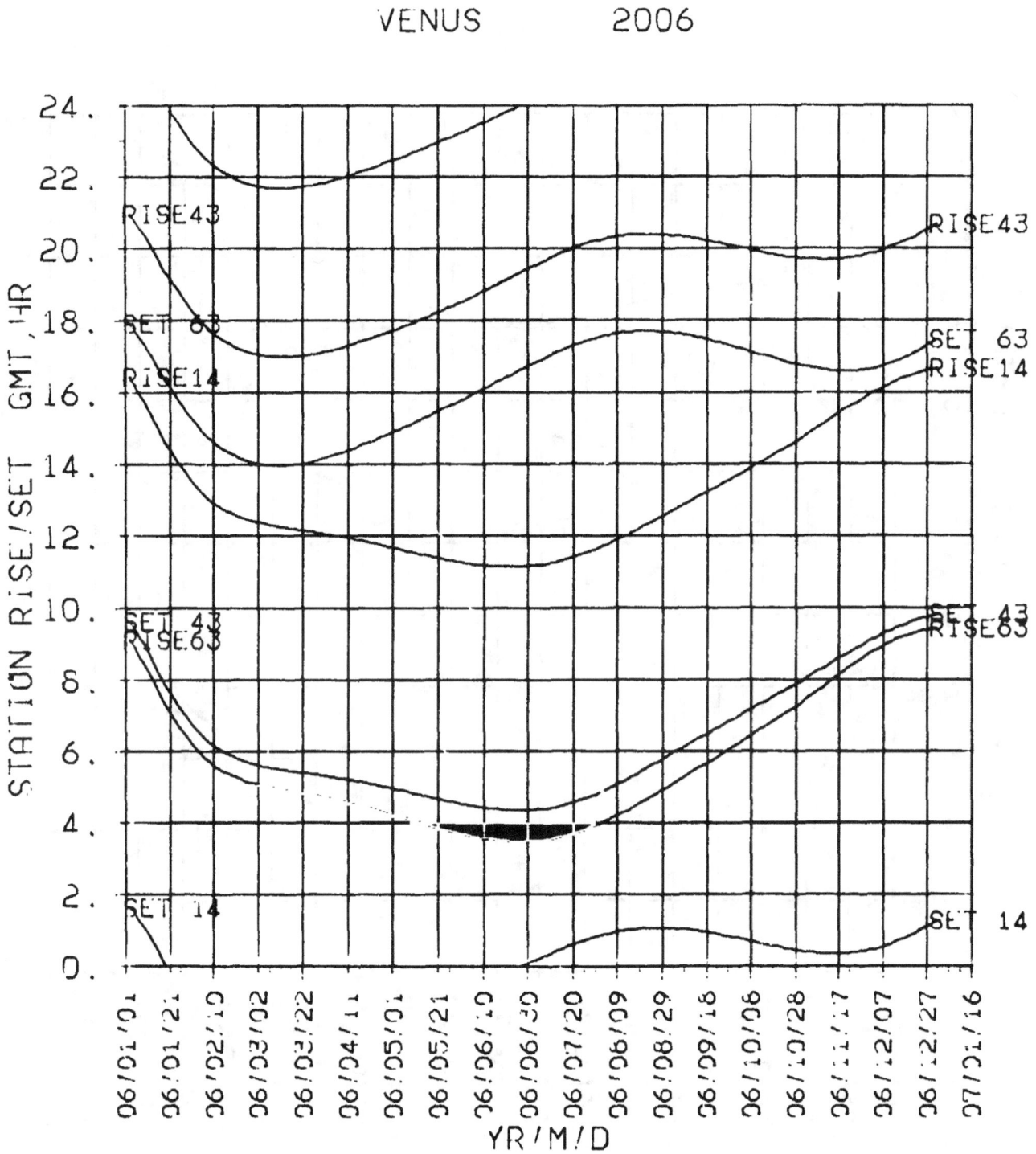

STA R/S
NON-DSN
2006

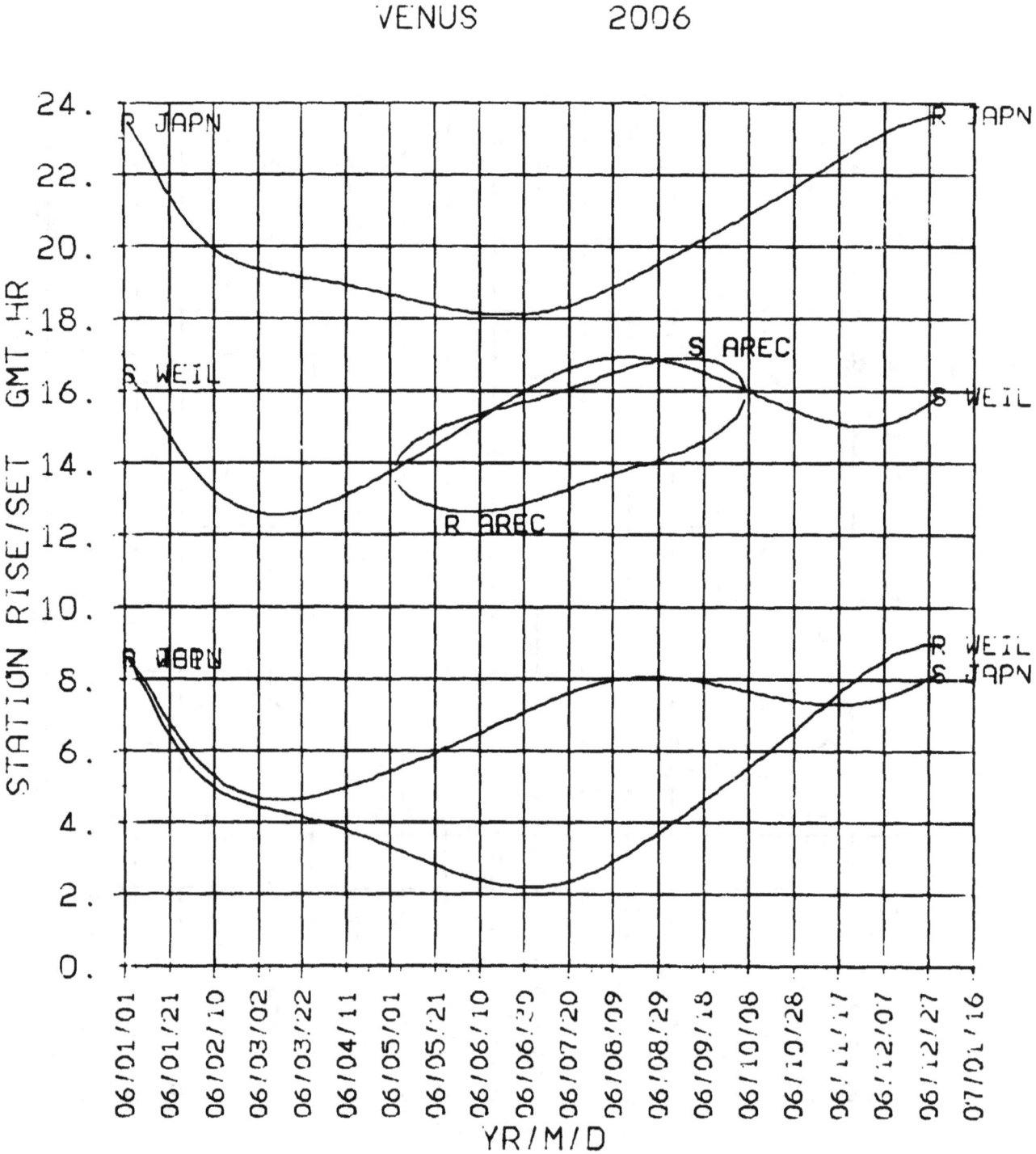

Venus

2007

DECLIN RT.ASC 2007

VENUS 2007

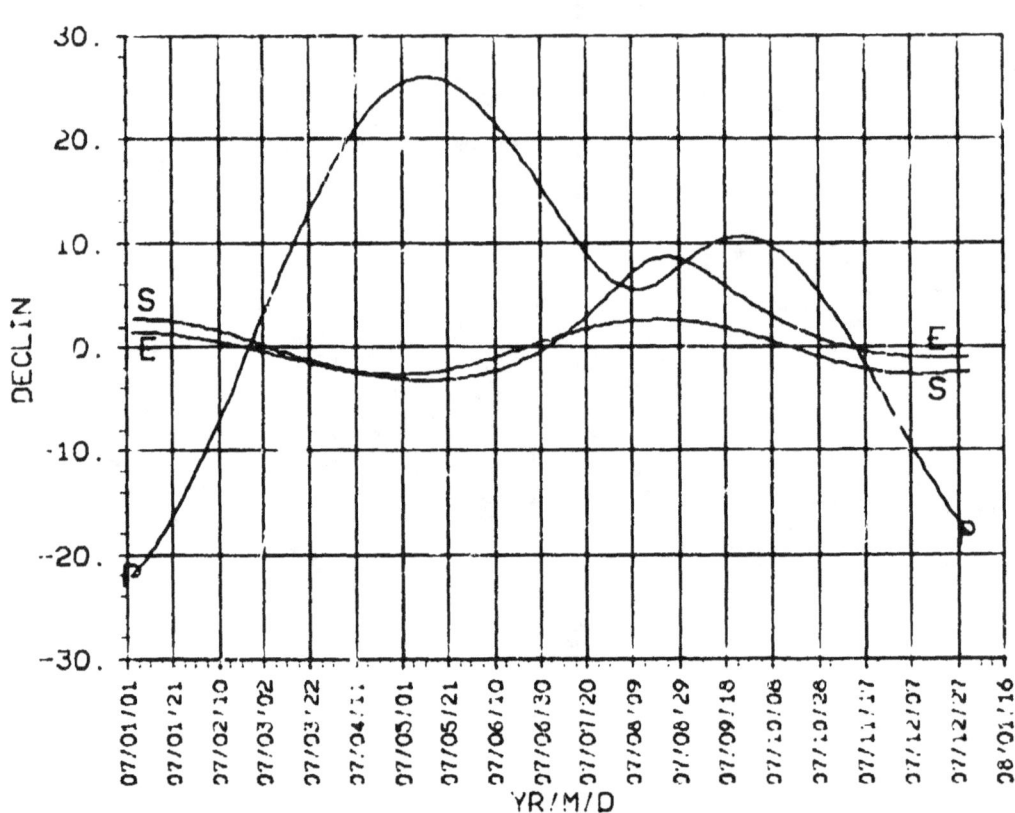

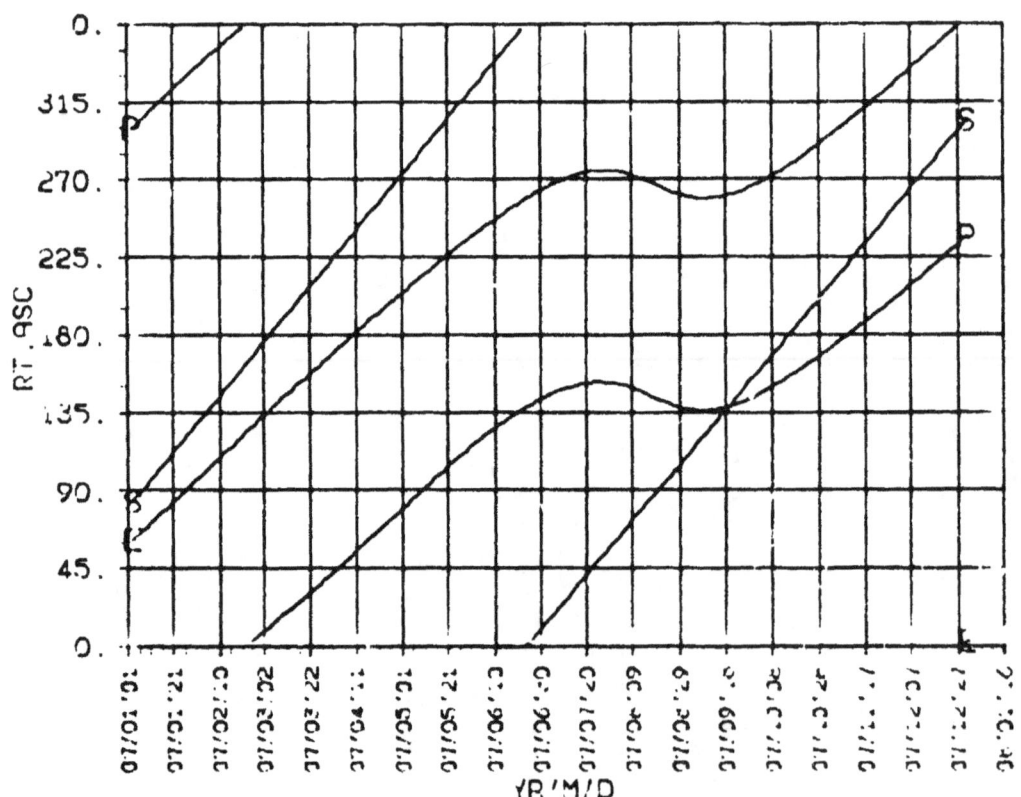

VENUS 2007

DISTANCE EC.LON 2007

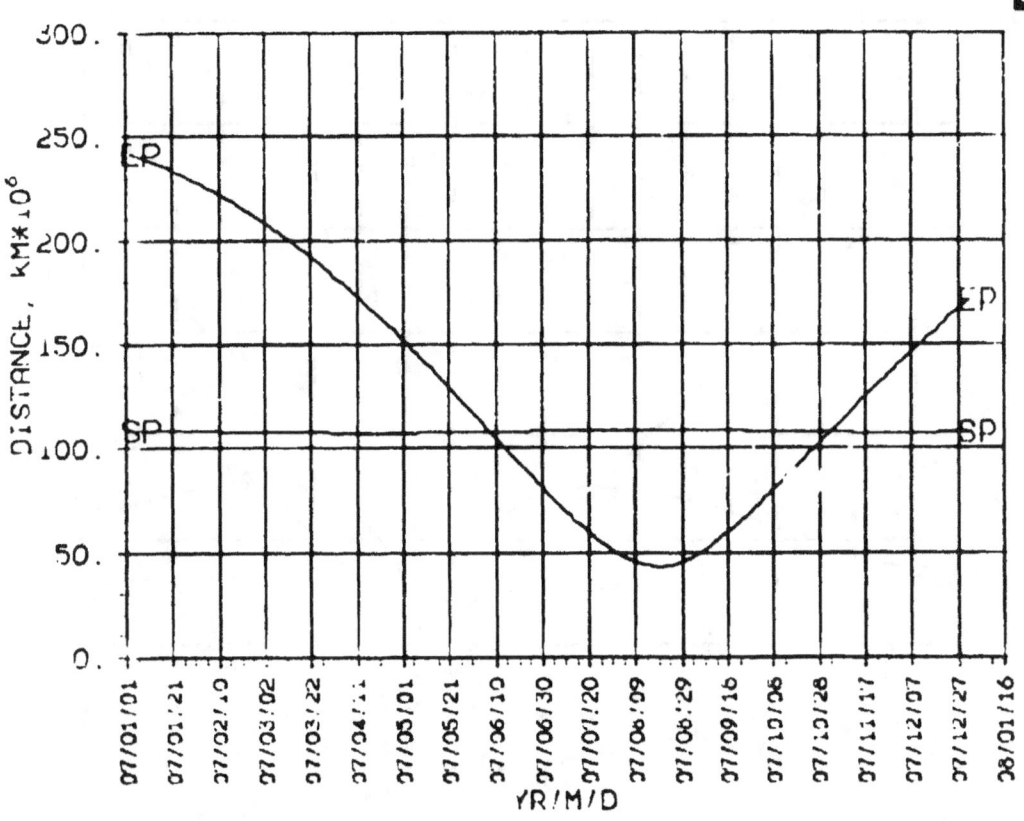
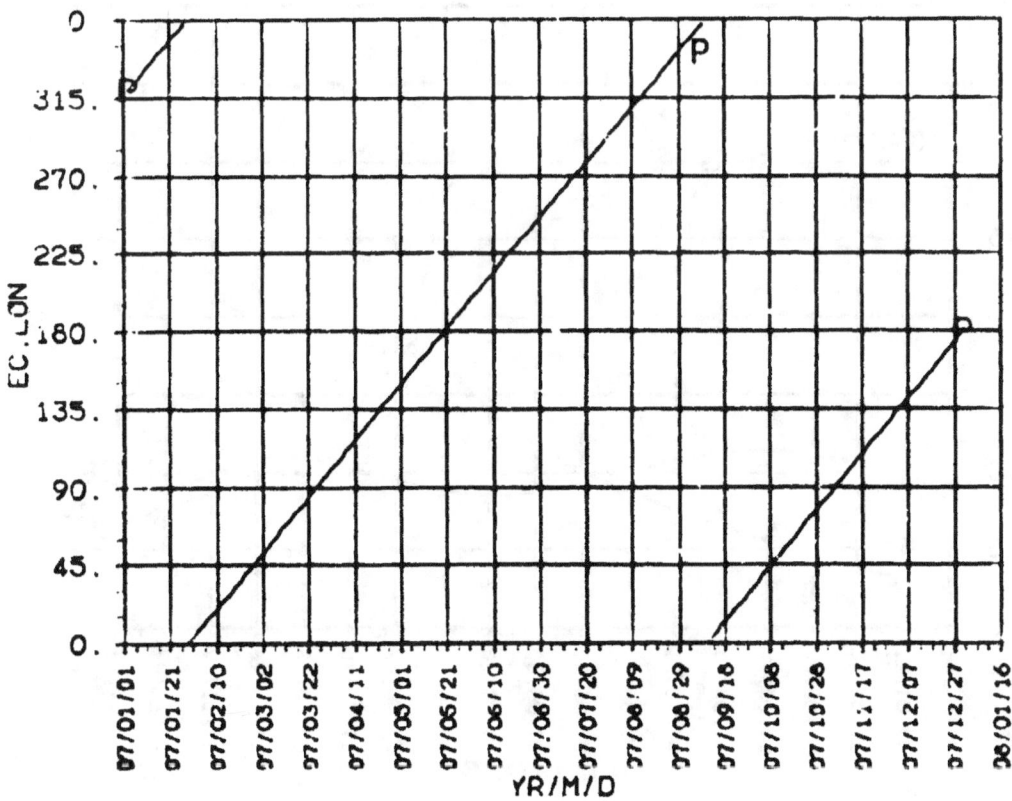

VENUS 2007
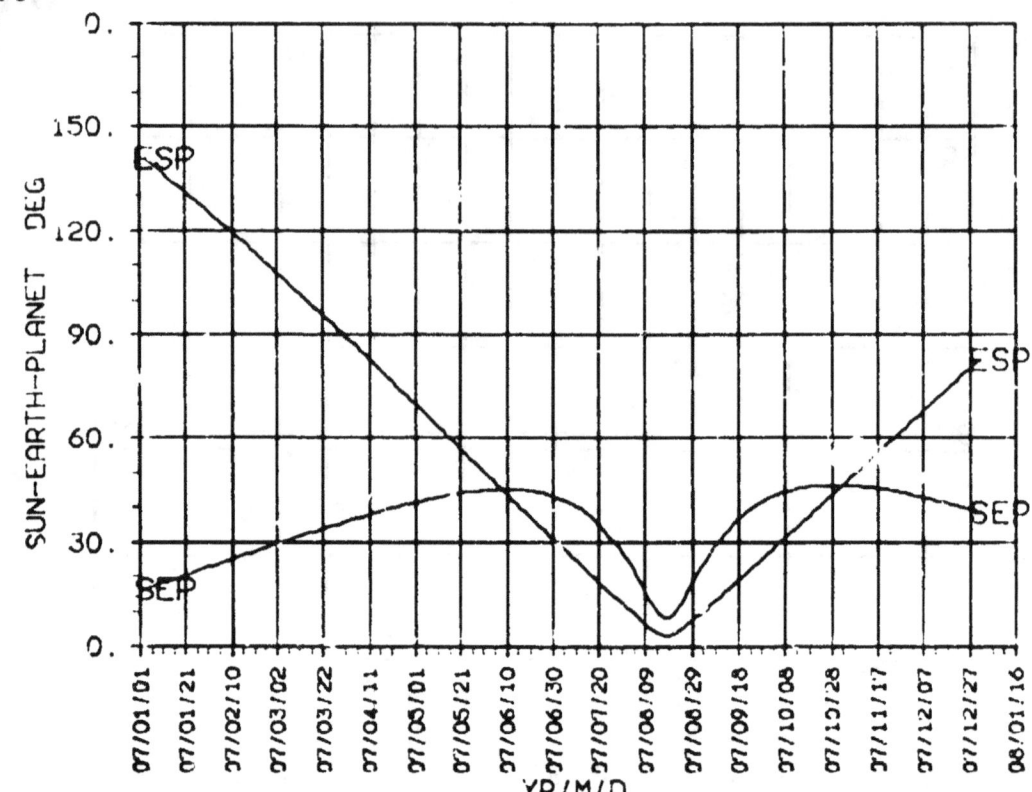
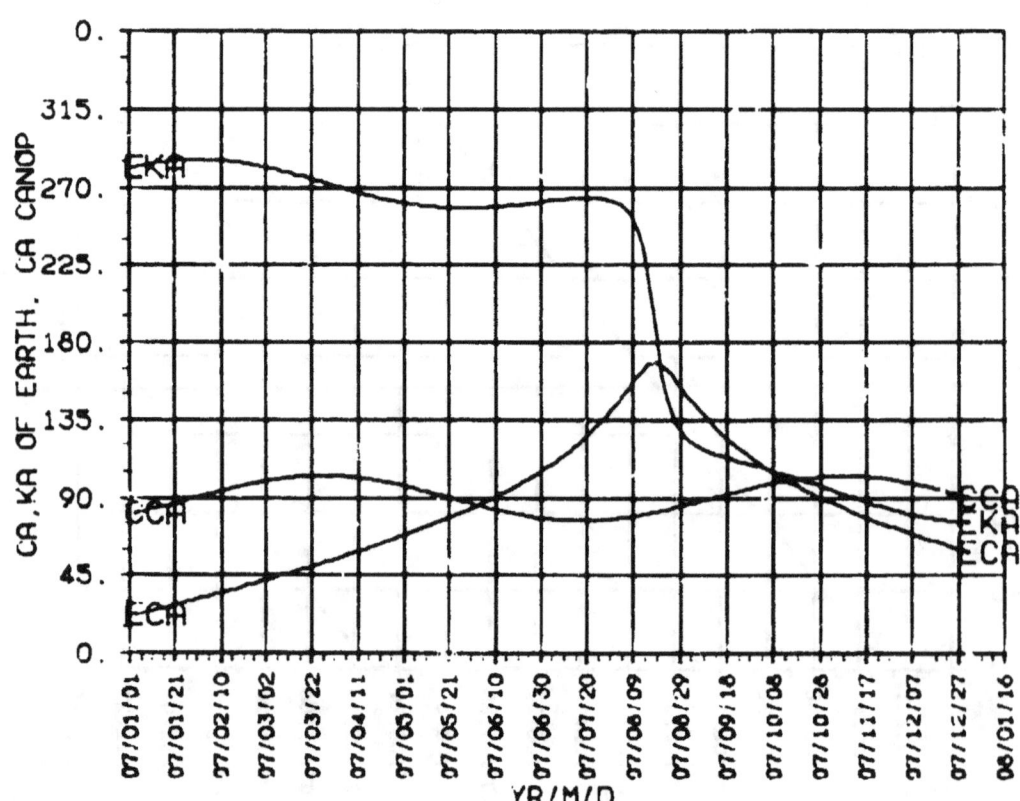

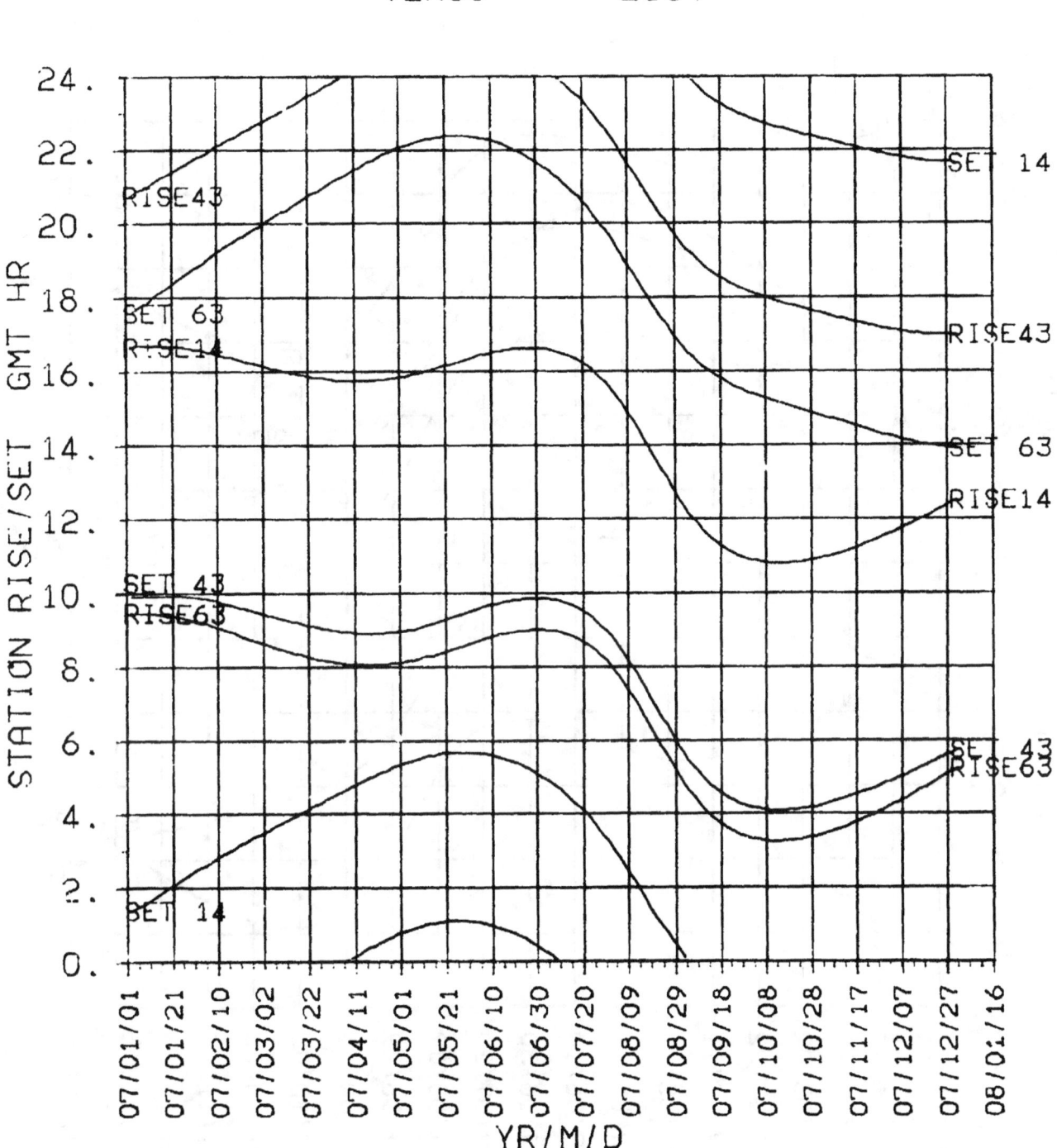

STA R/S
NON-DSN
2007

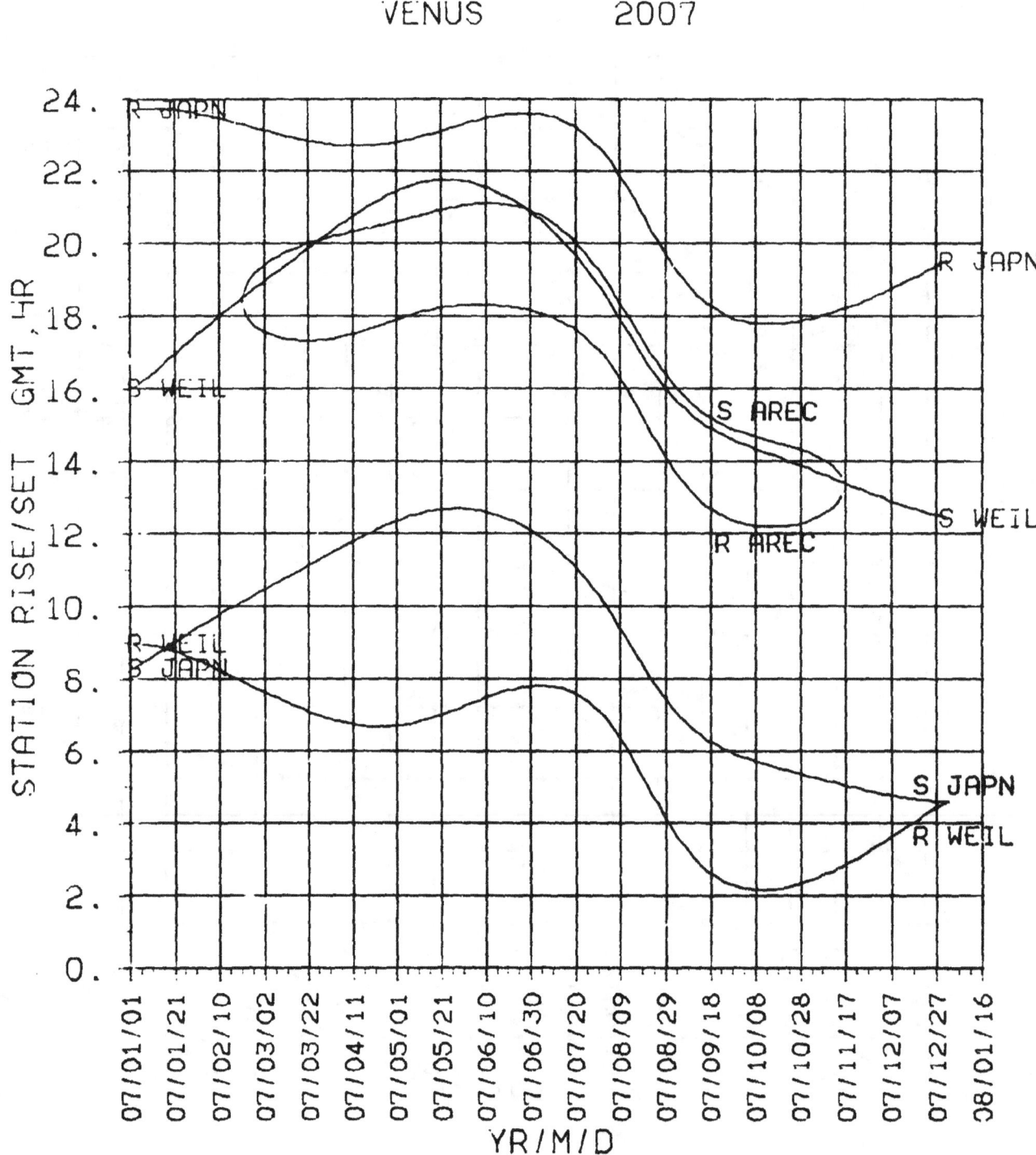

Venus

2008

DECLIN RT.ASC 2008

VENUS 2008

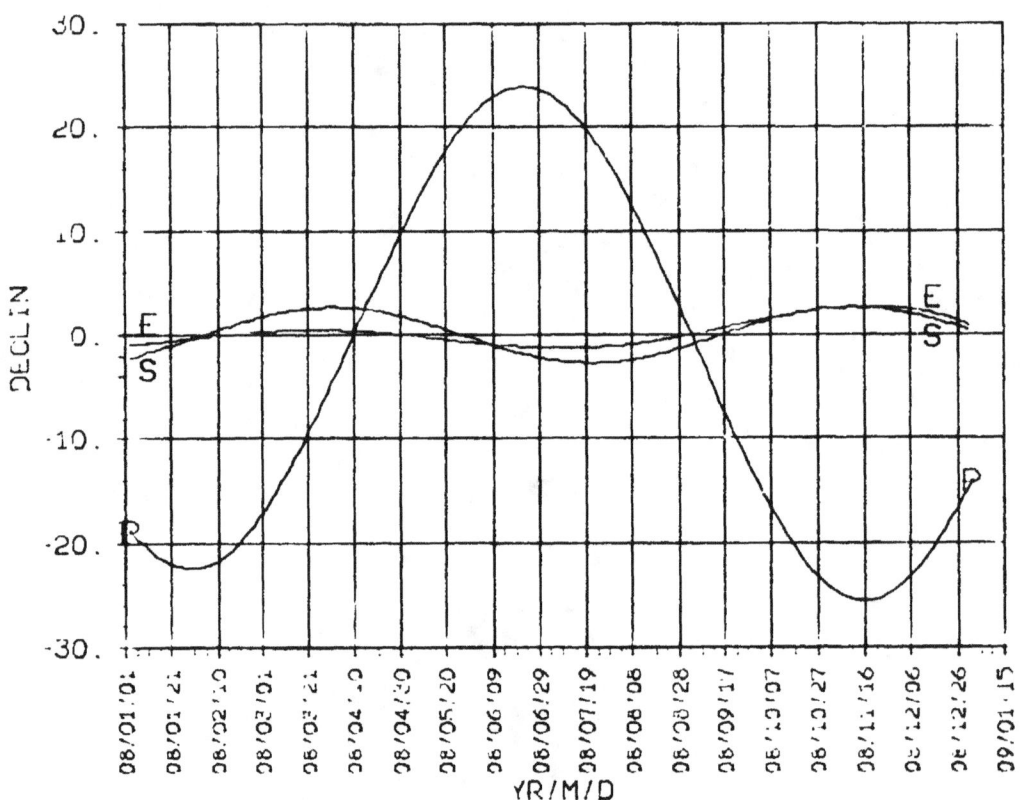

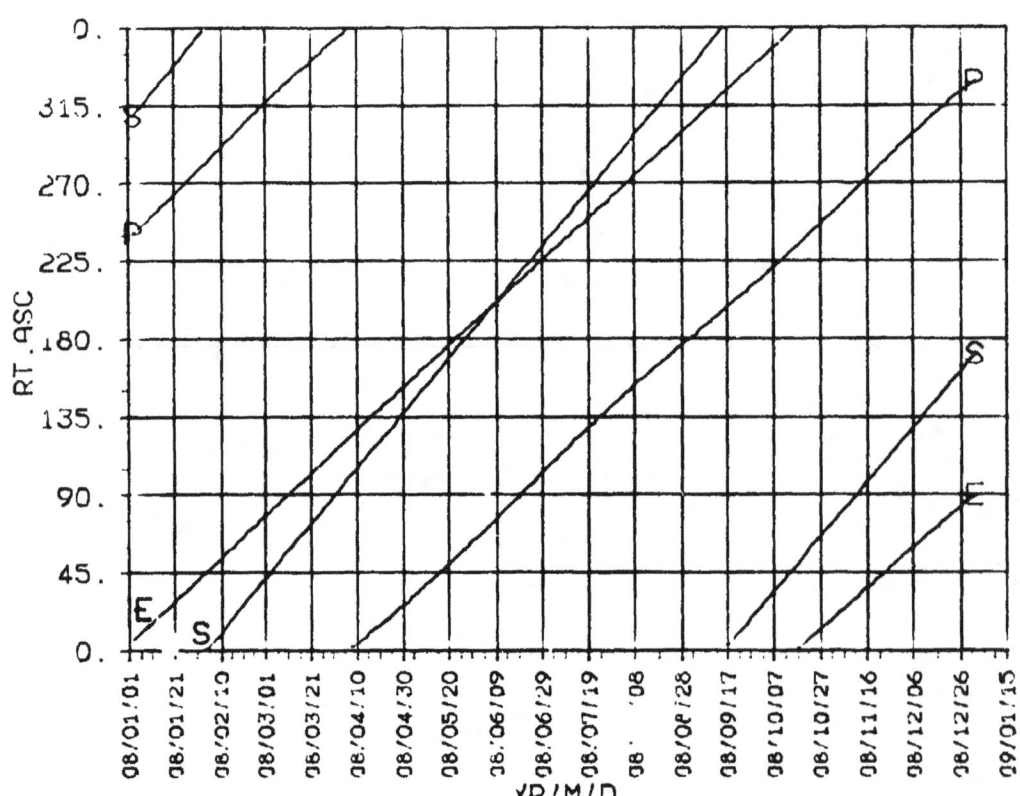

DISTANCE EC.LON 2008

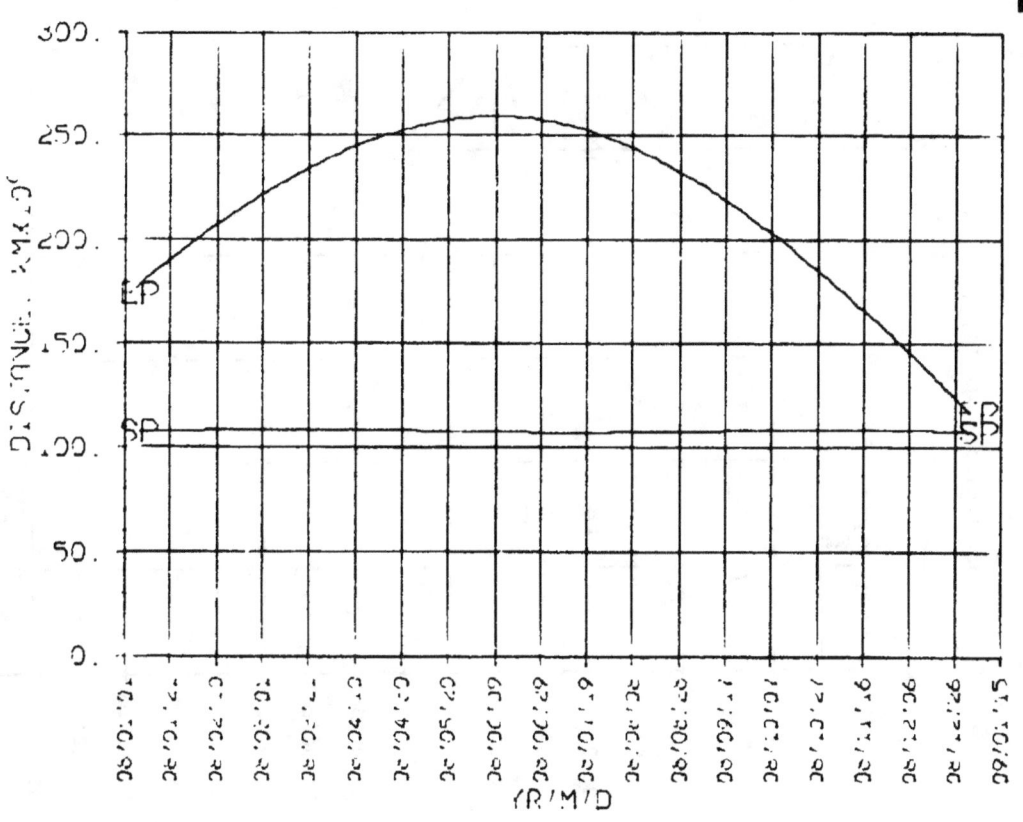

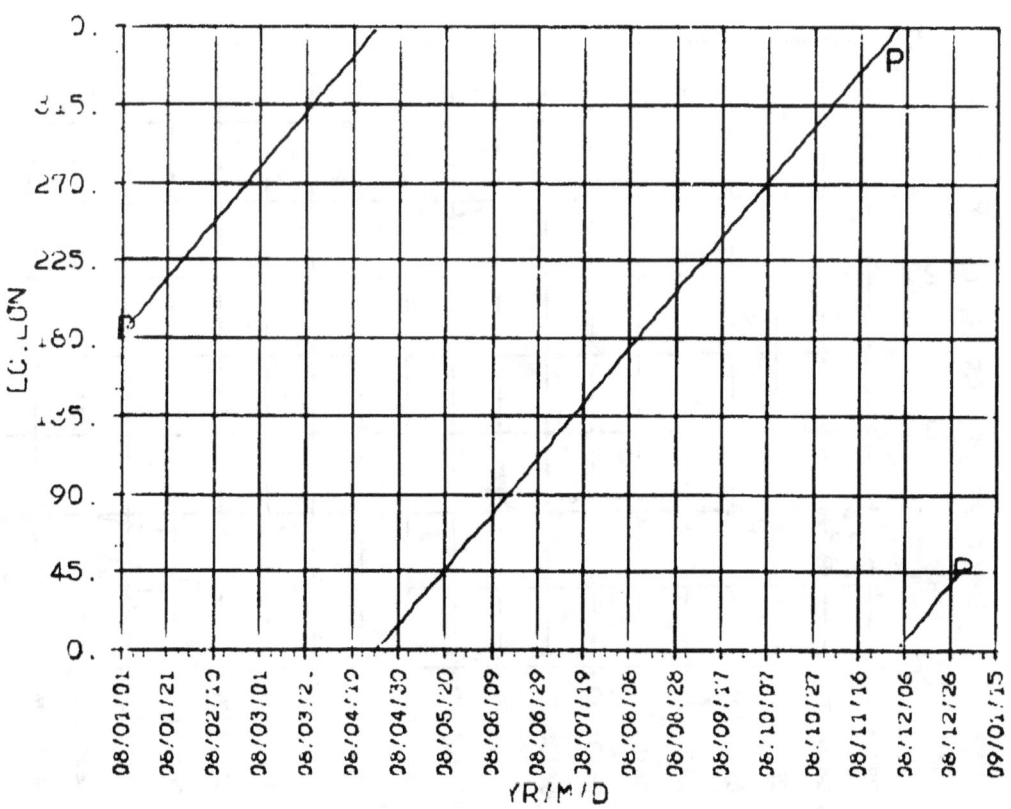

**SEP, ESP
CA, KA
2008**

VENUS 2008

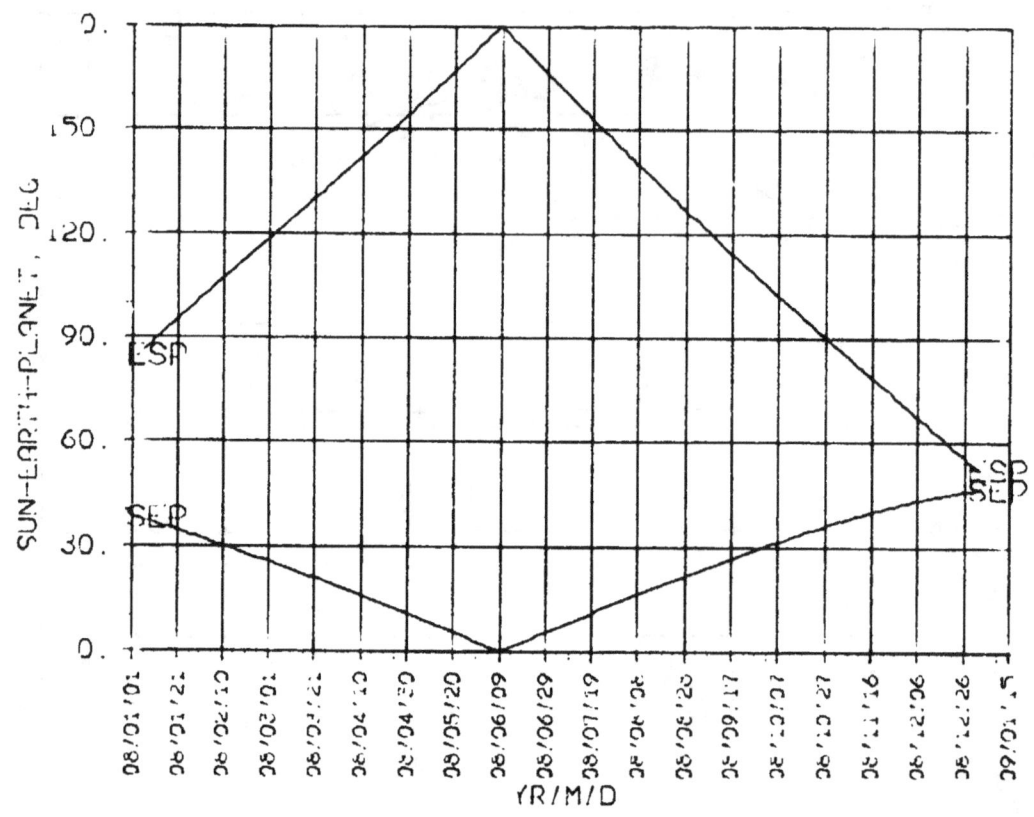

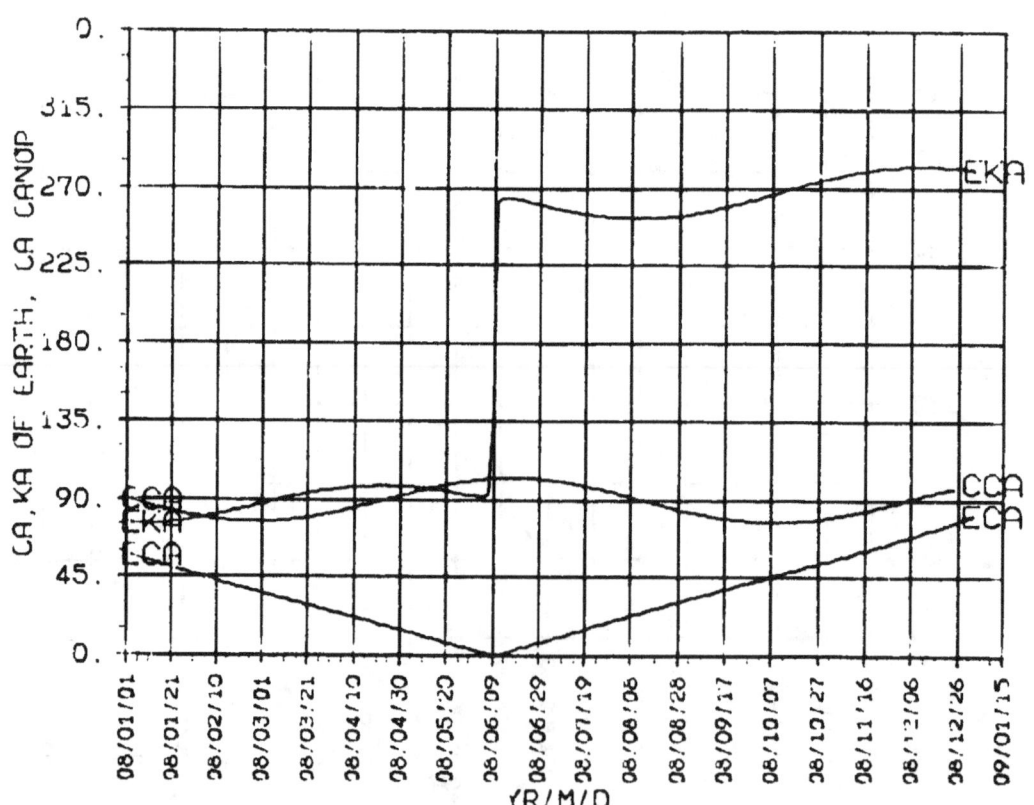

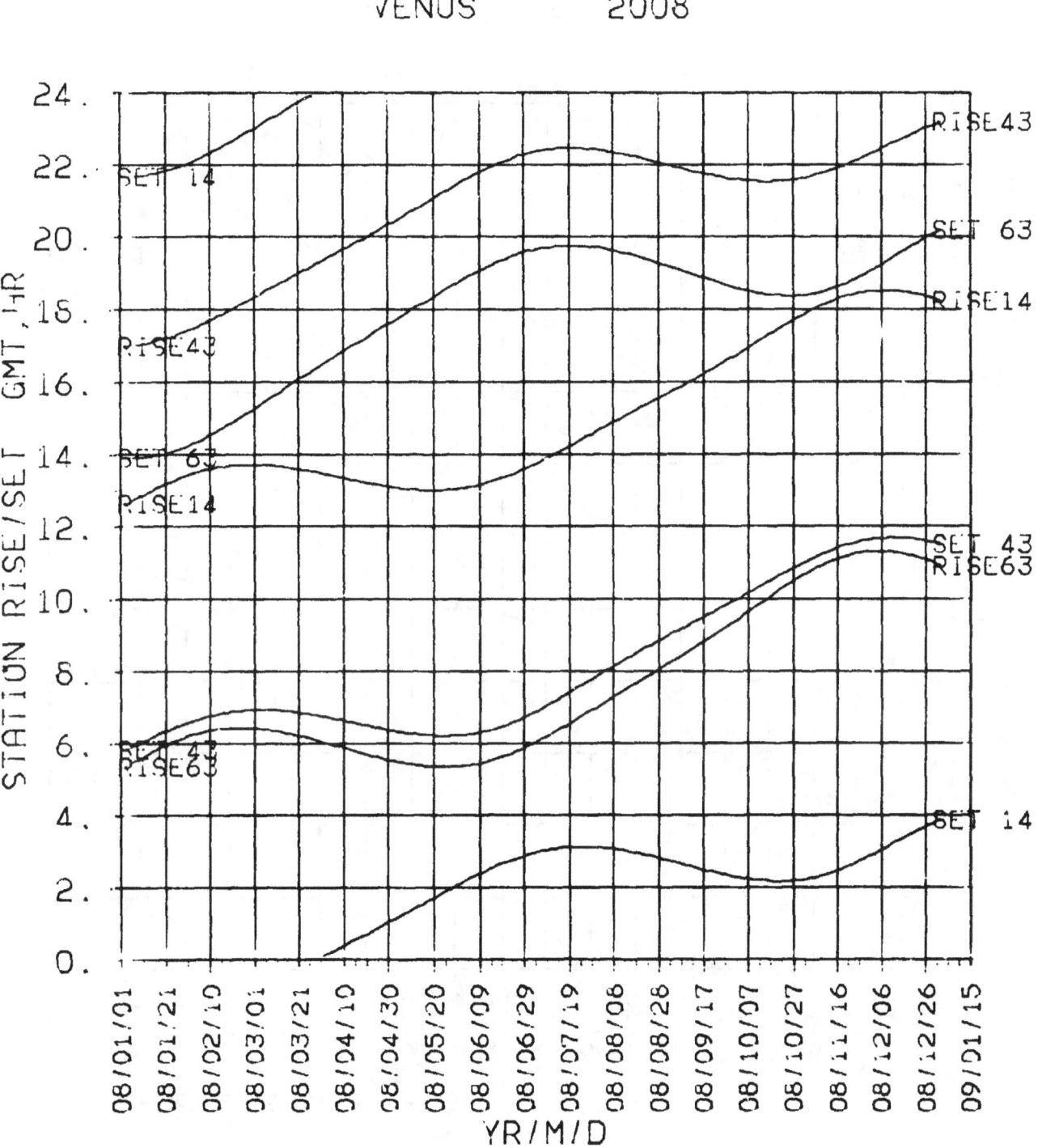

STA R/S
NON-DSN
2008

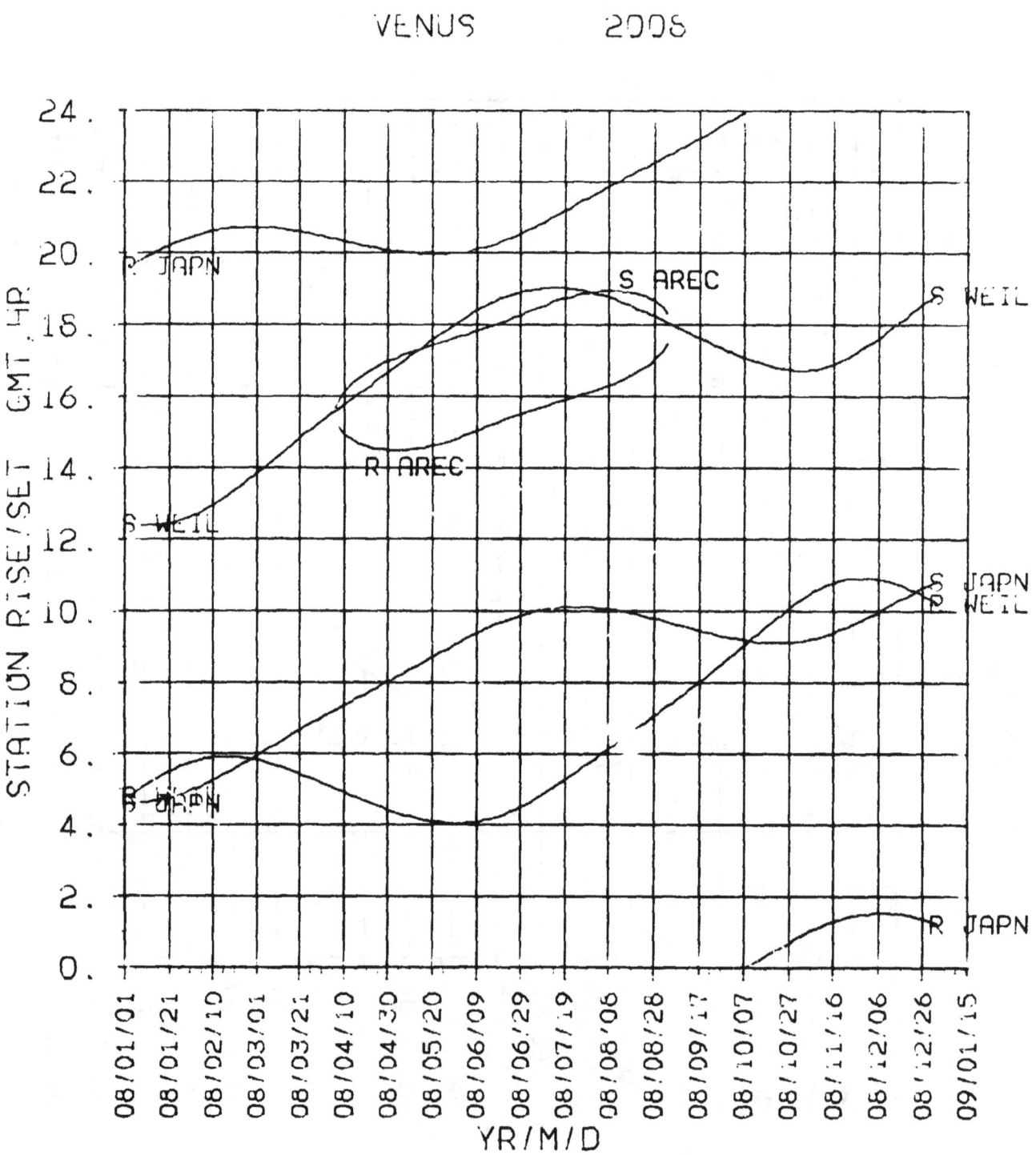

Venus

2009

DECLIN RT.ASC 2009

VENUS 2009

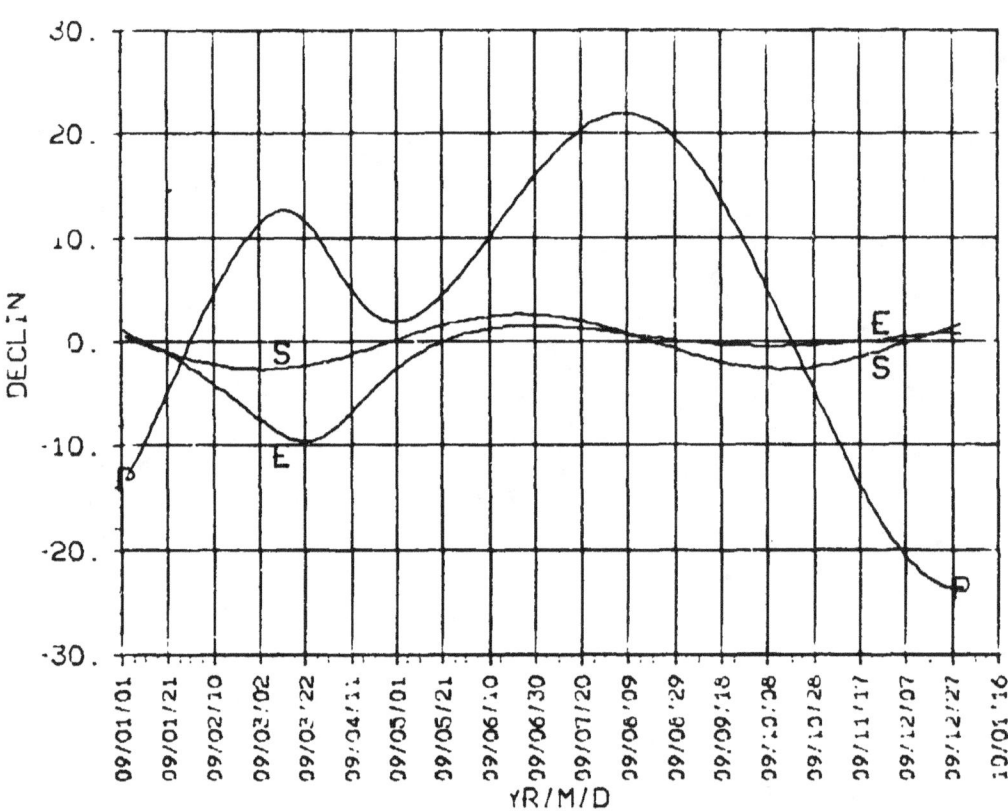

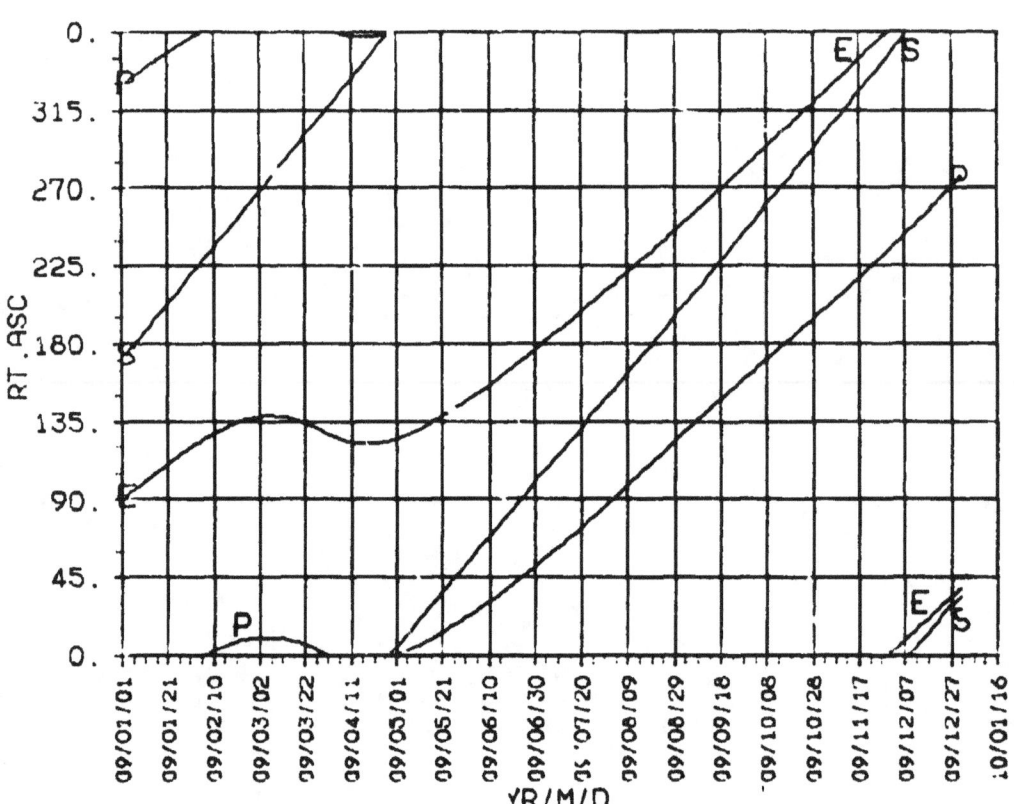

VENUS 2009

DISTANCE
EC.LON
2009

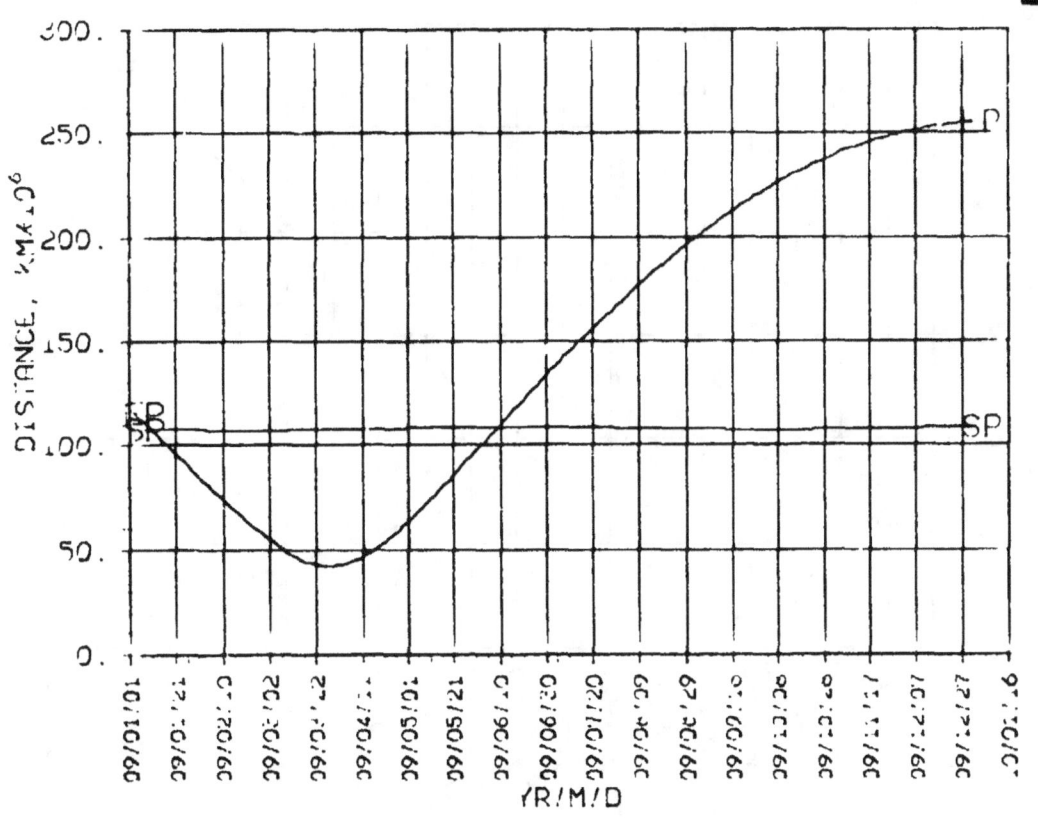

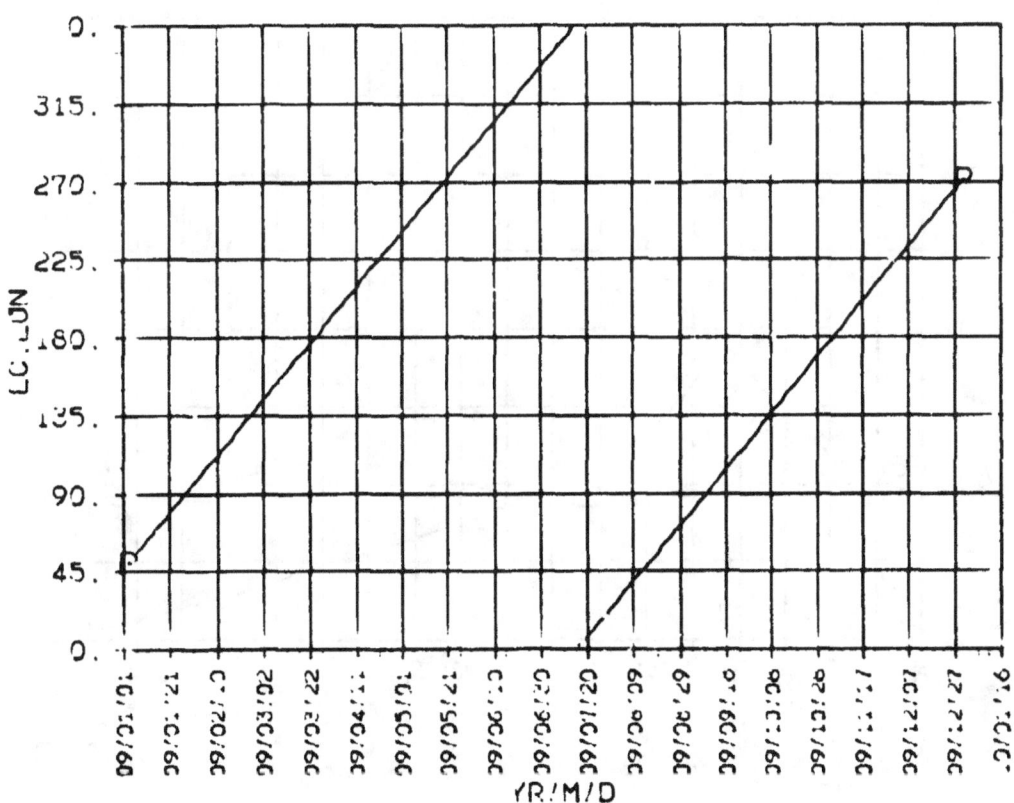

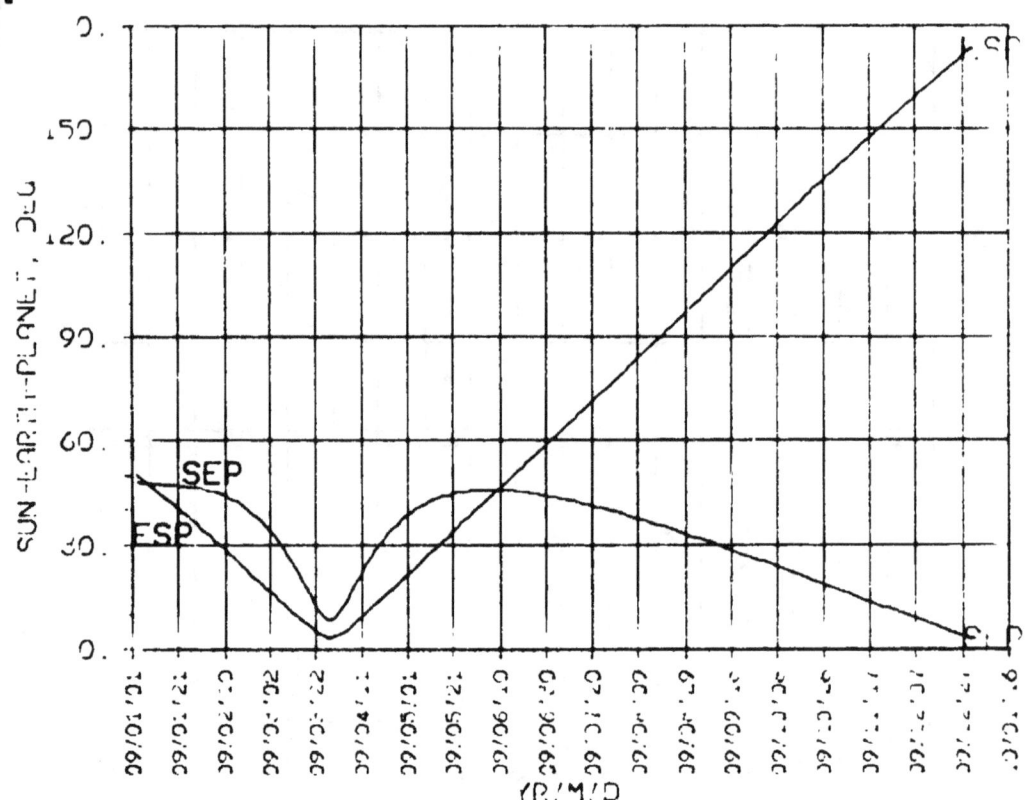
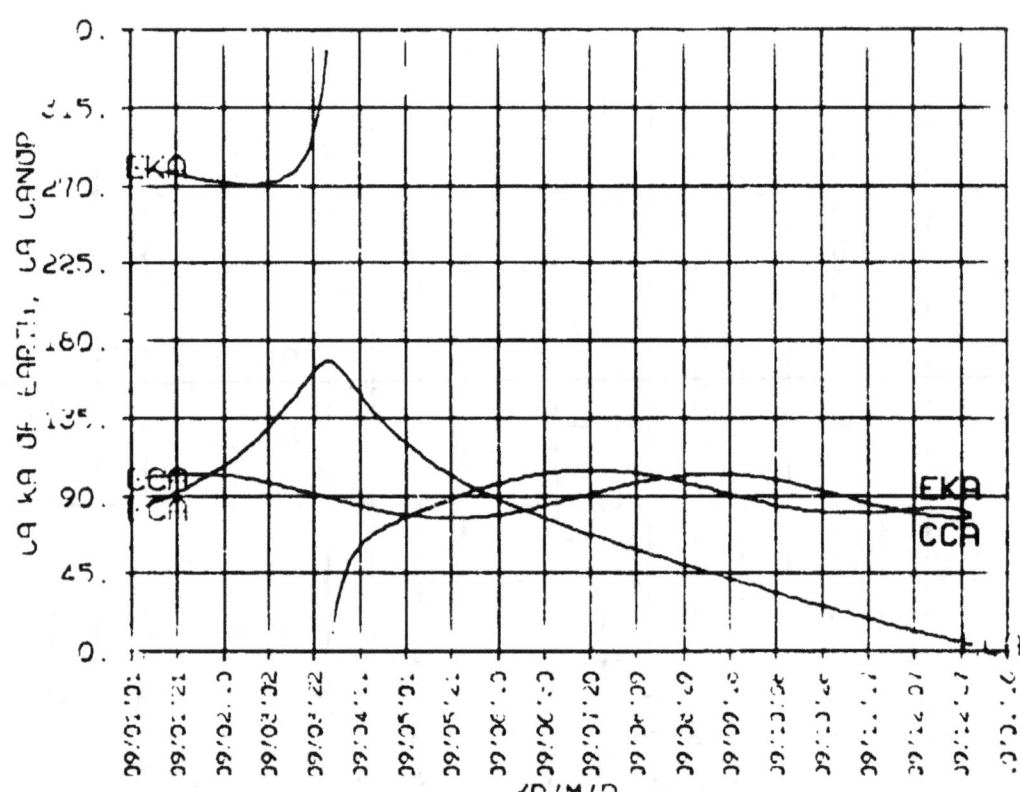

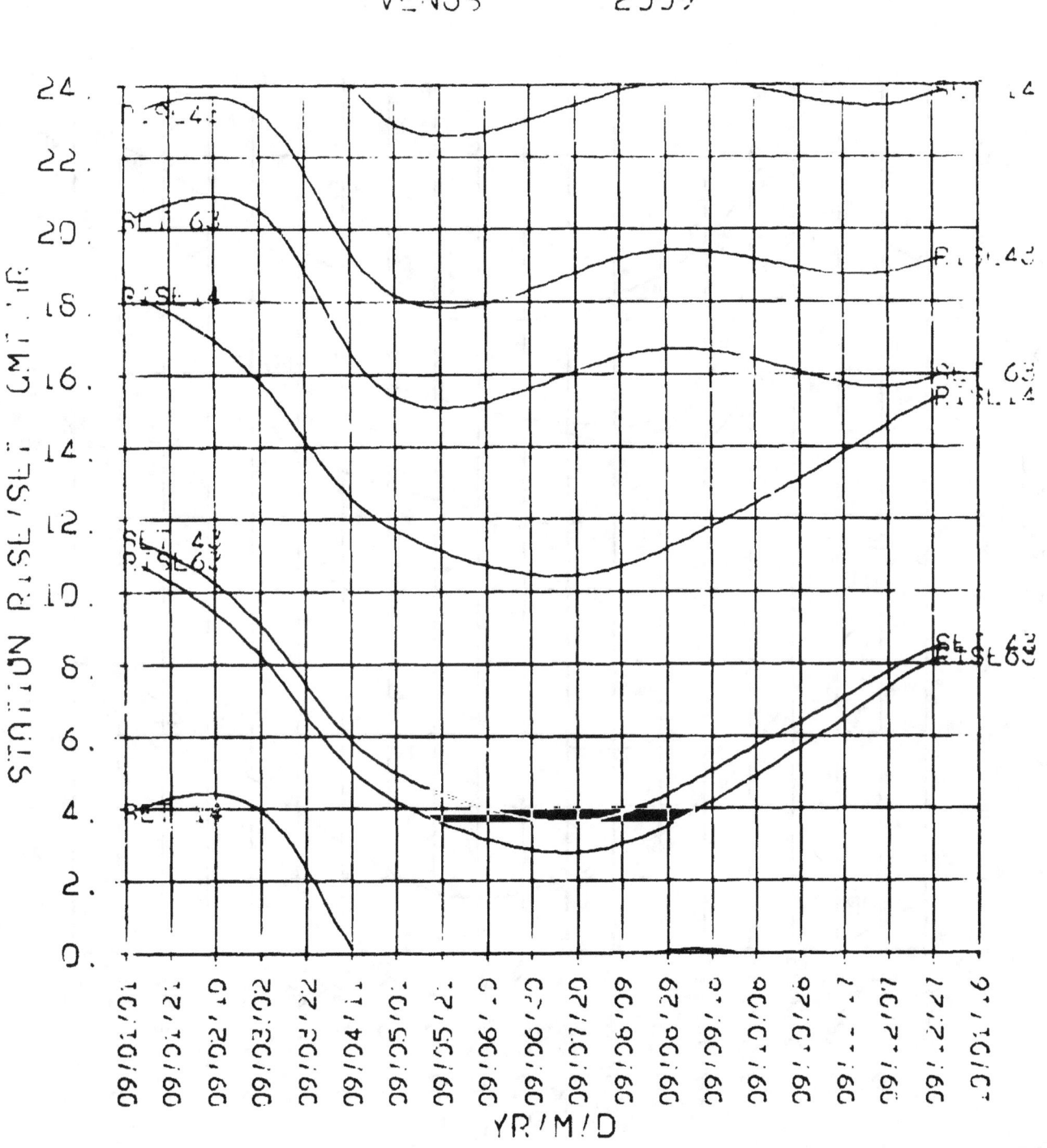

STA R/S
NON-DSN
2009

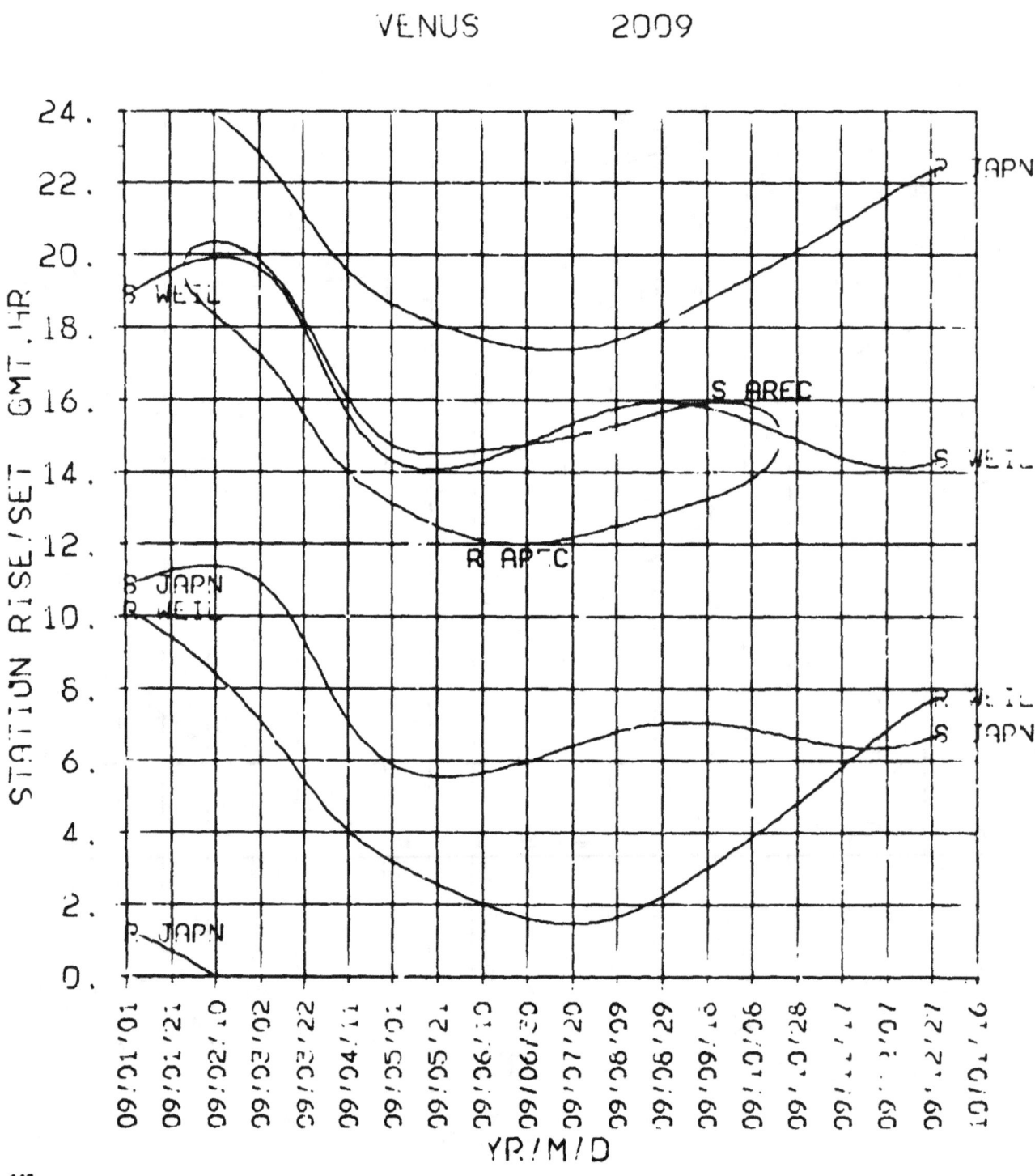

Venus

2010

DECLIN RT.ASC 2010

VENUS 2010

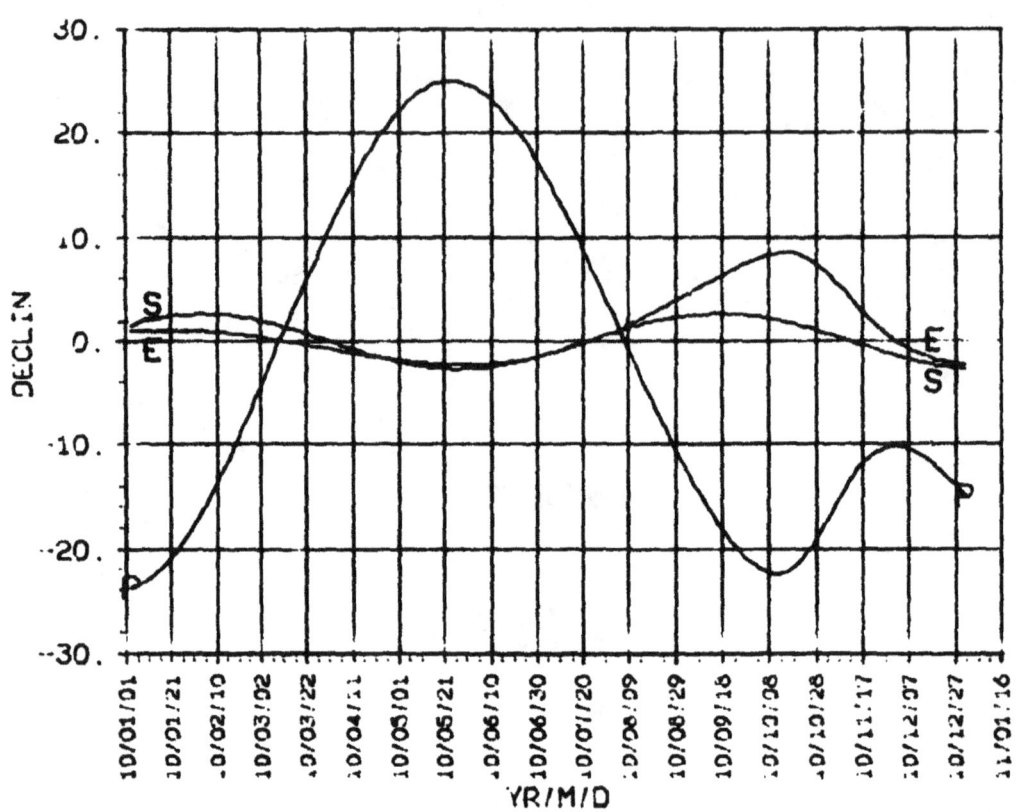

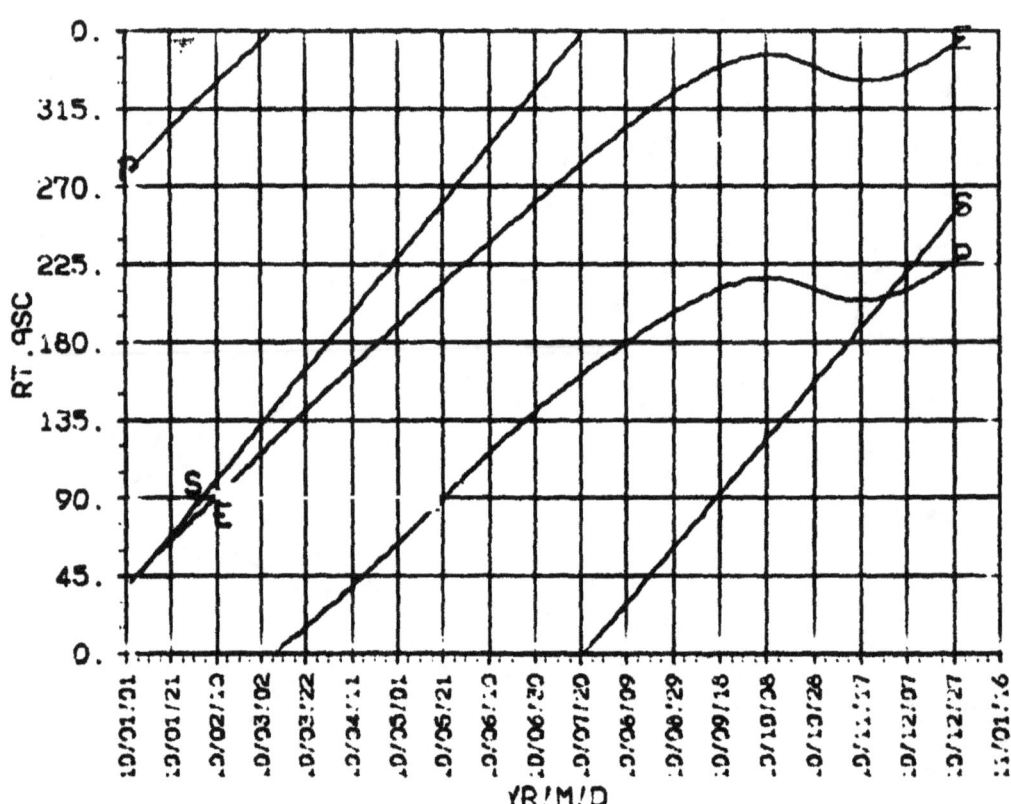

VENUS 2010

DISTANCE
EC.LON
2010

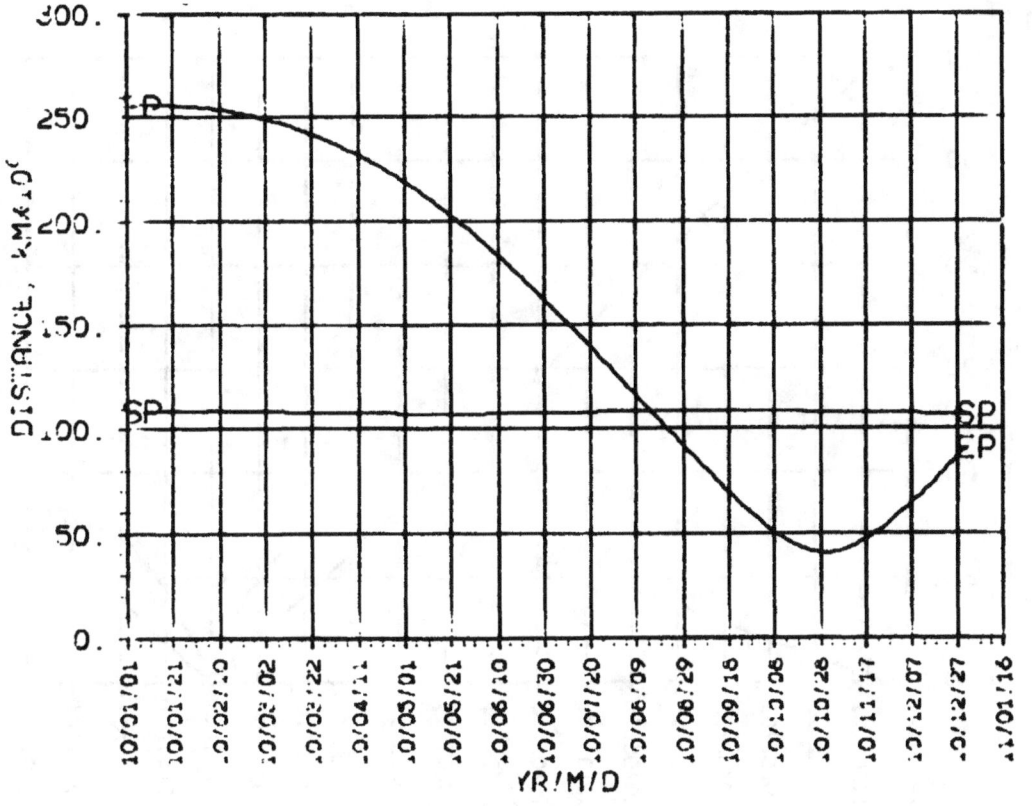

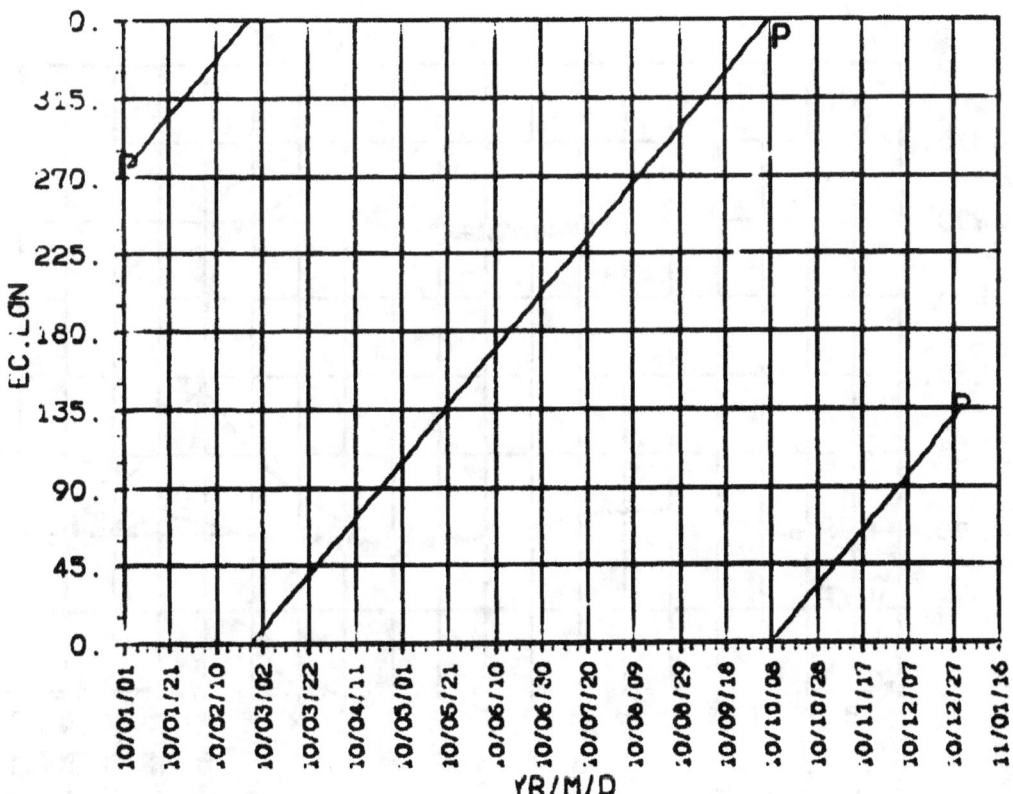

SEP, ESP CA, KA 2010

VENUS 2010

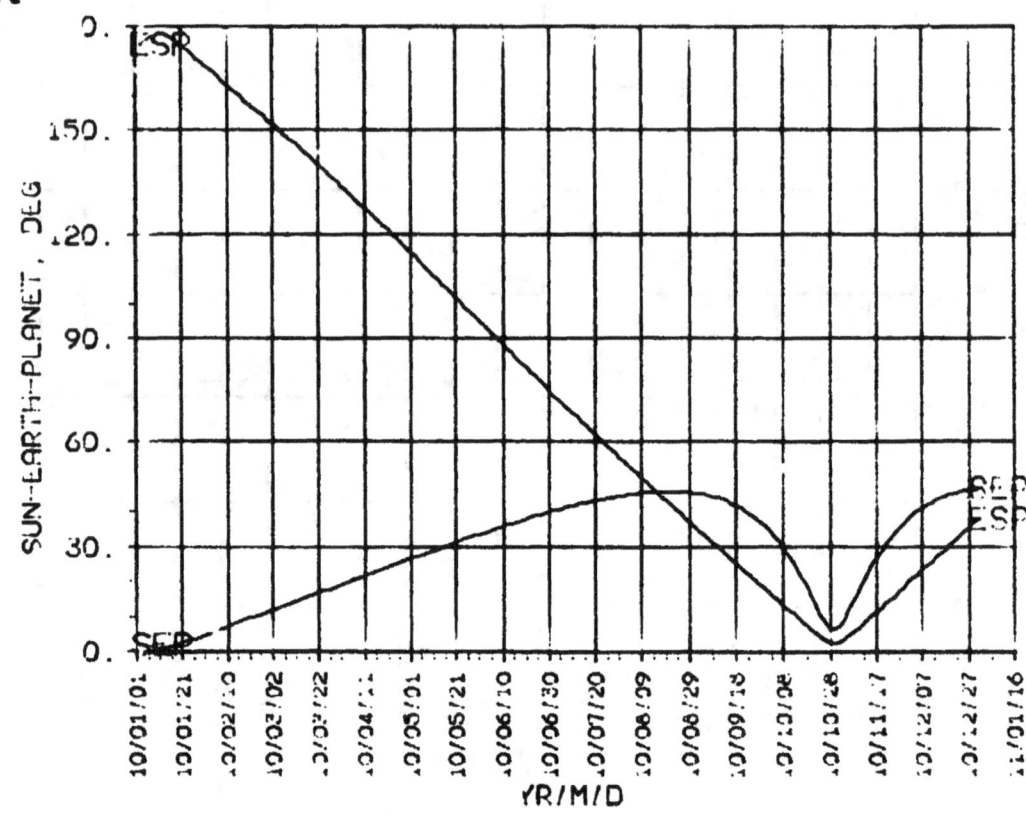

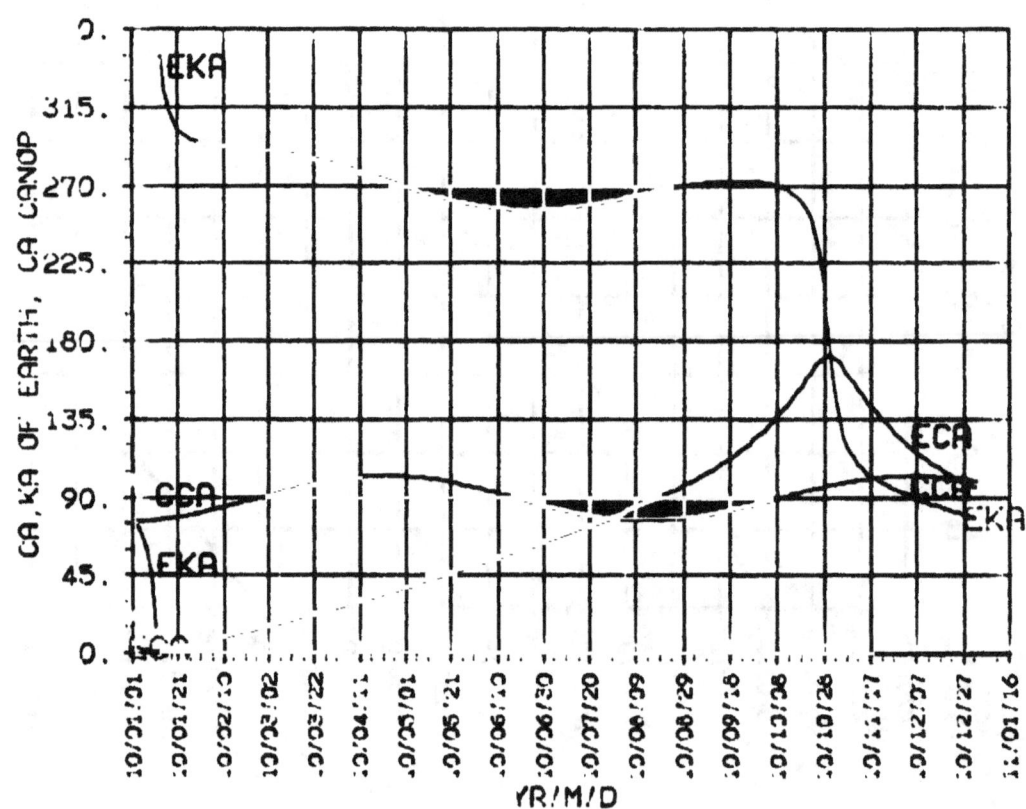

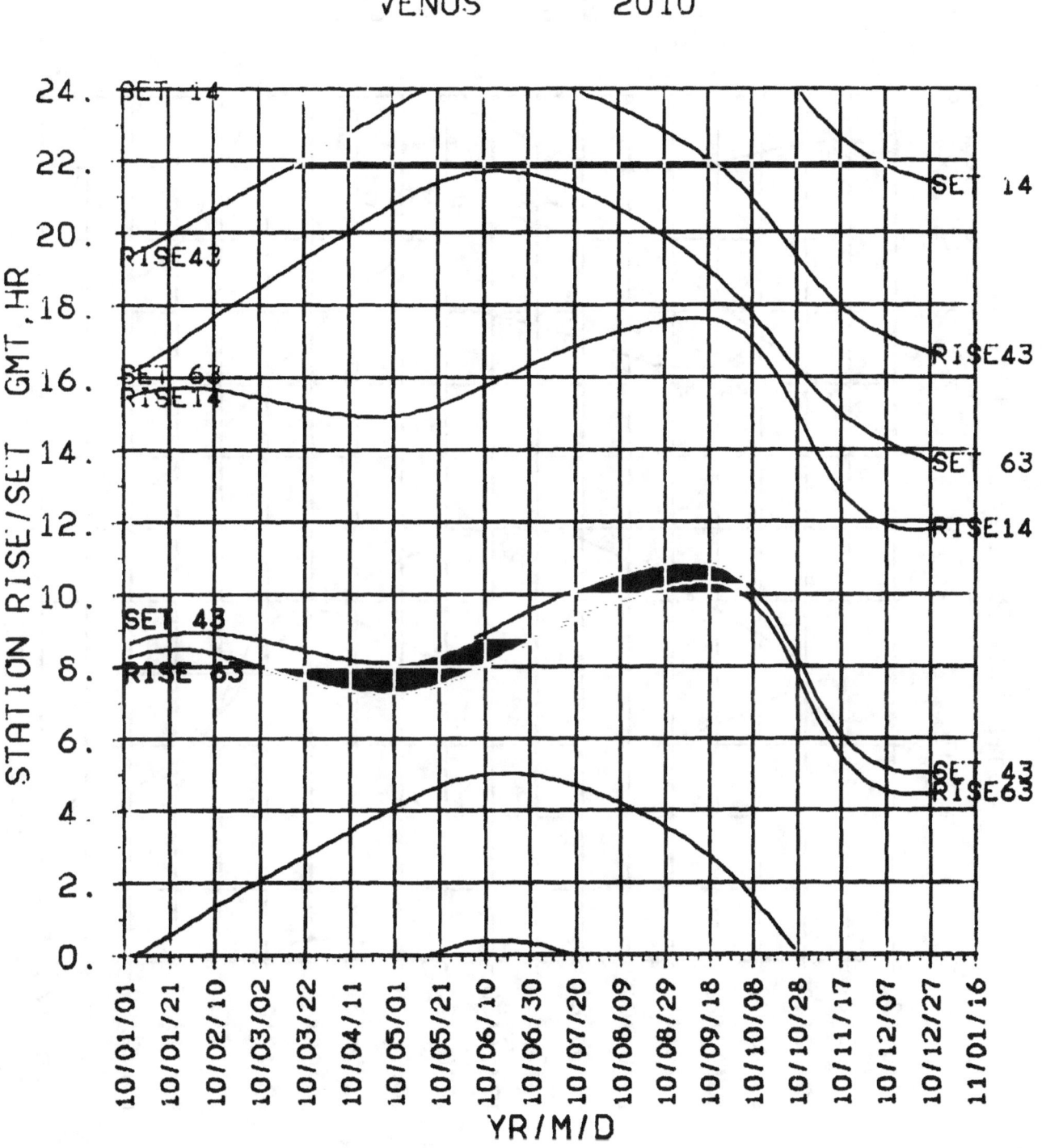

STA R/S NON-DSN 2010

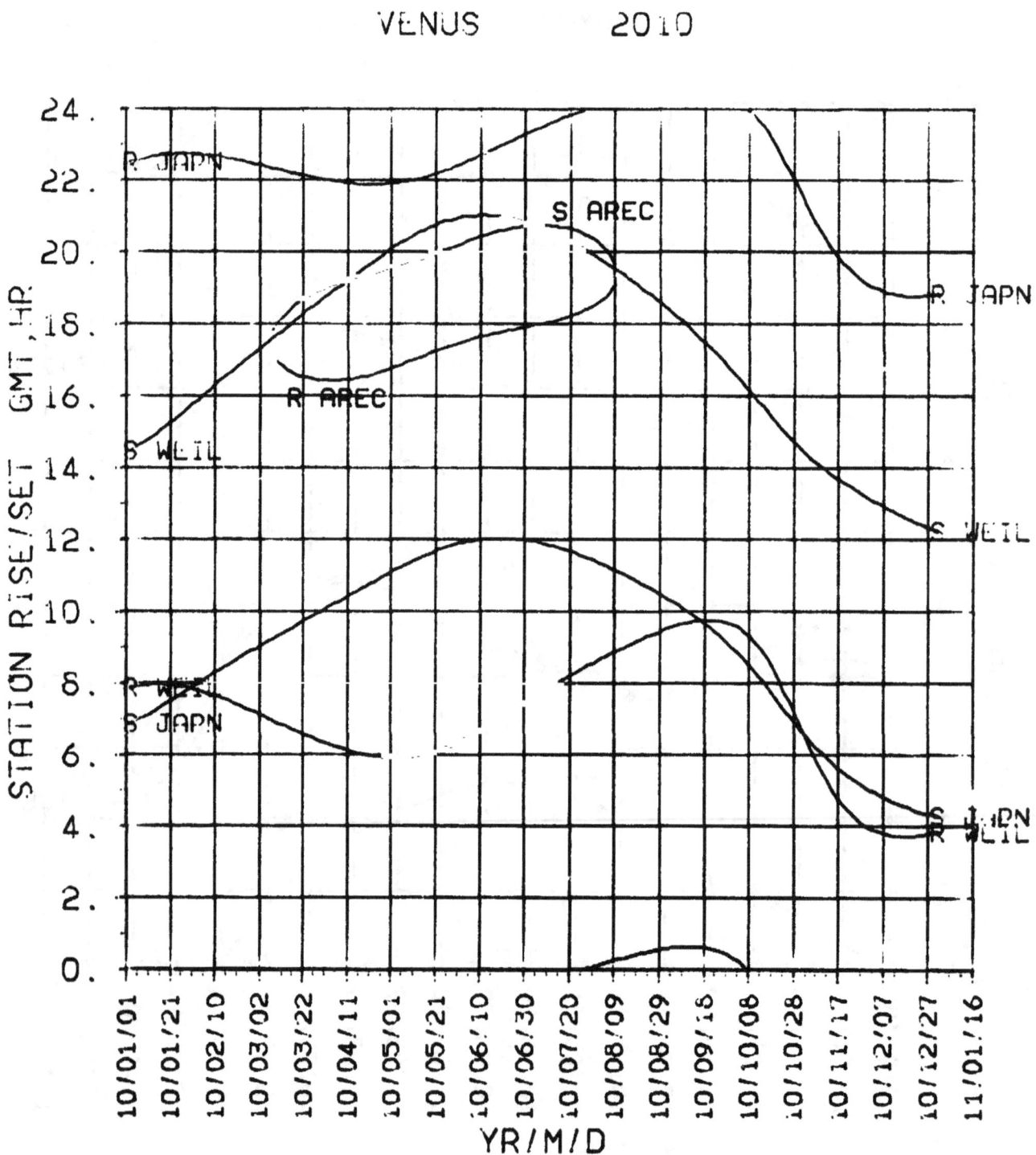

Venus

2011

DECLIN RT.ASC 2011

VENUS 2011

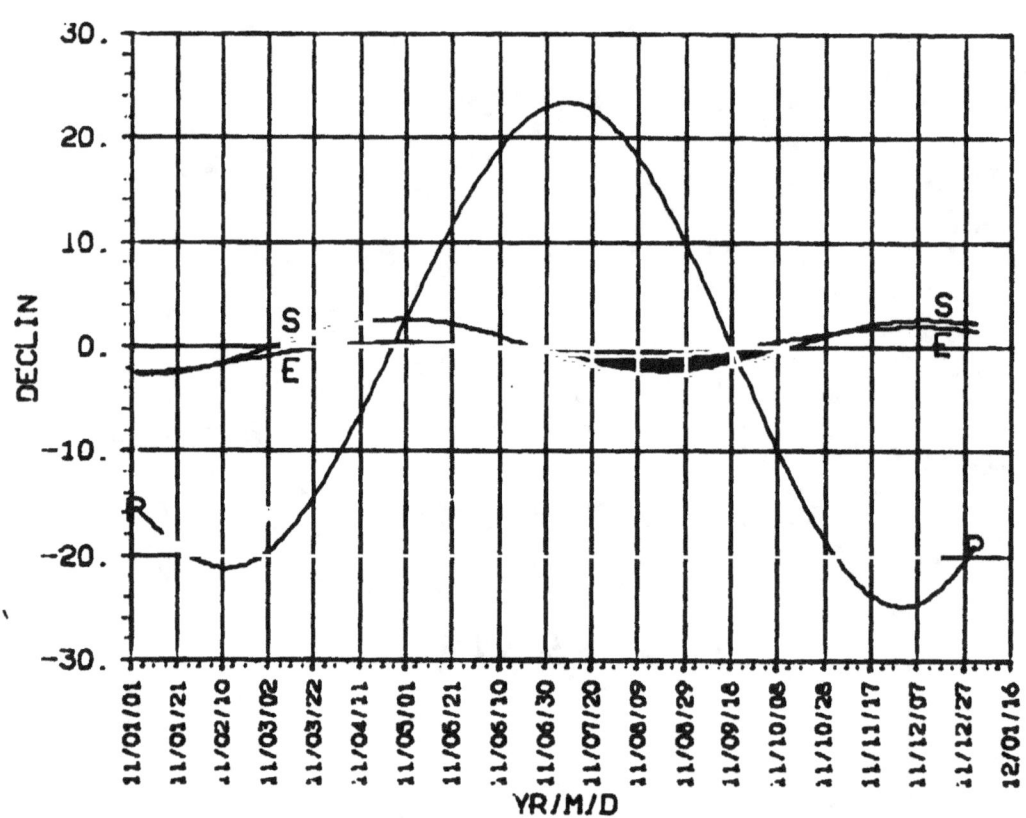

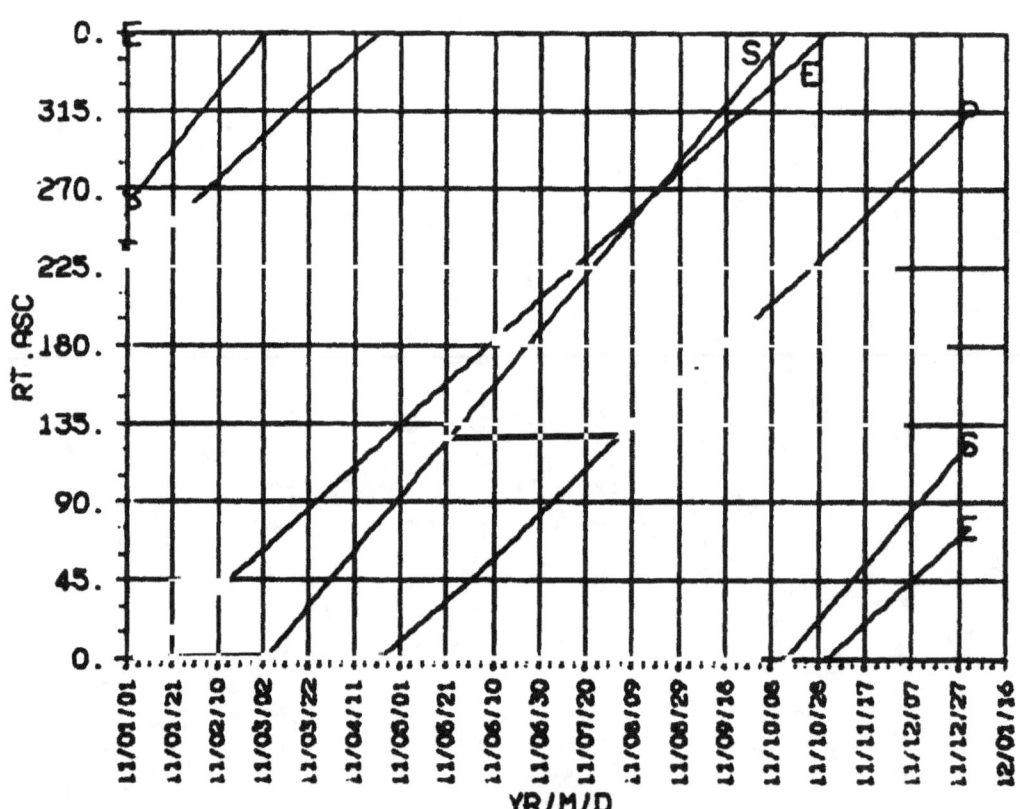

VENUS 2011

DISTANCE EC.LON 2011

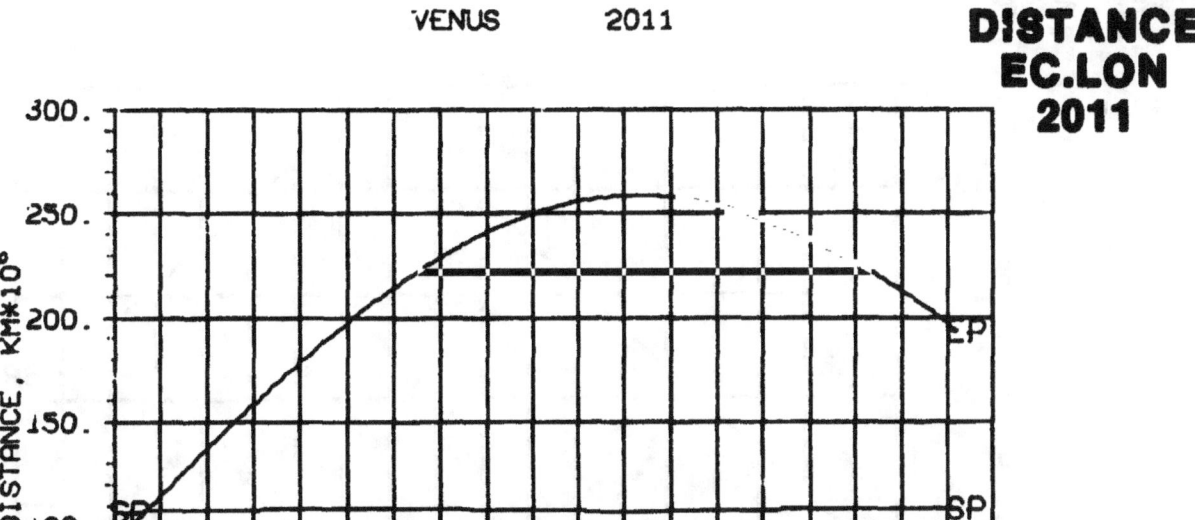

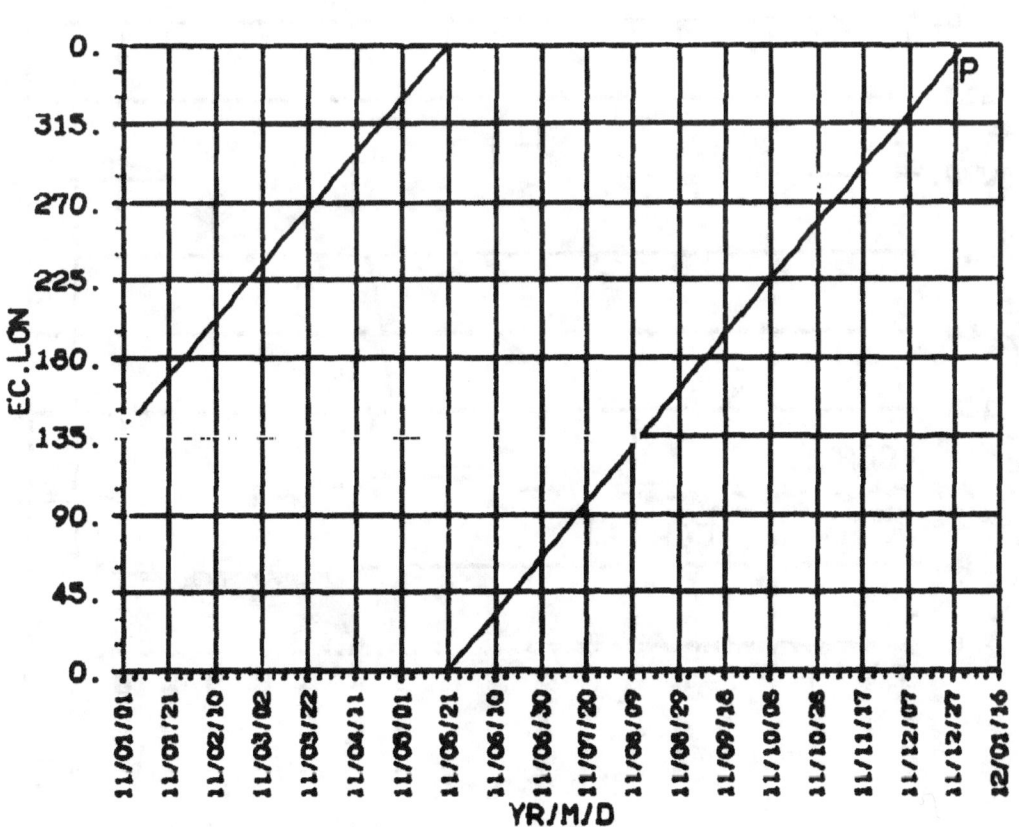

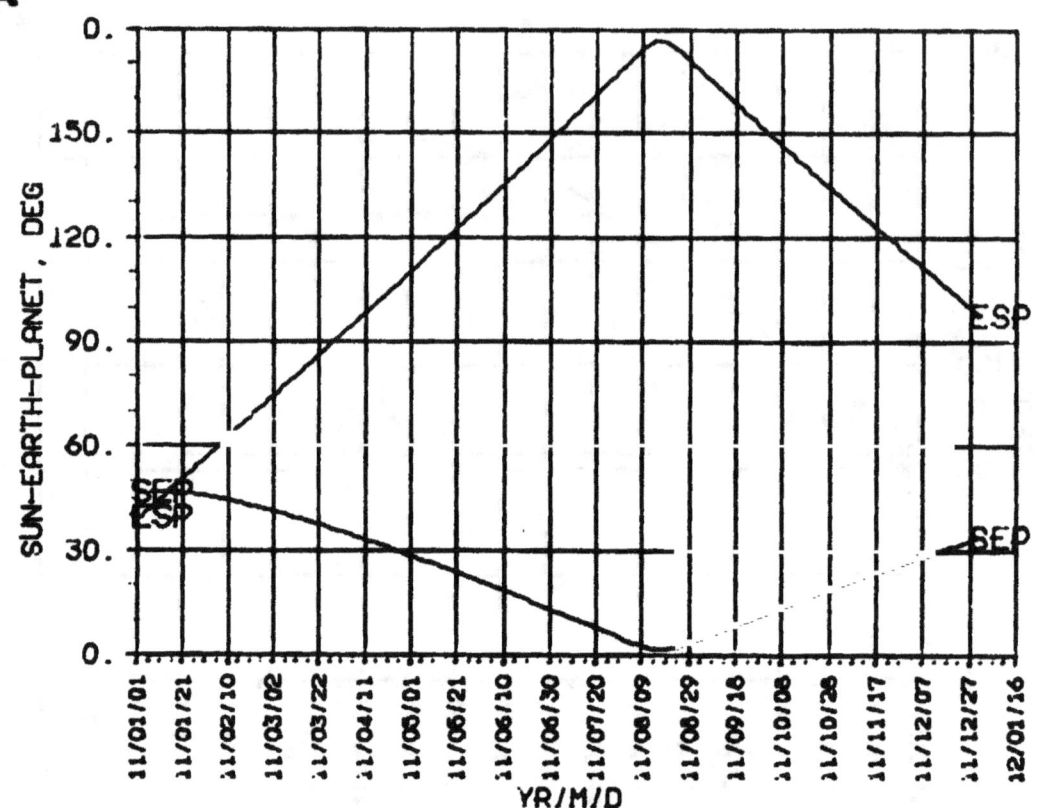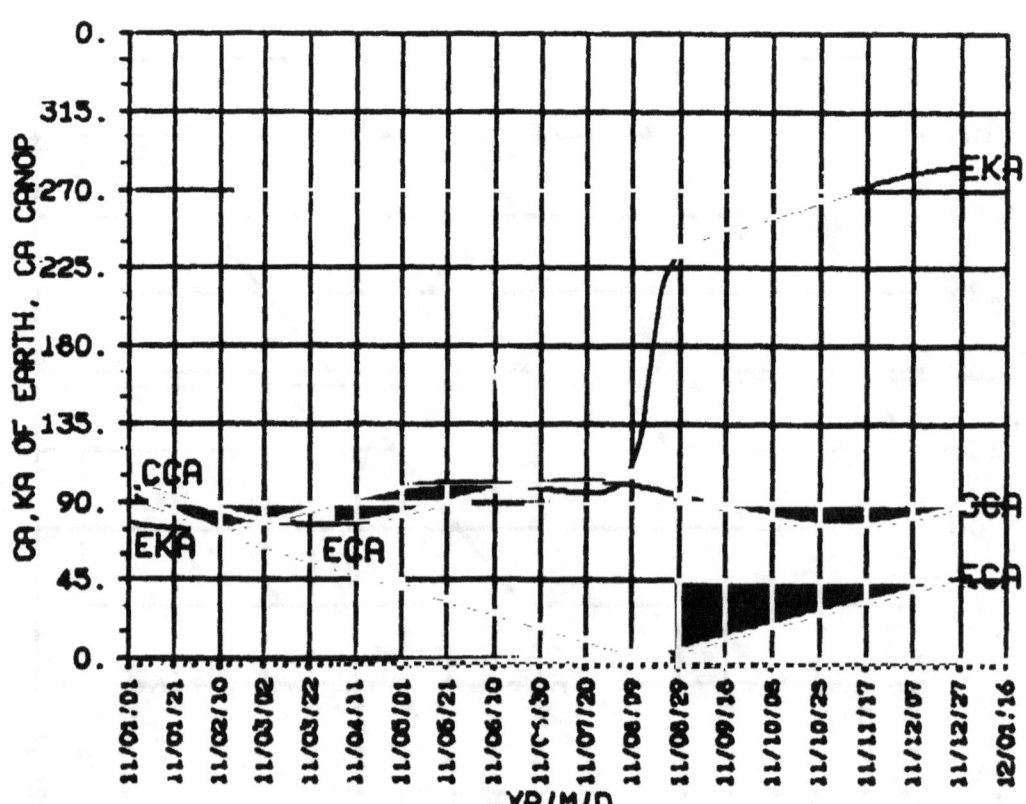

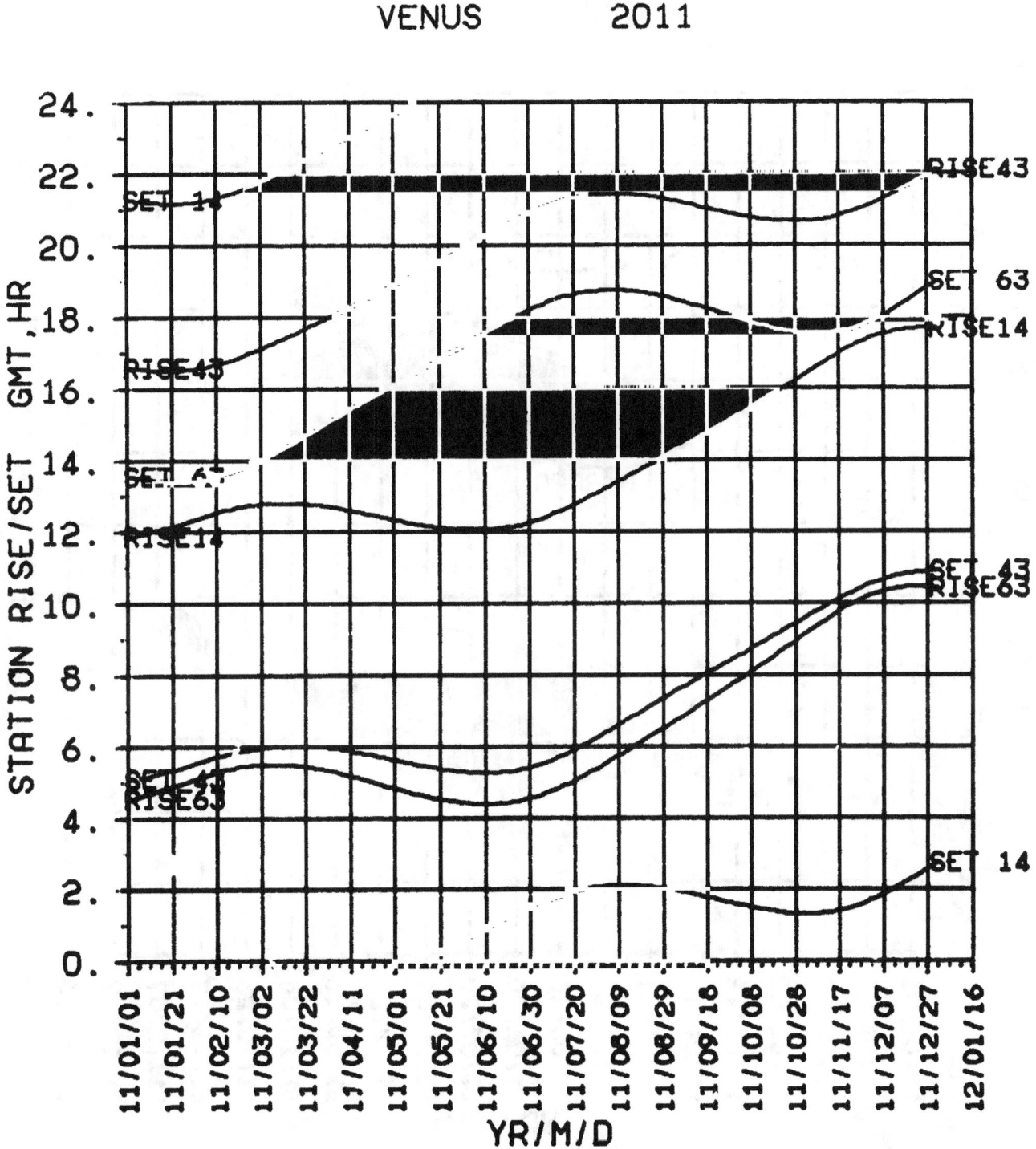

STA R/S NON-DSN 2011

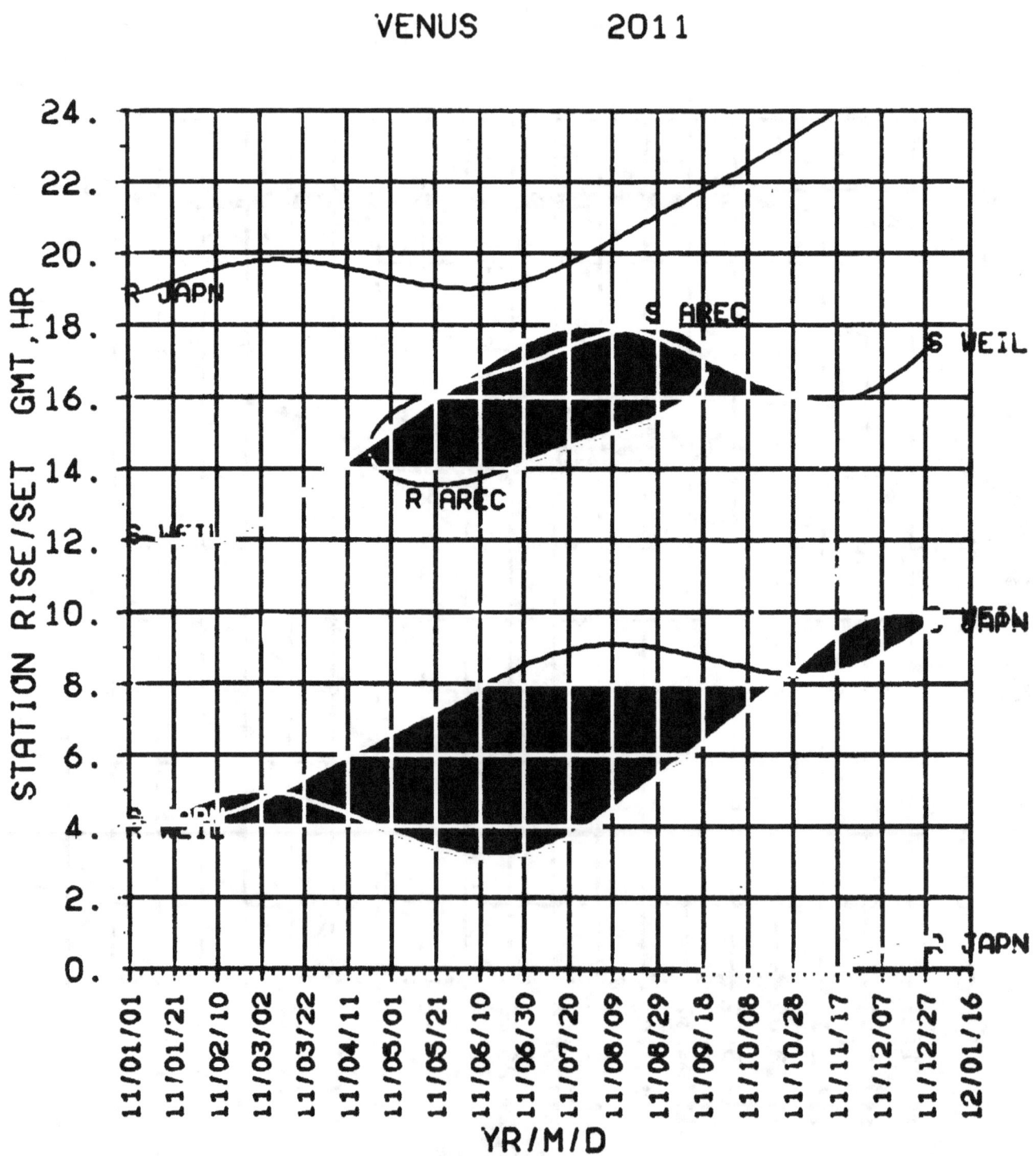

Venus

2012

DECLIN RT.ASC 2012

VENUS 2012

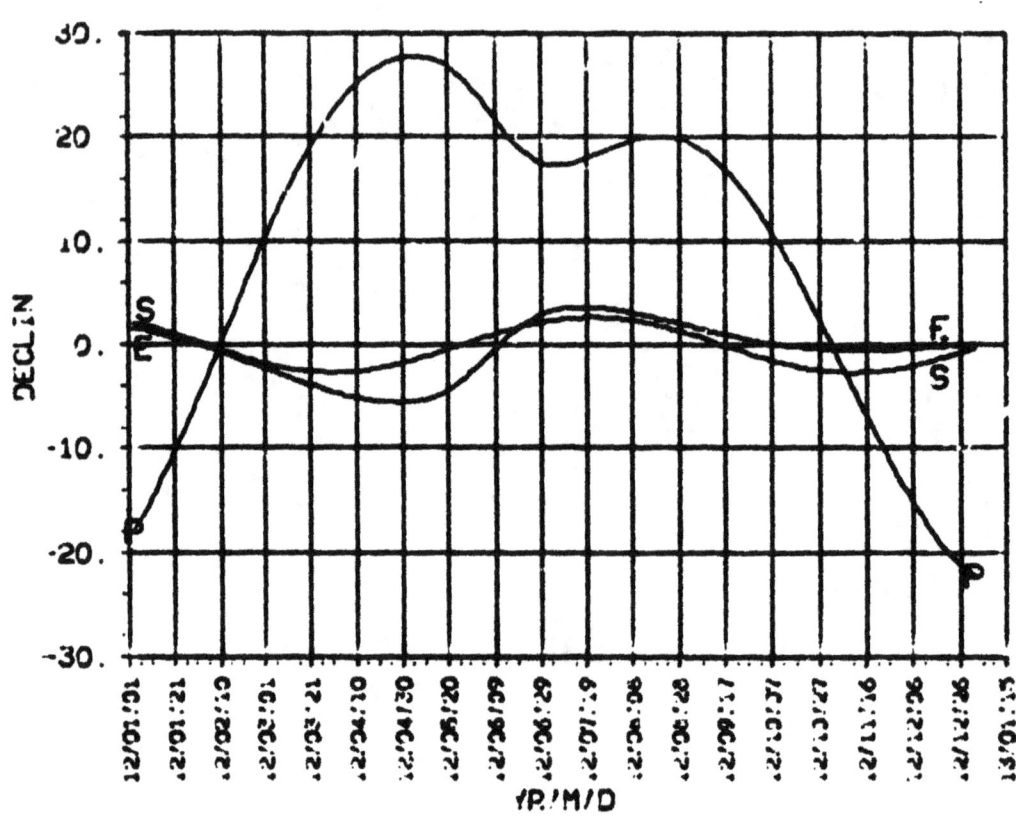

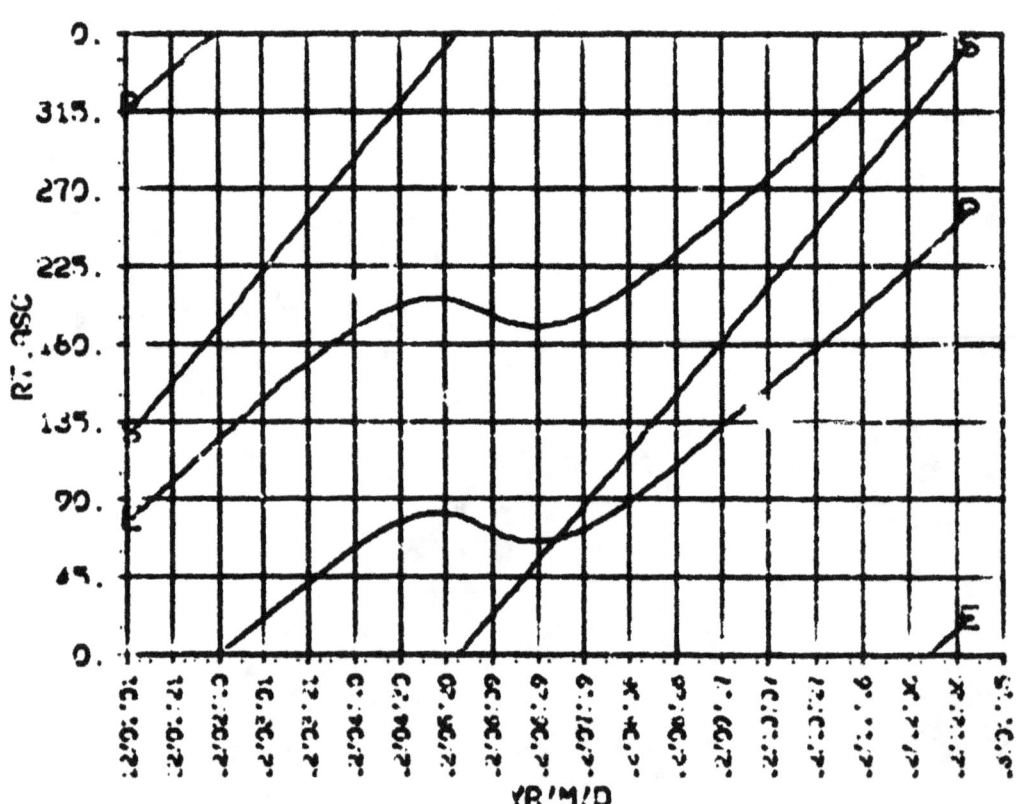

DISTANCE EC.LON 2012

VENUS 2012

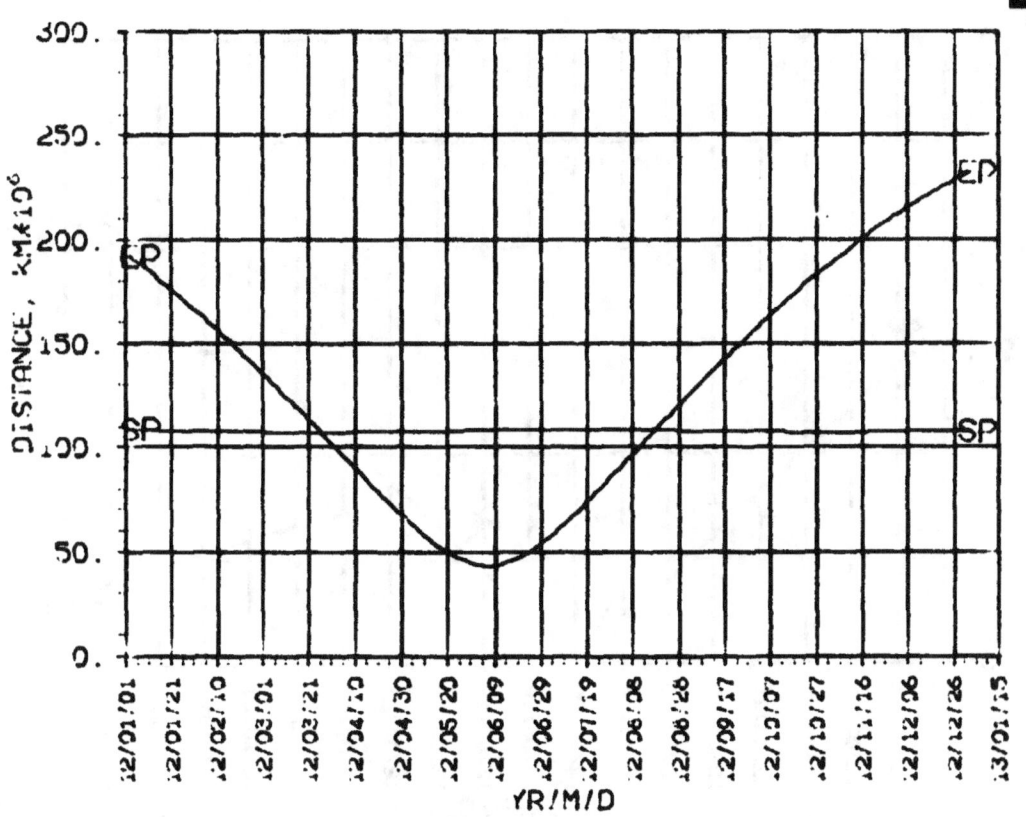

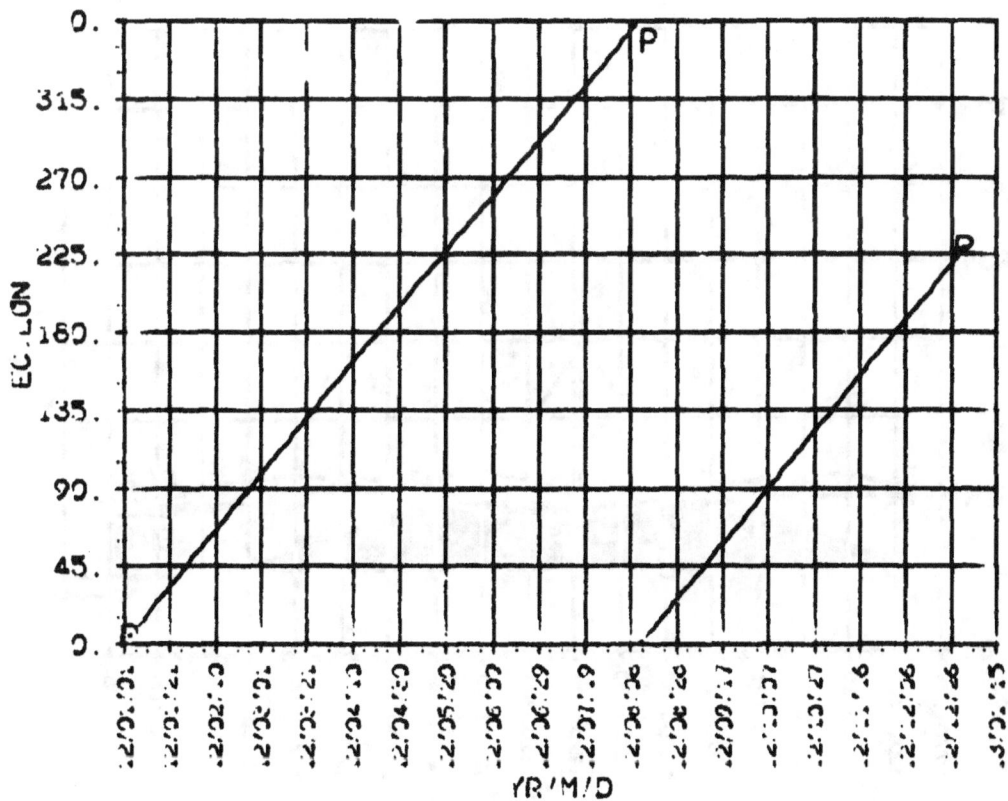

SEP, ESP CA, KA 2012

VENUS 2012

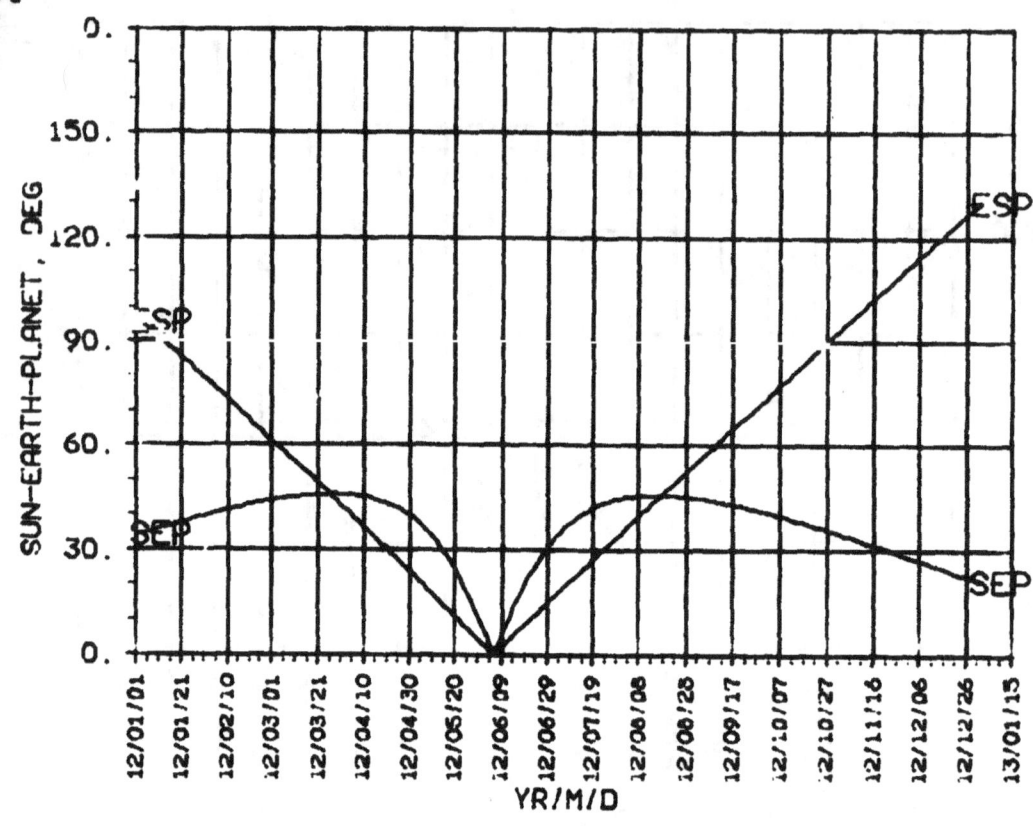

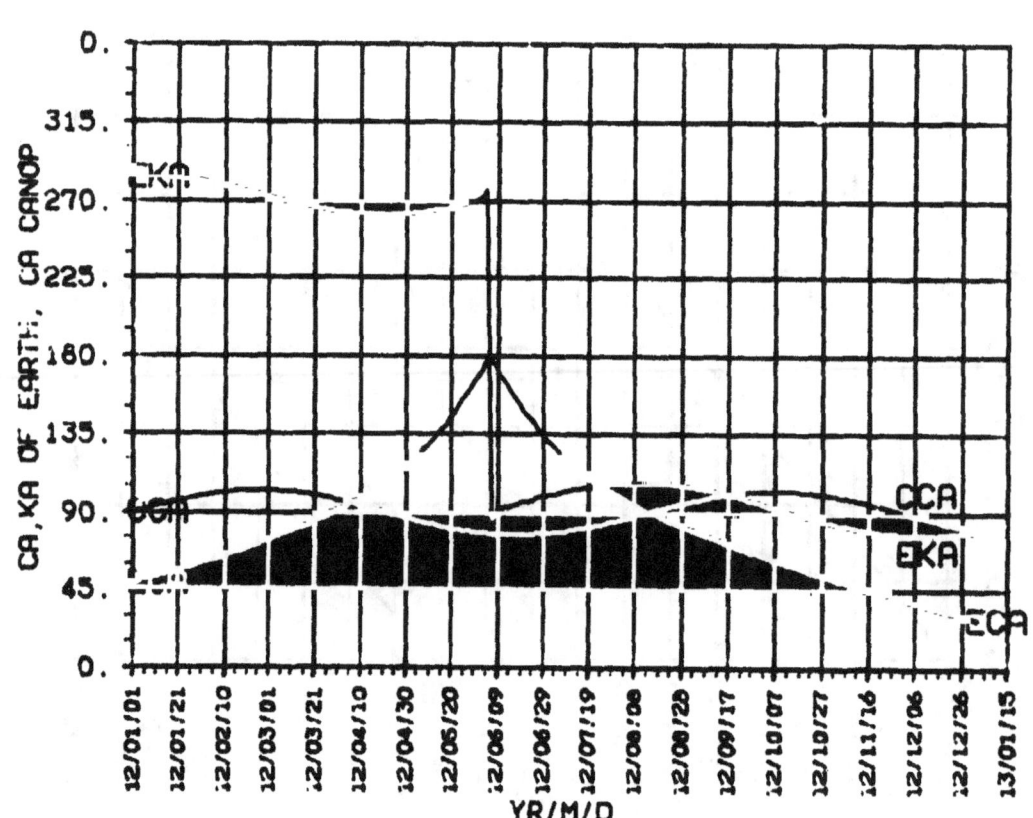

VENUS 2012

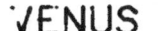

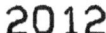

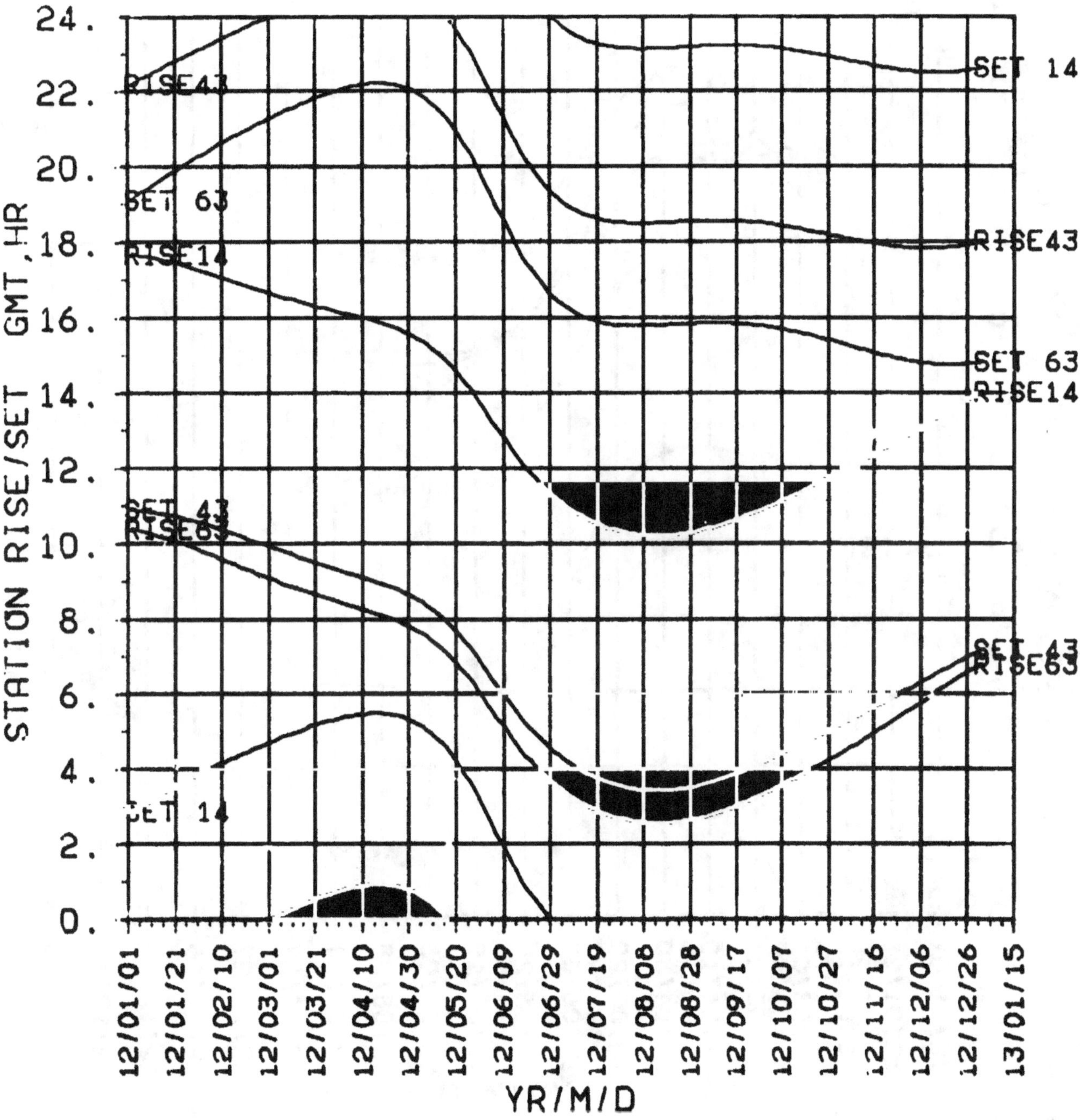
STATION RISE/SET GMT,HR
24.
22.
20.
18.
16.
14.
12.
10.
8.
6.
4.
2.
0.
RISE43
SET 63
RISE14
SET 43
RISE63
SET 14
SET 14
RISE43
SET 63
RISE14
SET 43
RISE63
12/01/01
12/01/21
12/02/10
12/03/01
12/03/21
12/04/10
12/04/30
12/05/20
12/06/09
12/06/29
12/07/19
12/08/08
12/08/28
12/09/17
12/10/07
12/10/27
12/11/16
12/12/06
12/12/26
13/01/15
YR/M/D

STA R/S NON-DSN 2012

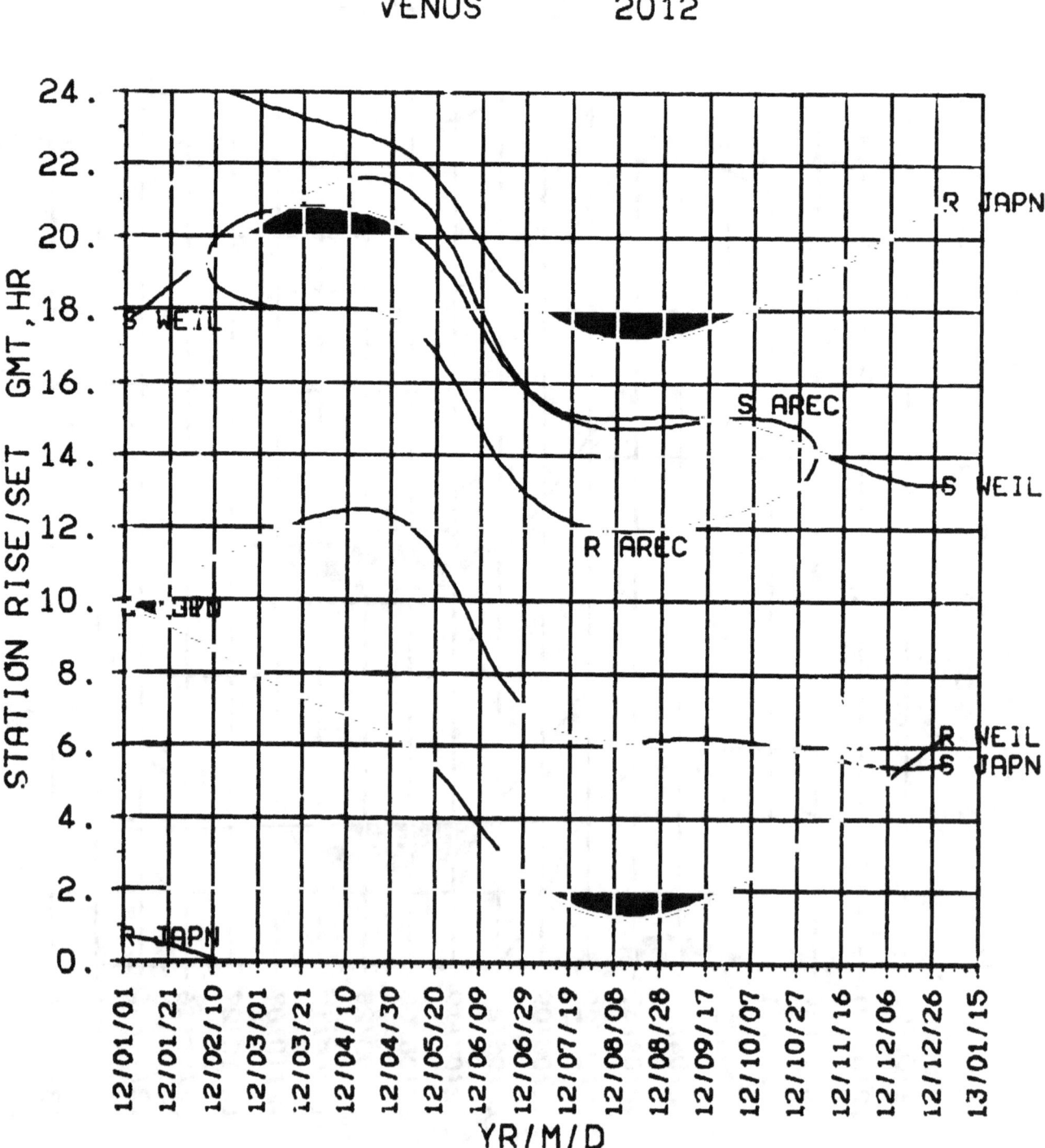

Venus

2013

DECLIN RT.ASC 2013

VENUS 2013

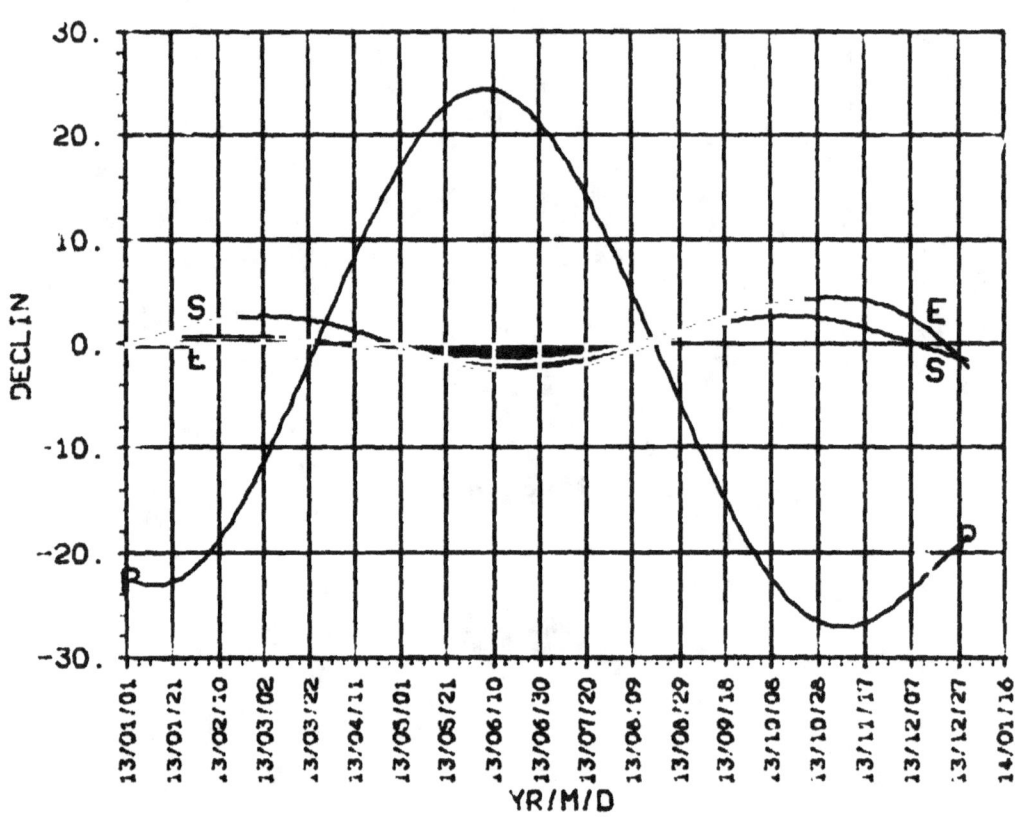

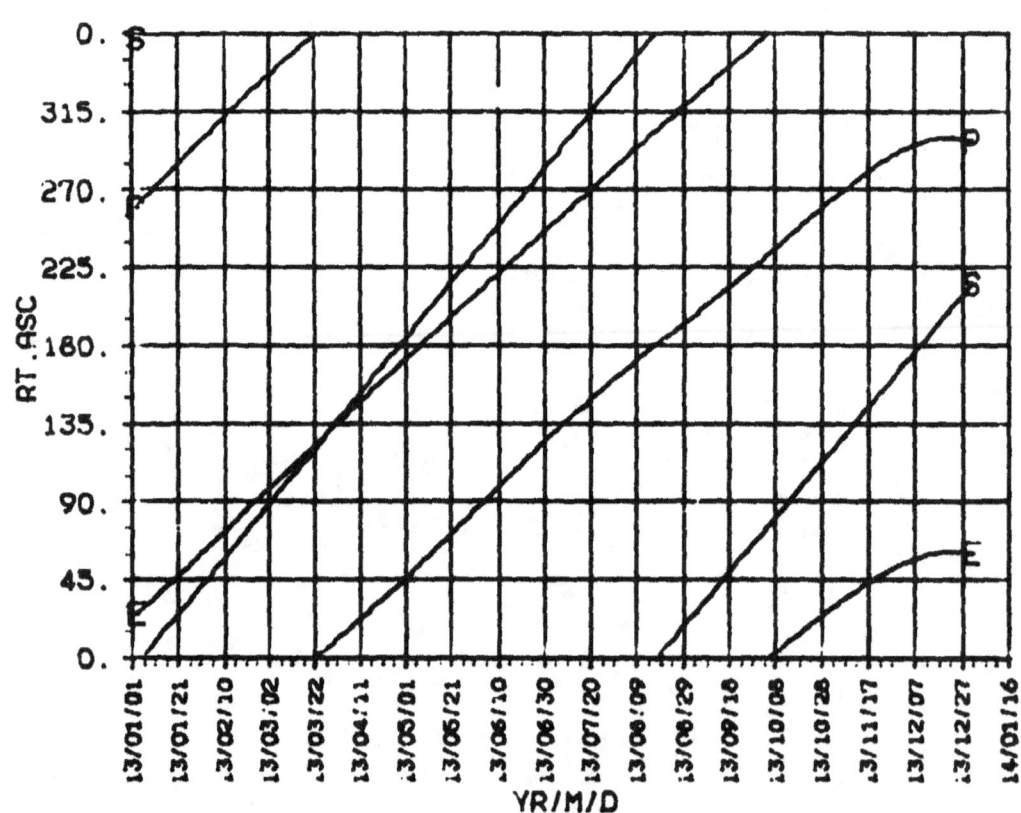

VENUS 2013

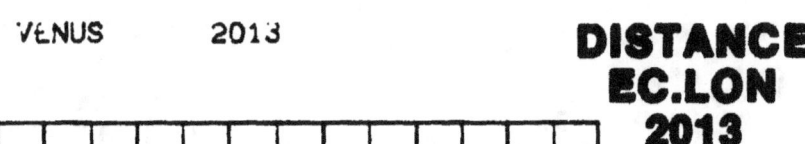

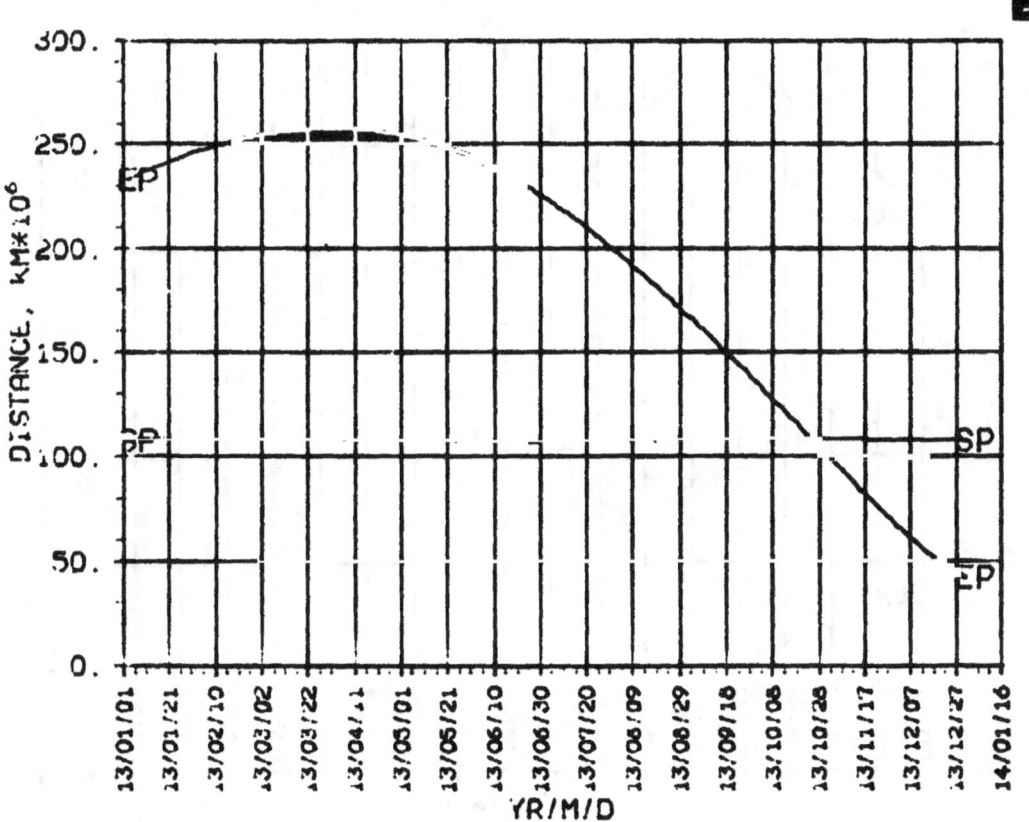

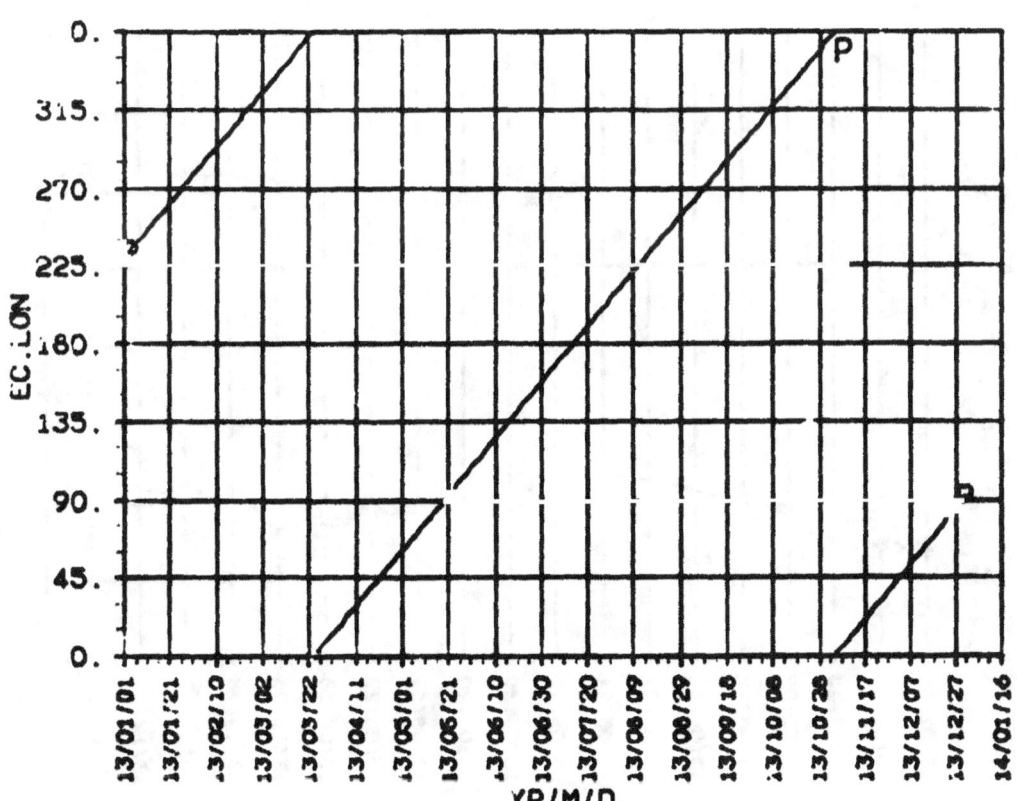

SEP, ESP CA, KA 2013

VENUS 2013

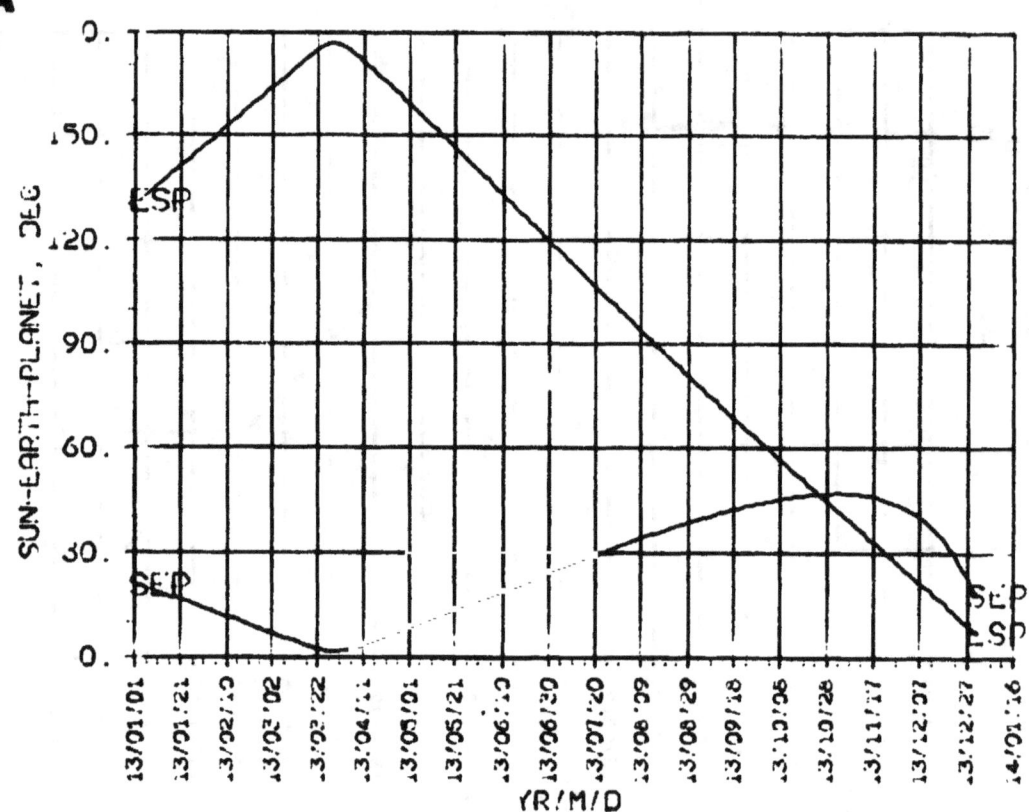

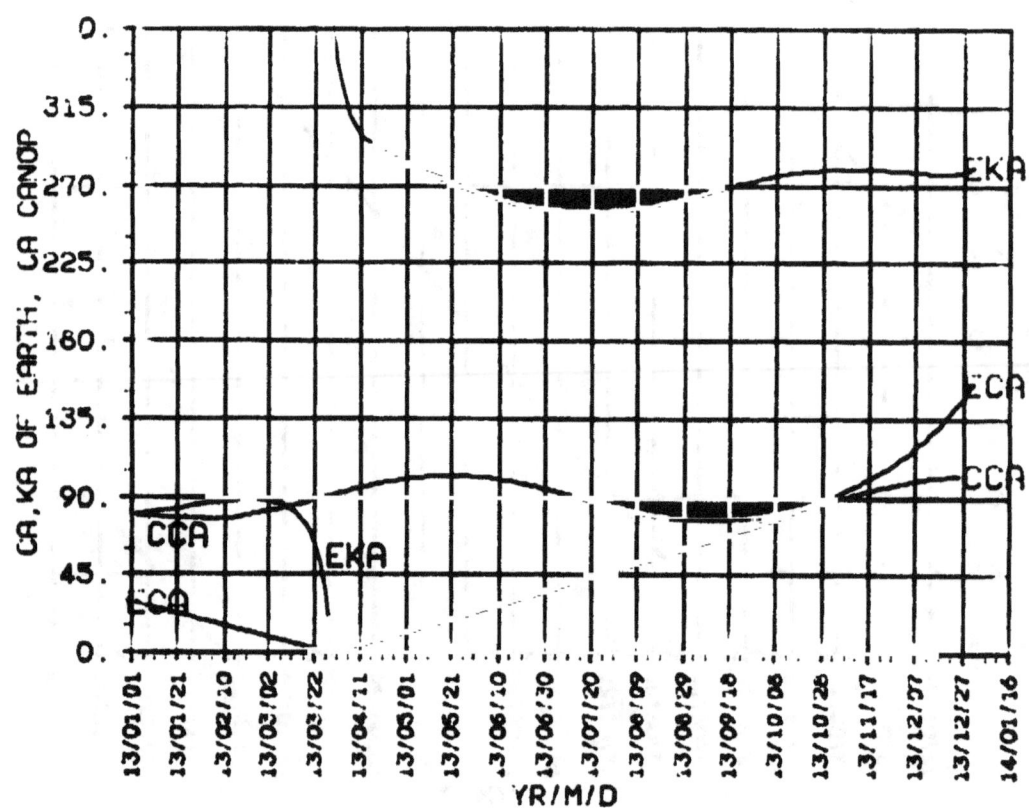

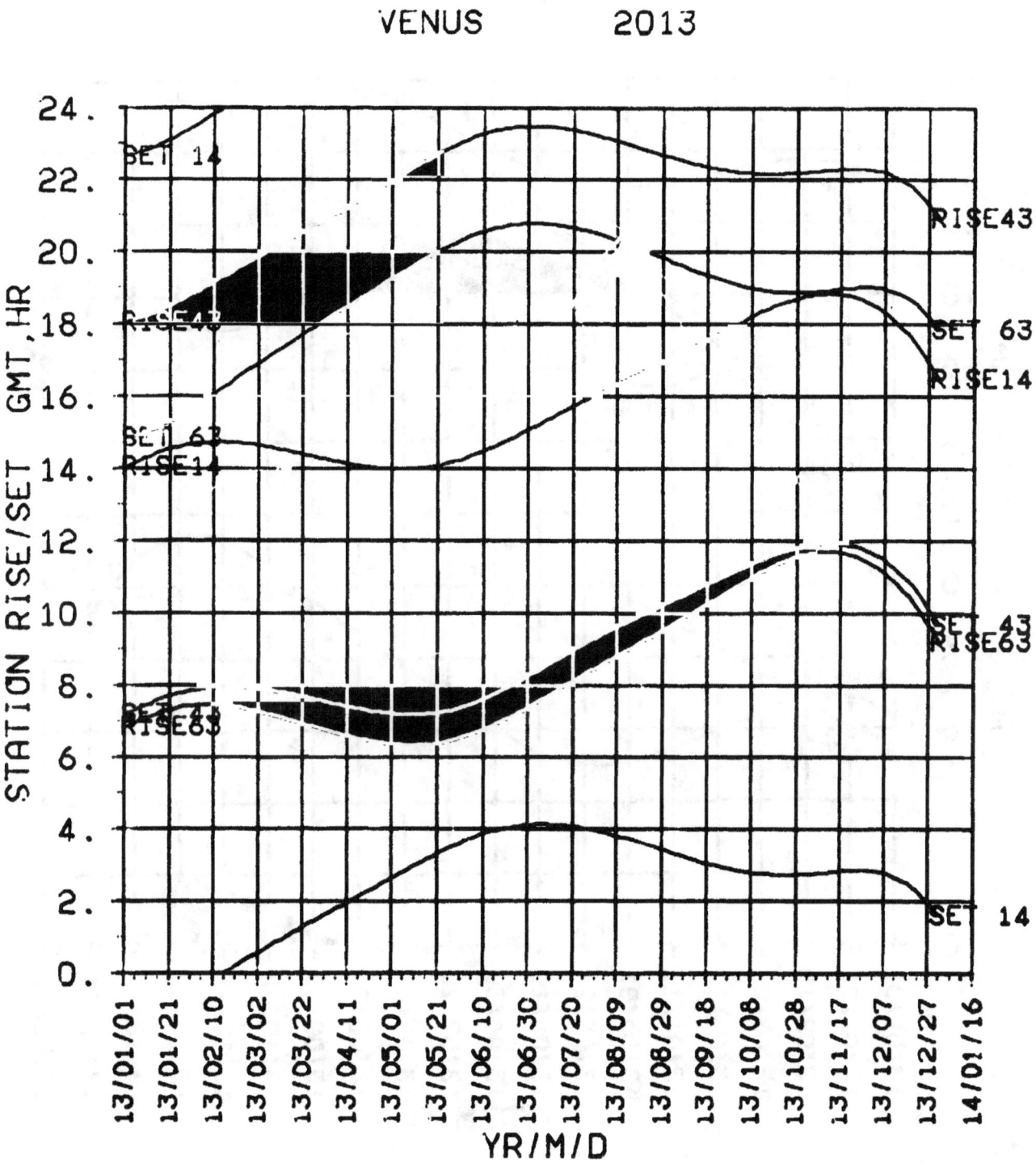

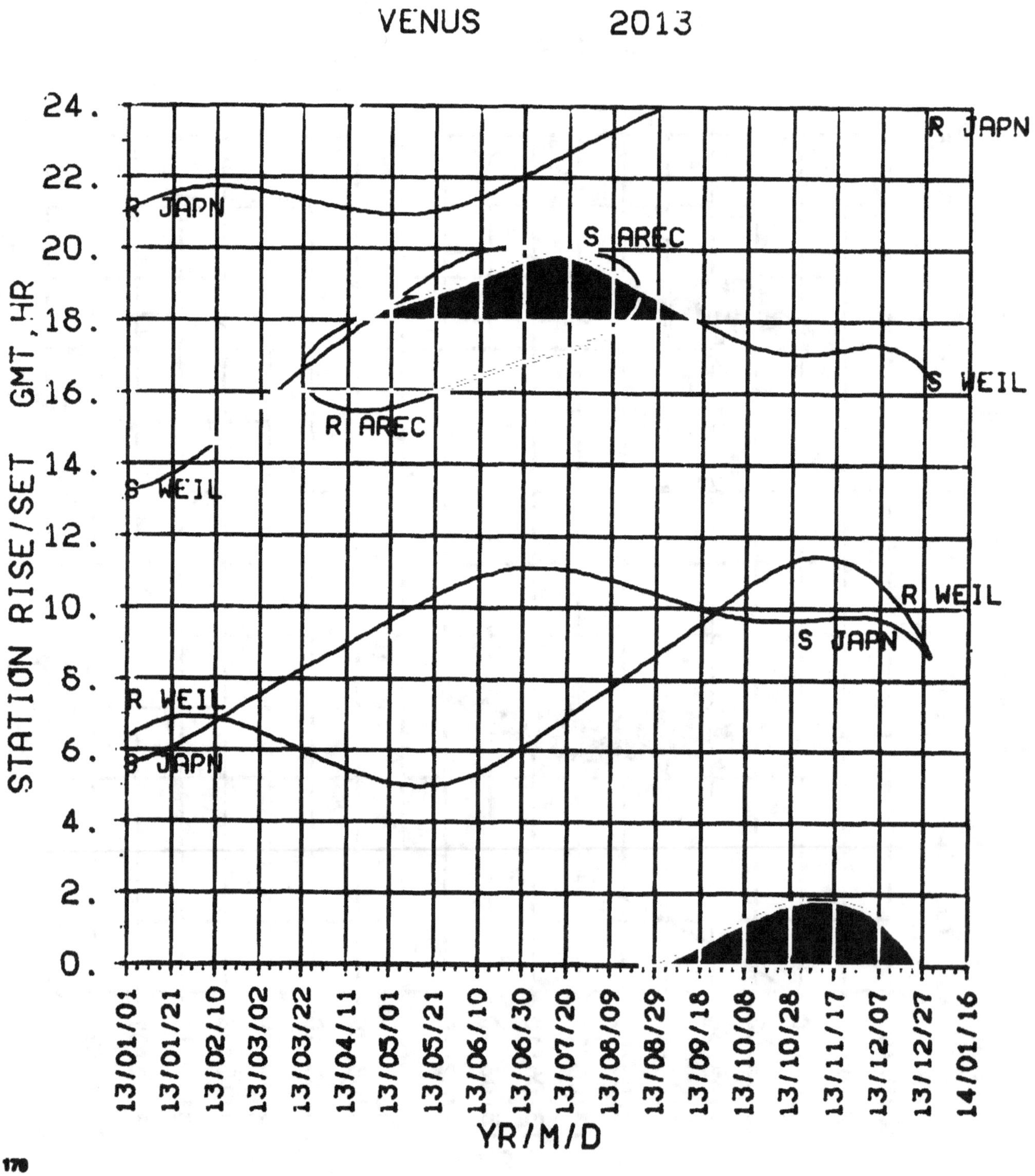

STA R/S NON-DSN 2013

VENUS 2013

Venus

2014

DECLIN RT.ASC 2014

VENUS 2014

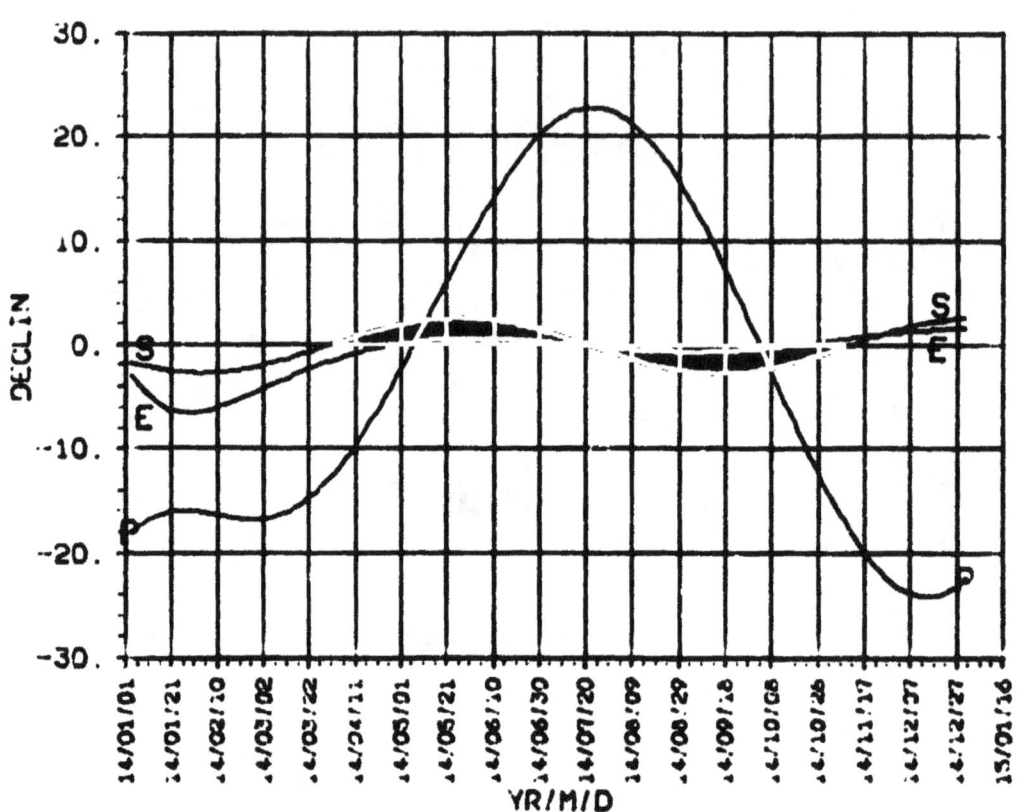

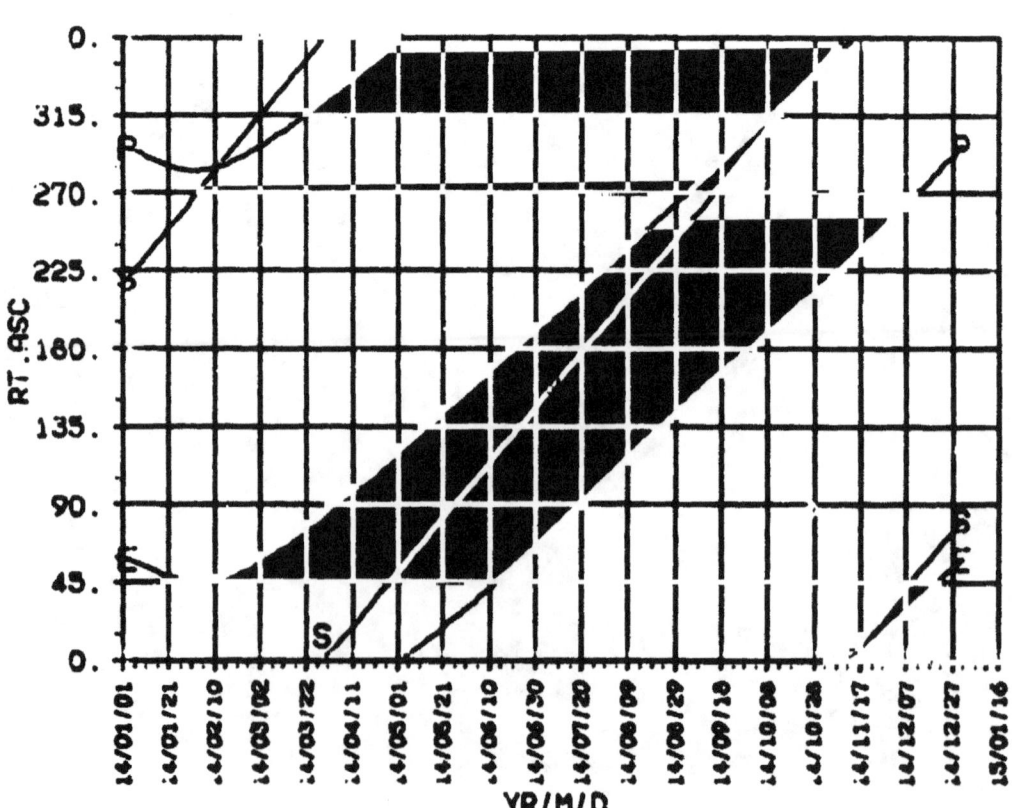

DISTANCE EC.LON 2014

VENUS 2014

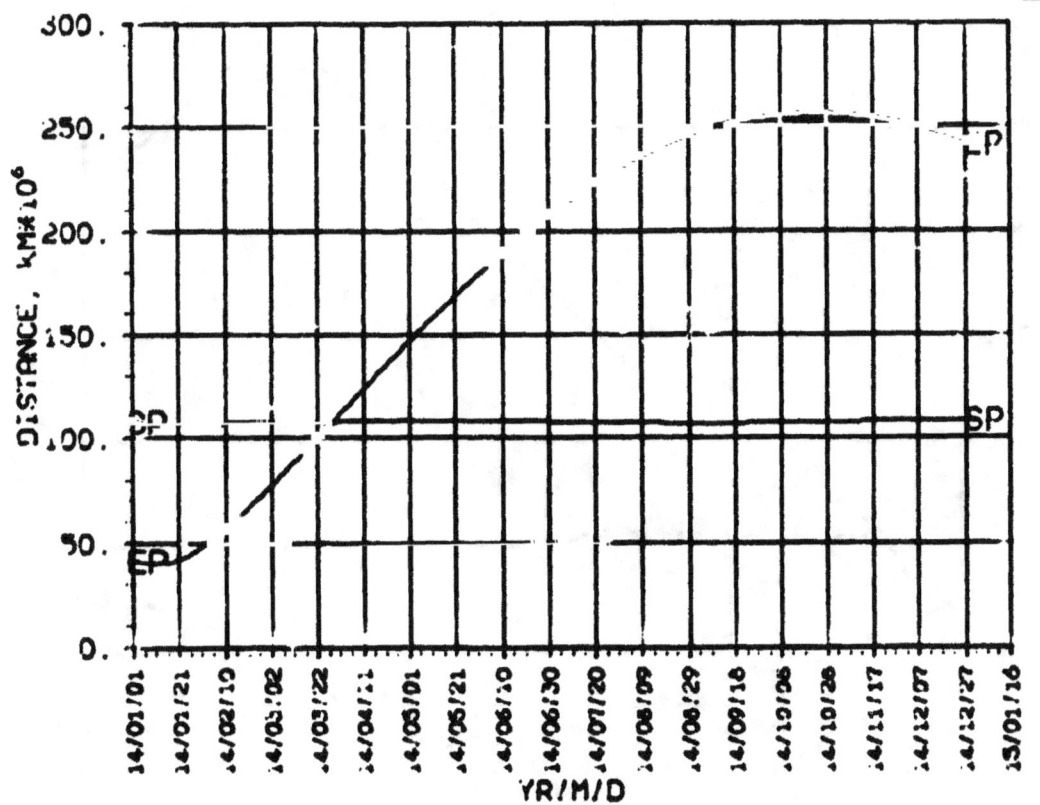

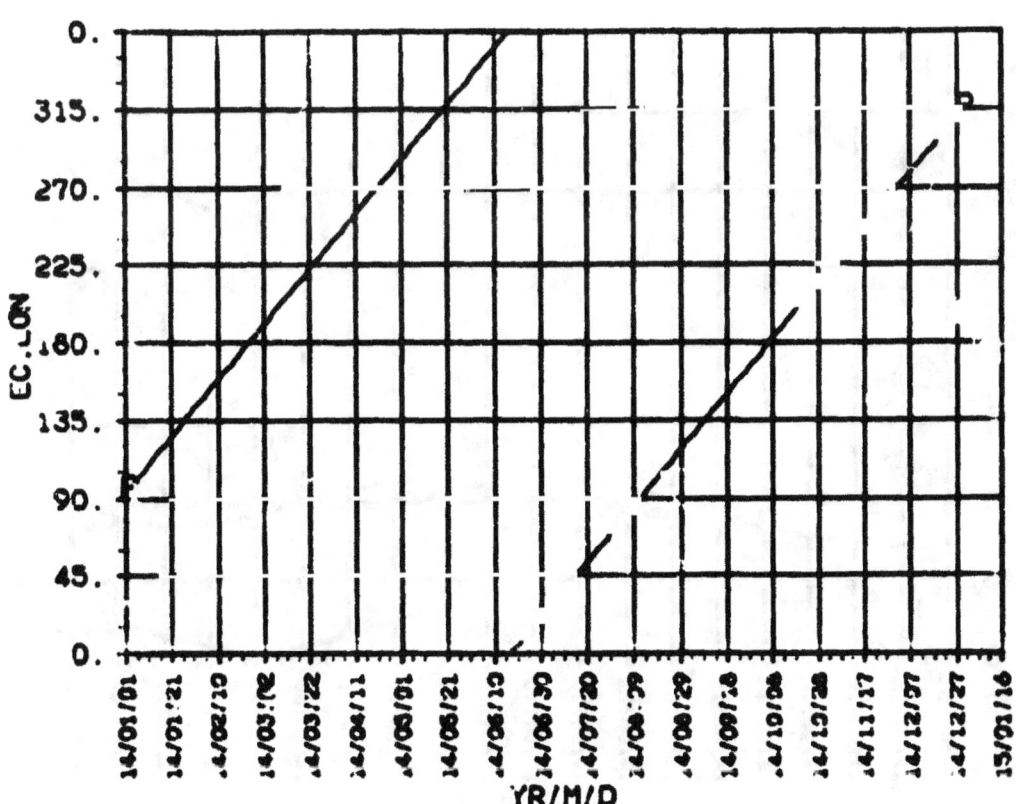

SEP, ESP CA, KA 2014

VENUS 2014

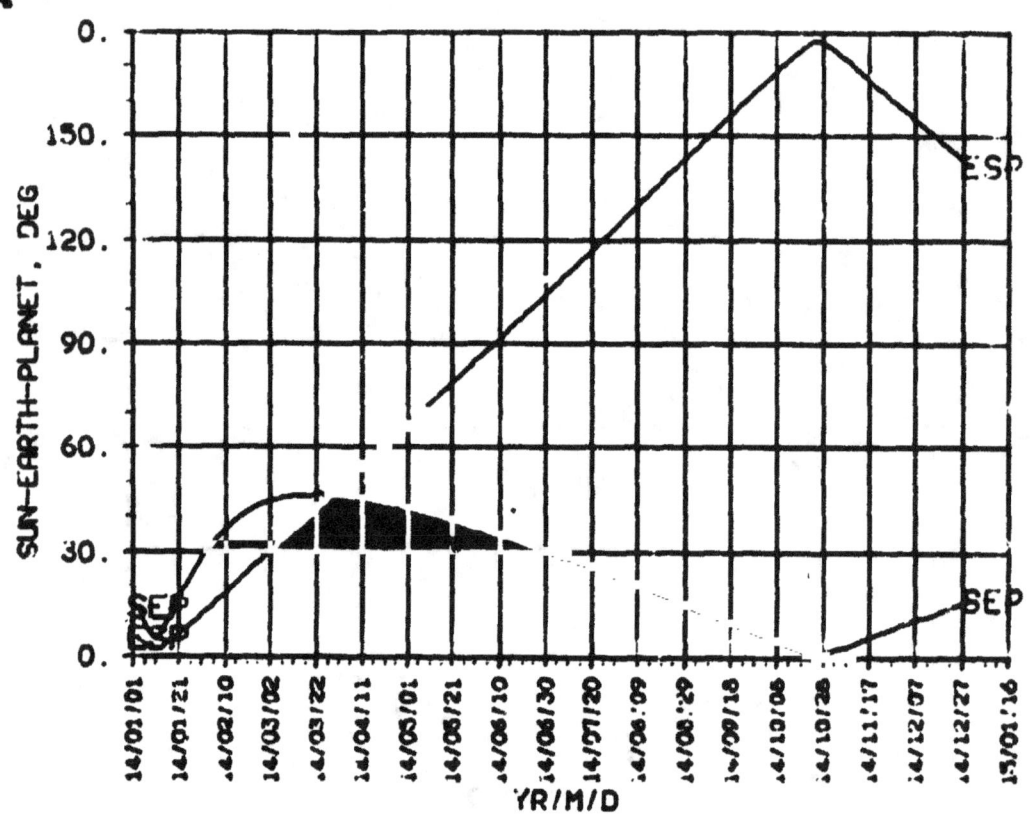

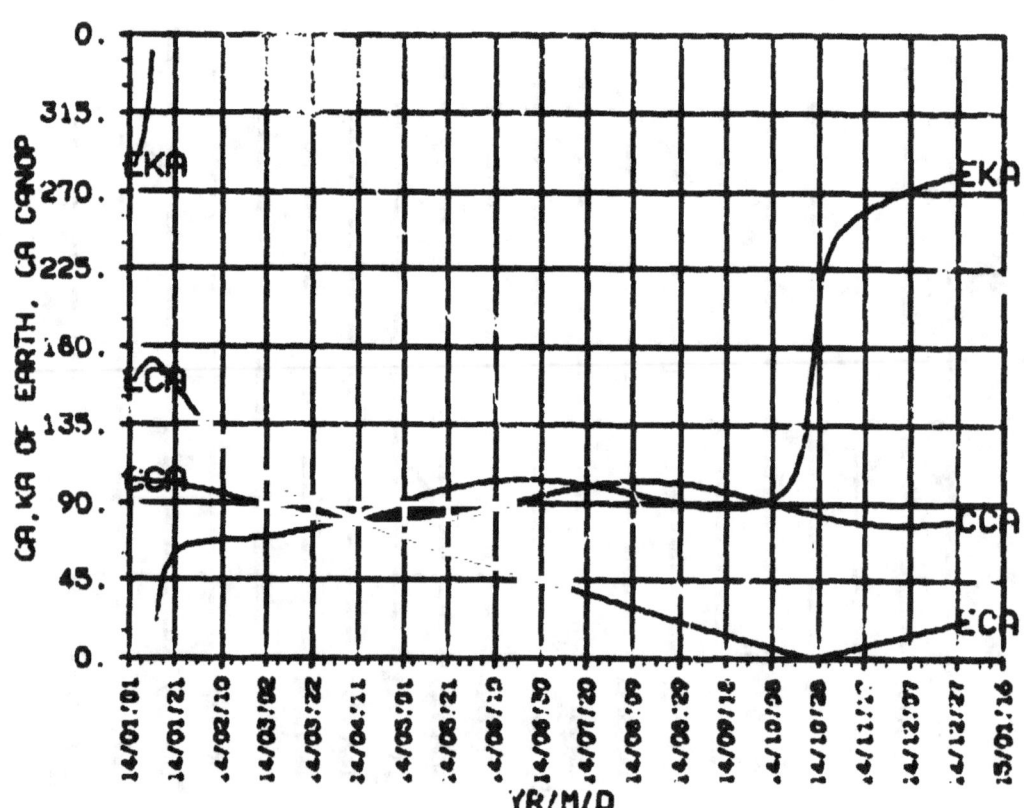

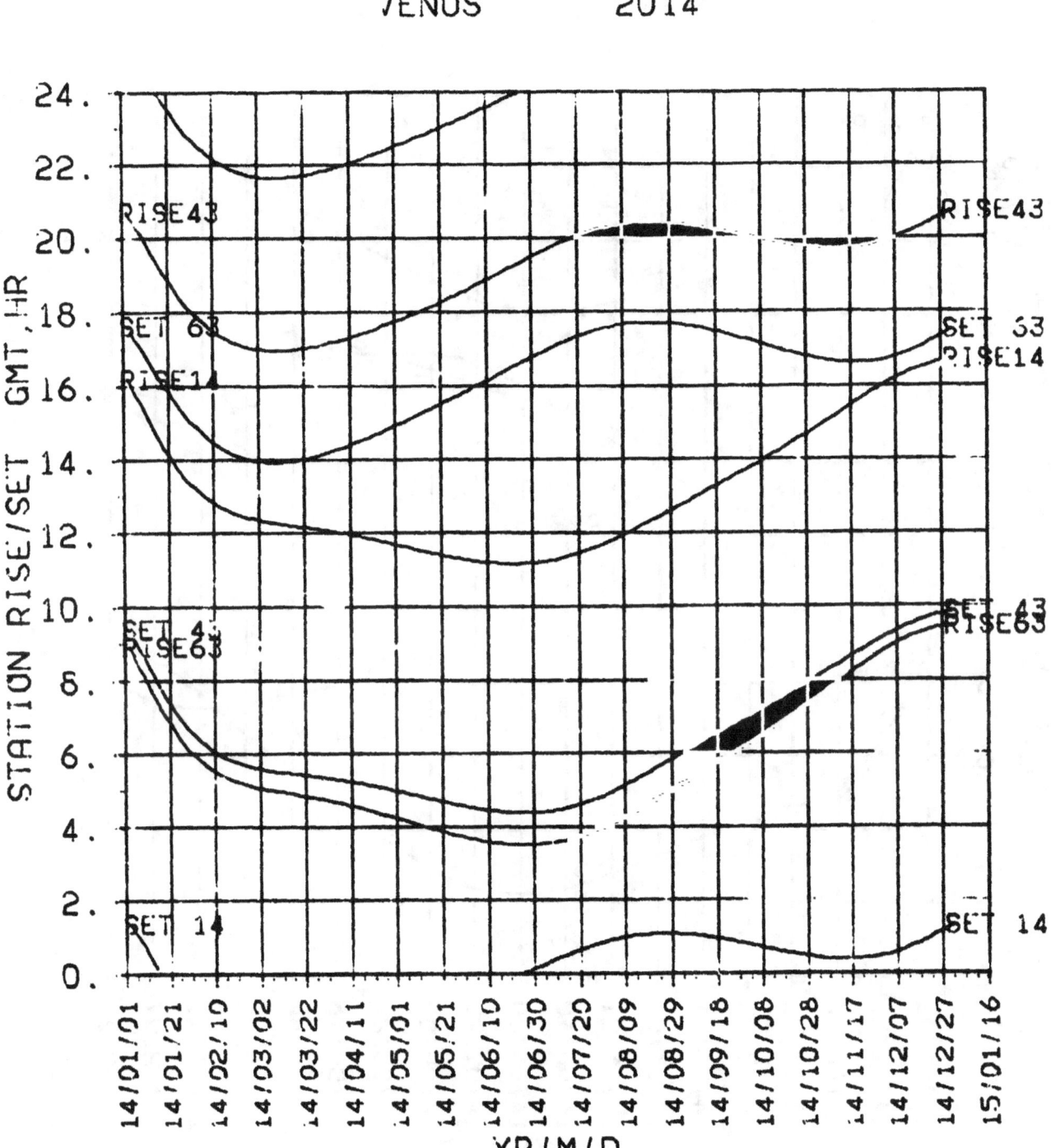

**STA R/S
NON-DSN
2014**

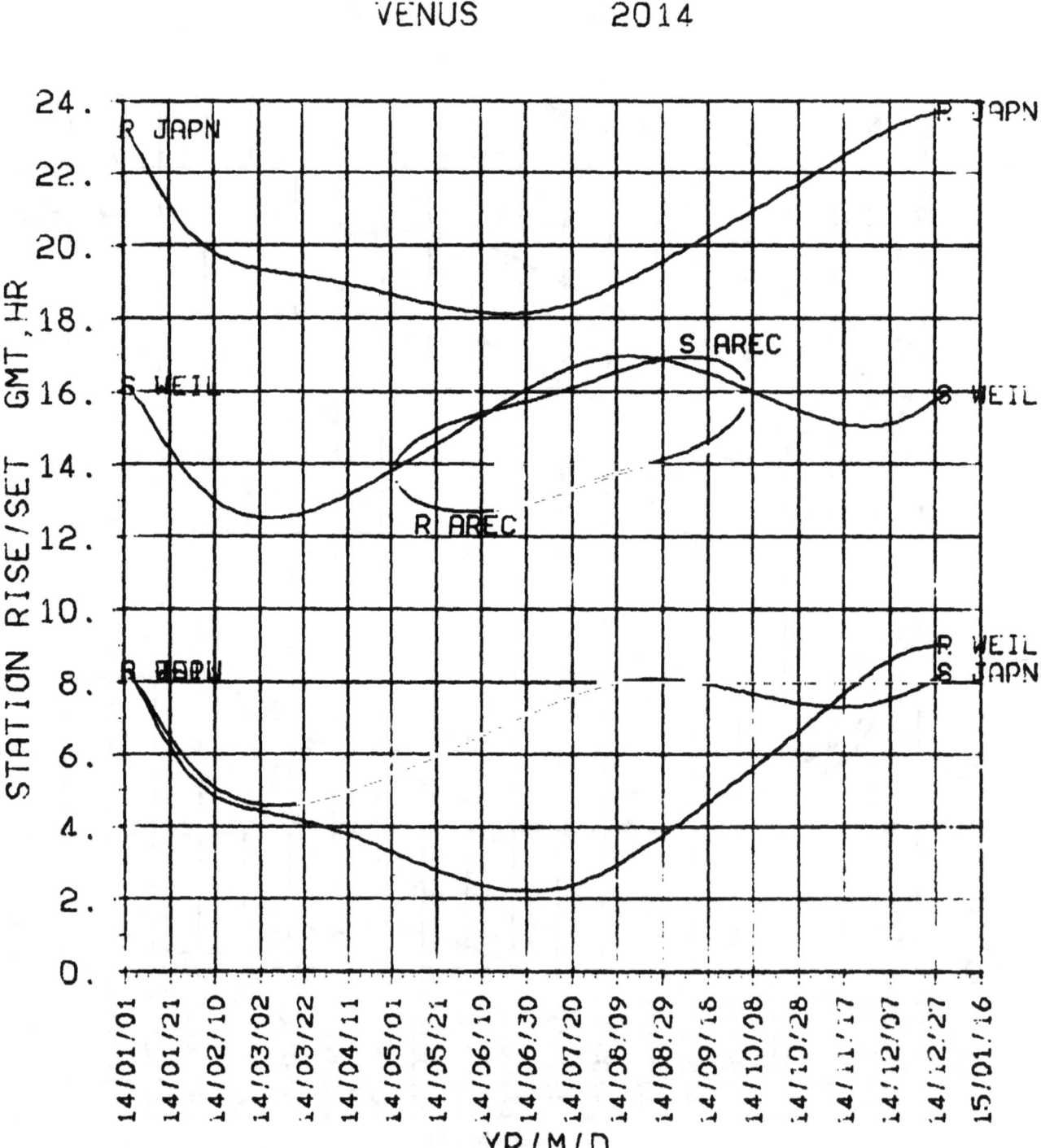

Venus

2015

DECLIN RT.ASC 2015

VENUS 2015

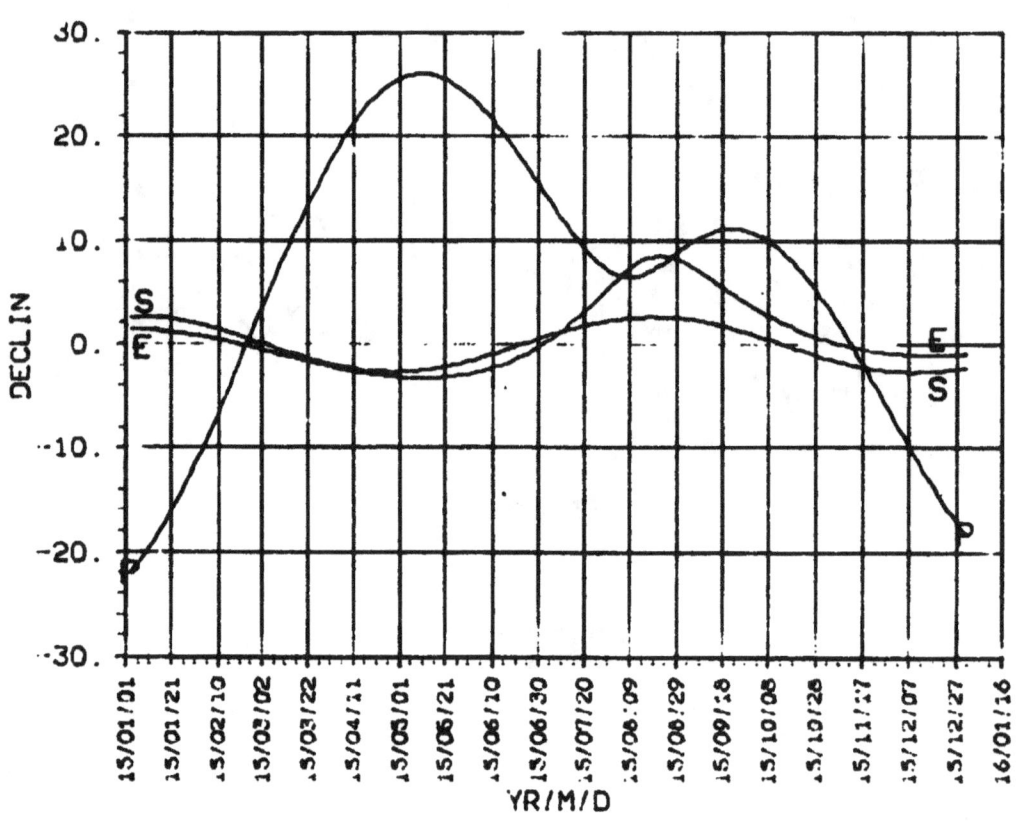

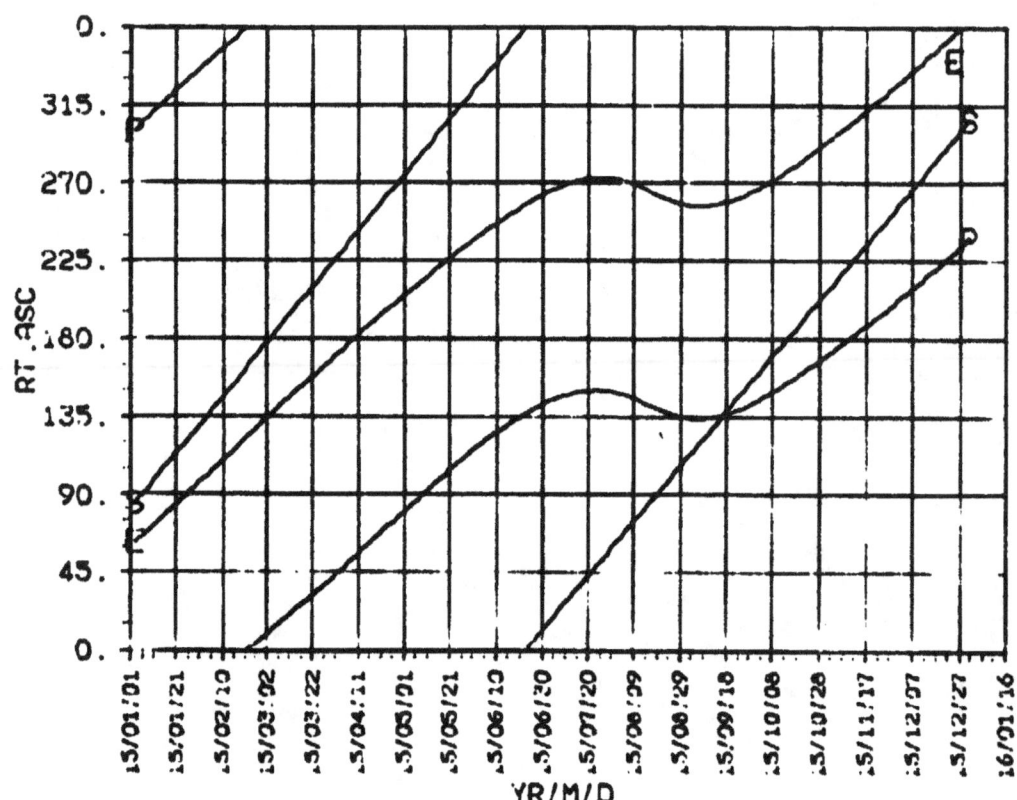

VENUS 2015

DISTANCE EC.LON 2015

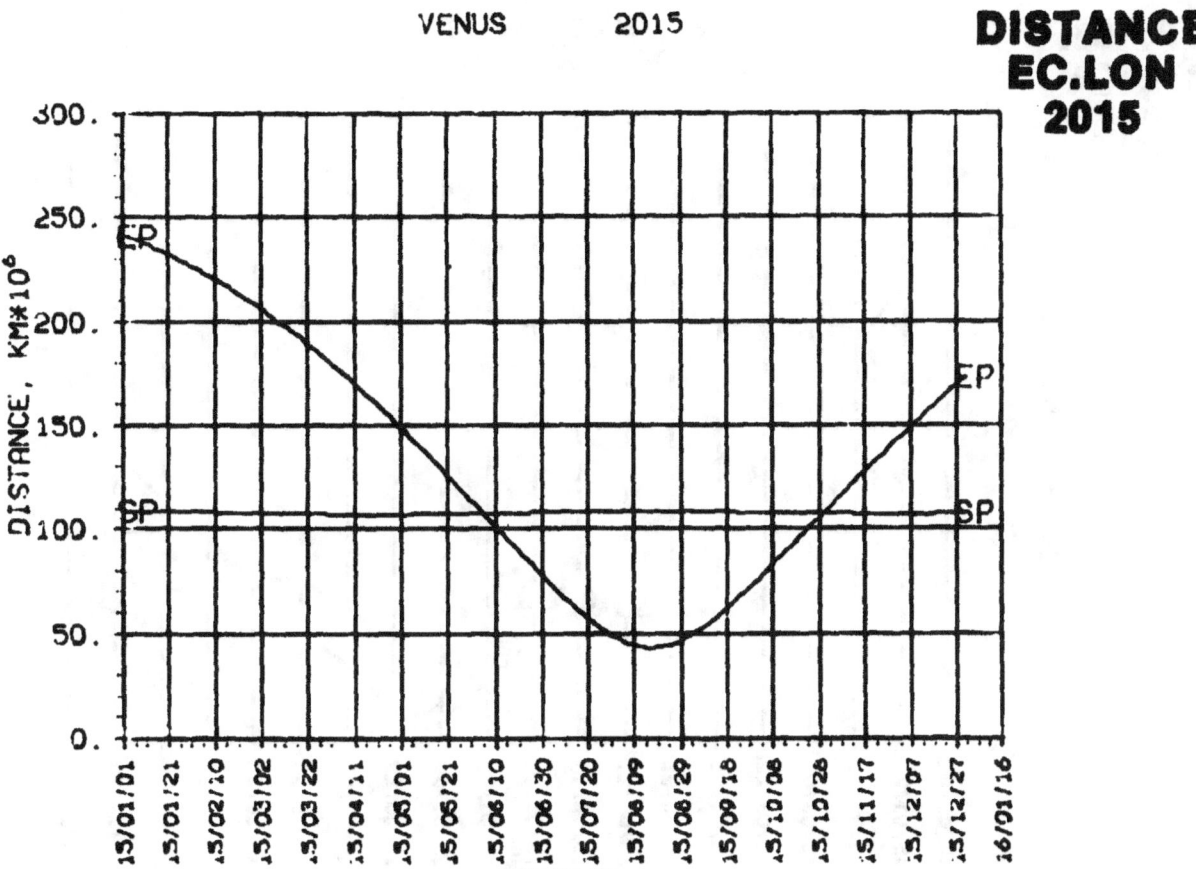

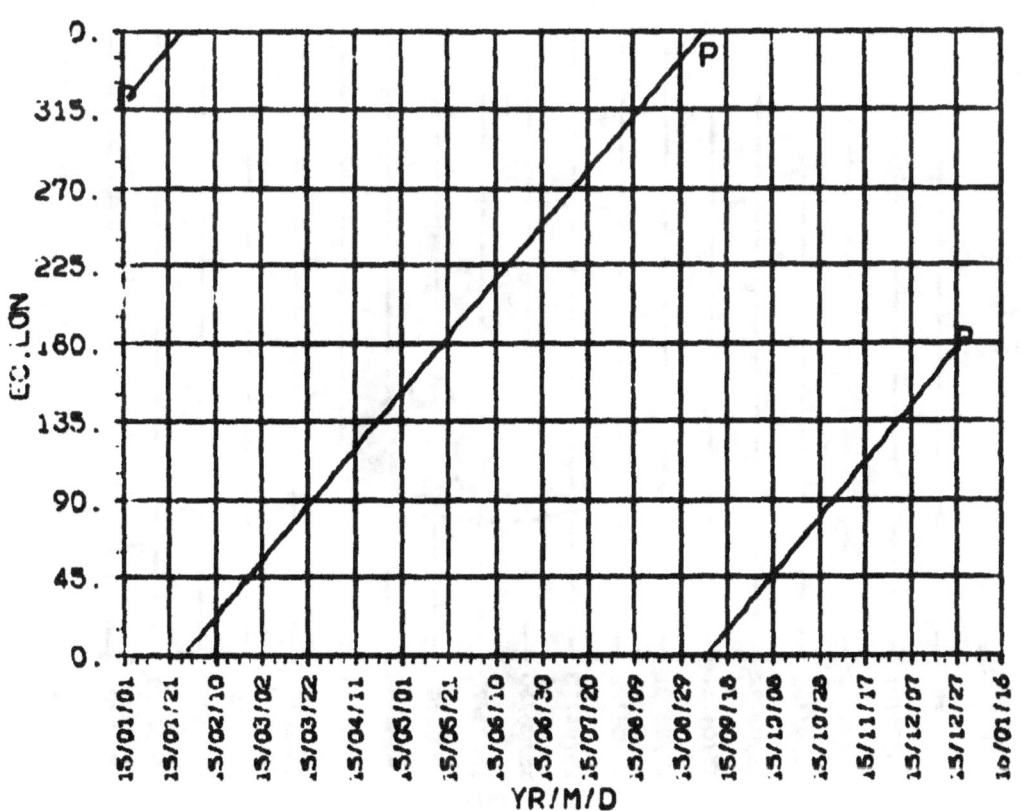

SEP, ESP CA, KA 2015

VENUS 2015

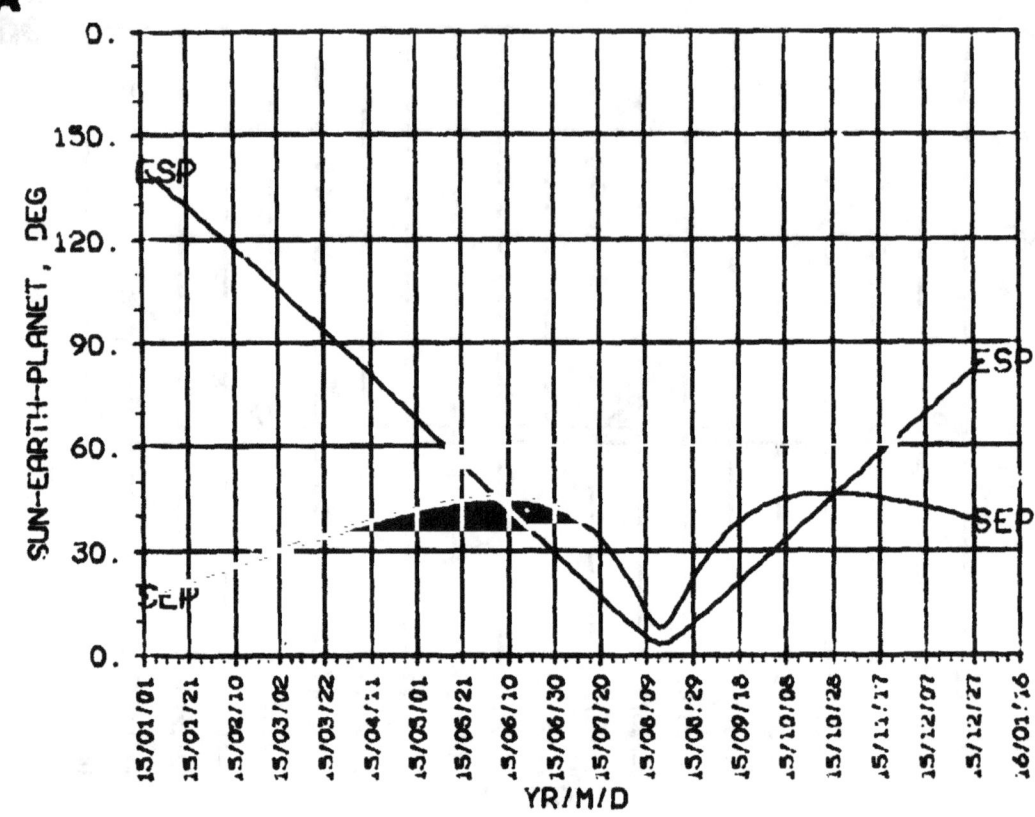

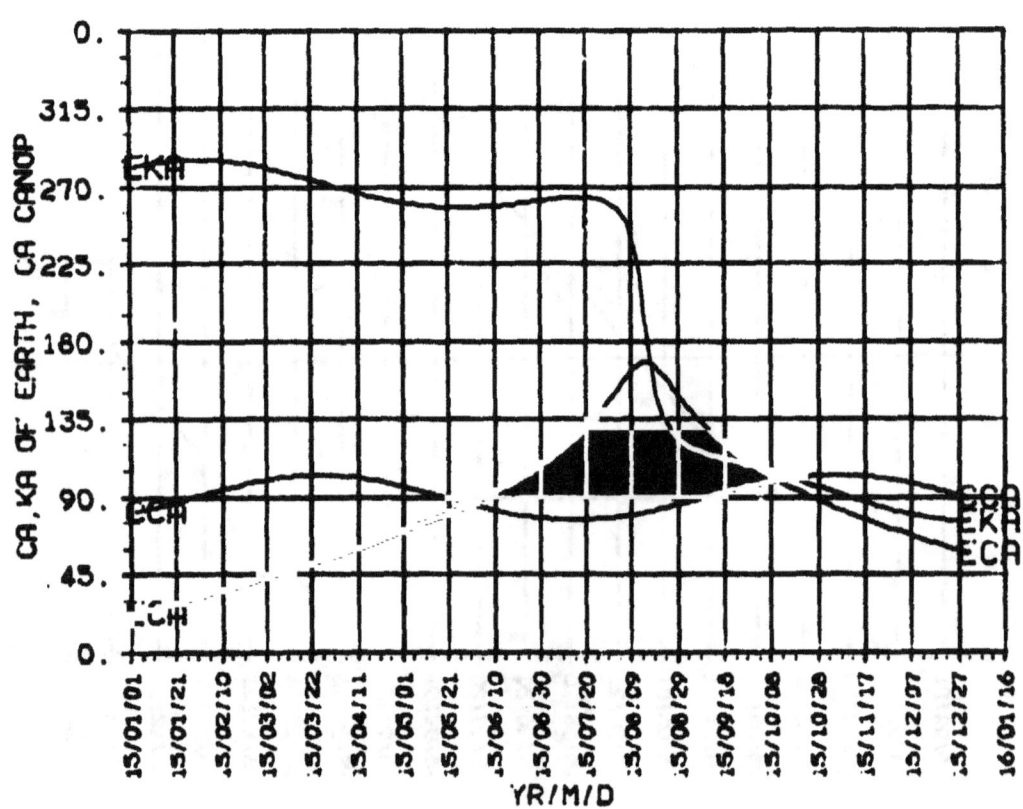

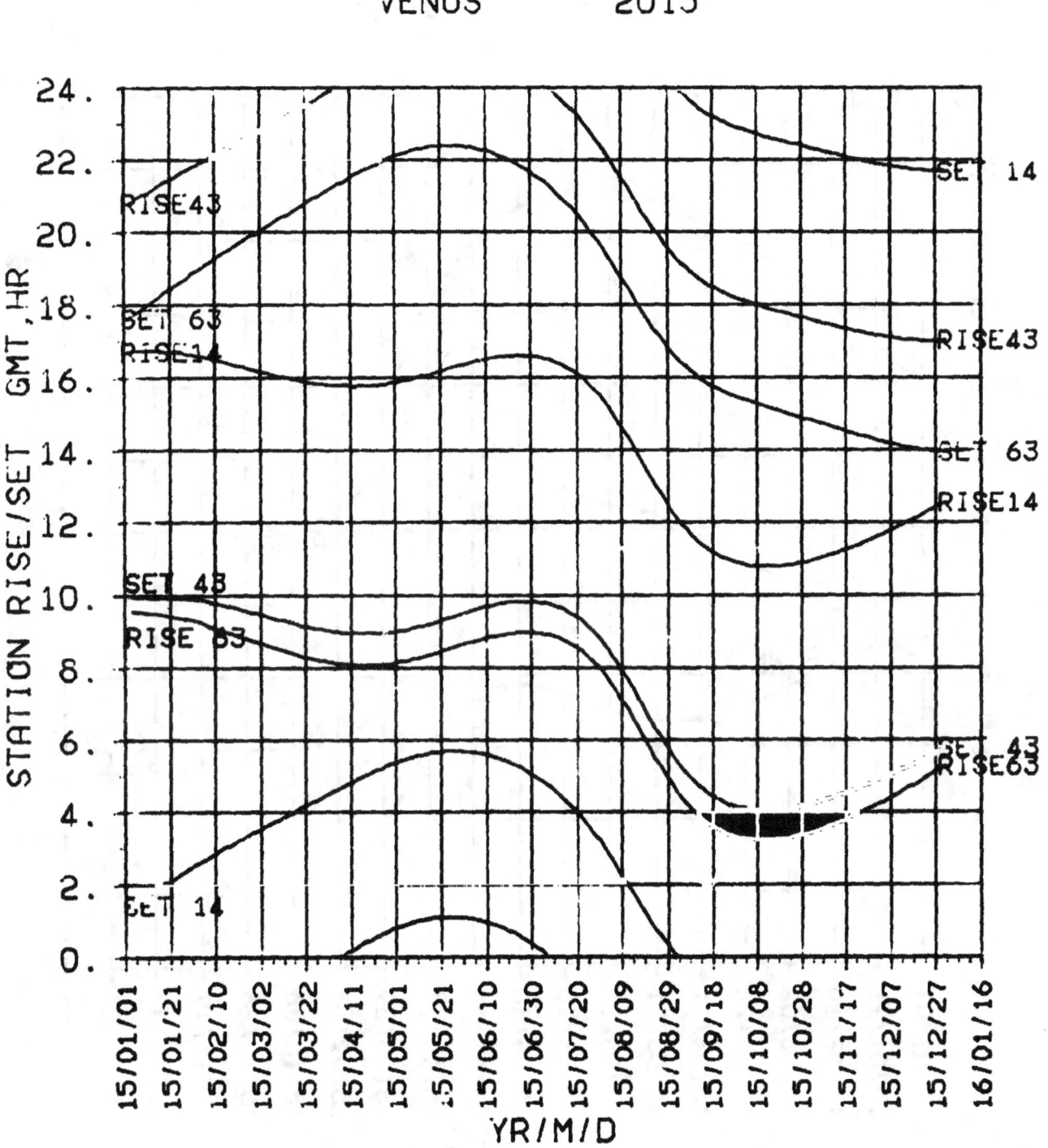

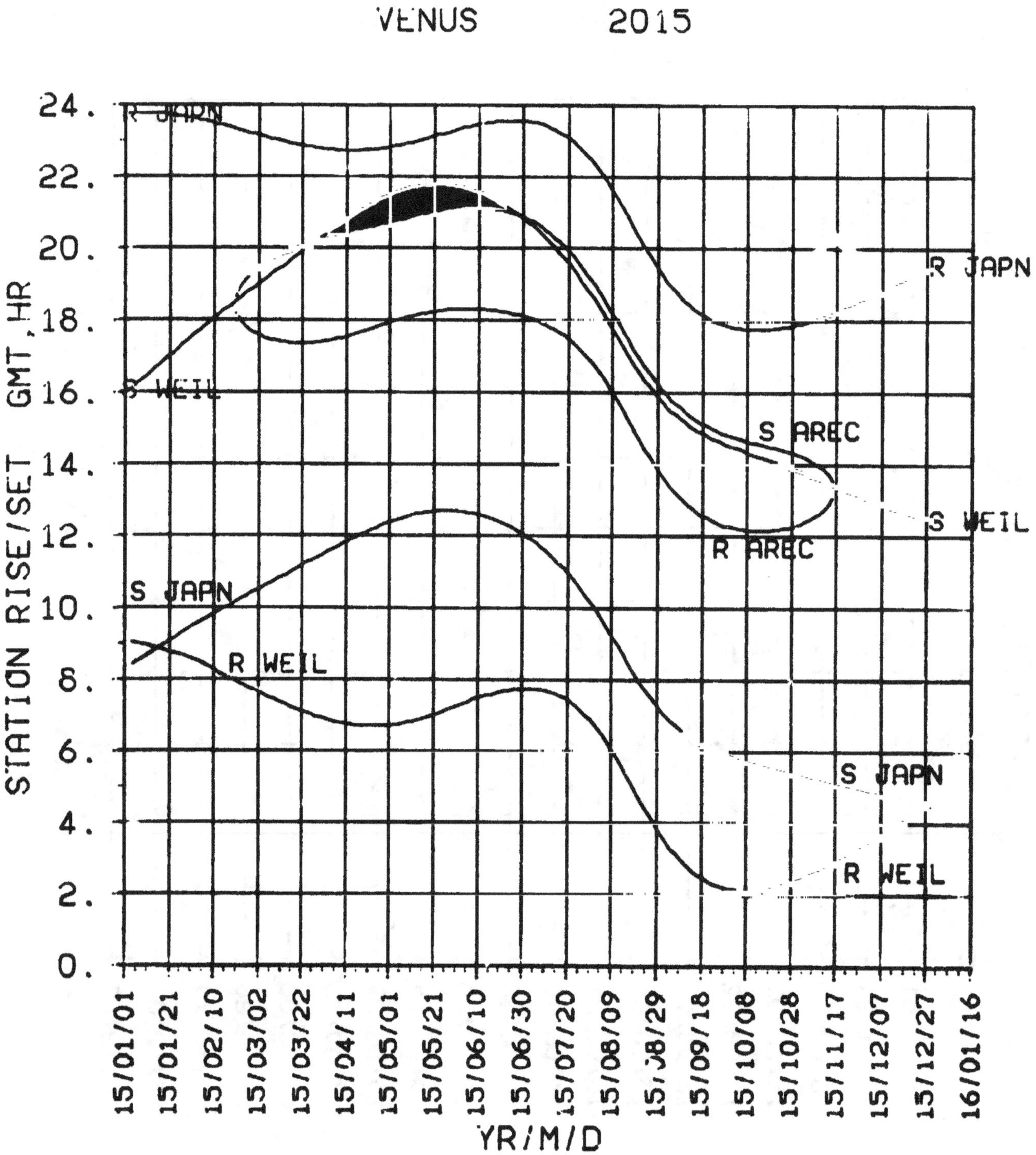
VENUS 2015
STATION RISE/SET GMT, HR
YR/M/D
24.
22.
20.
18.
16.
14.
12.
10.
8.
6.
4.
2.
0.
15/01/01
15/01/21
15/02/10
15/03/02
15/03/22
15/04/11
15/05/01
15/05/21
15/06/10
15/06/30
15/07/20
15/08/09
15/08/29
15/09/18
15/10/08
15/10/28
15/11/17
15/12/07
15/12/27
16/01/16
R JAPN
S WEIL
S AREC
R AREC
S JAPN
R WEIL

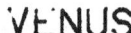

Venus

2016

DECLIN RT.ASC 2016

VENUS 2016

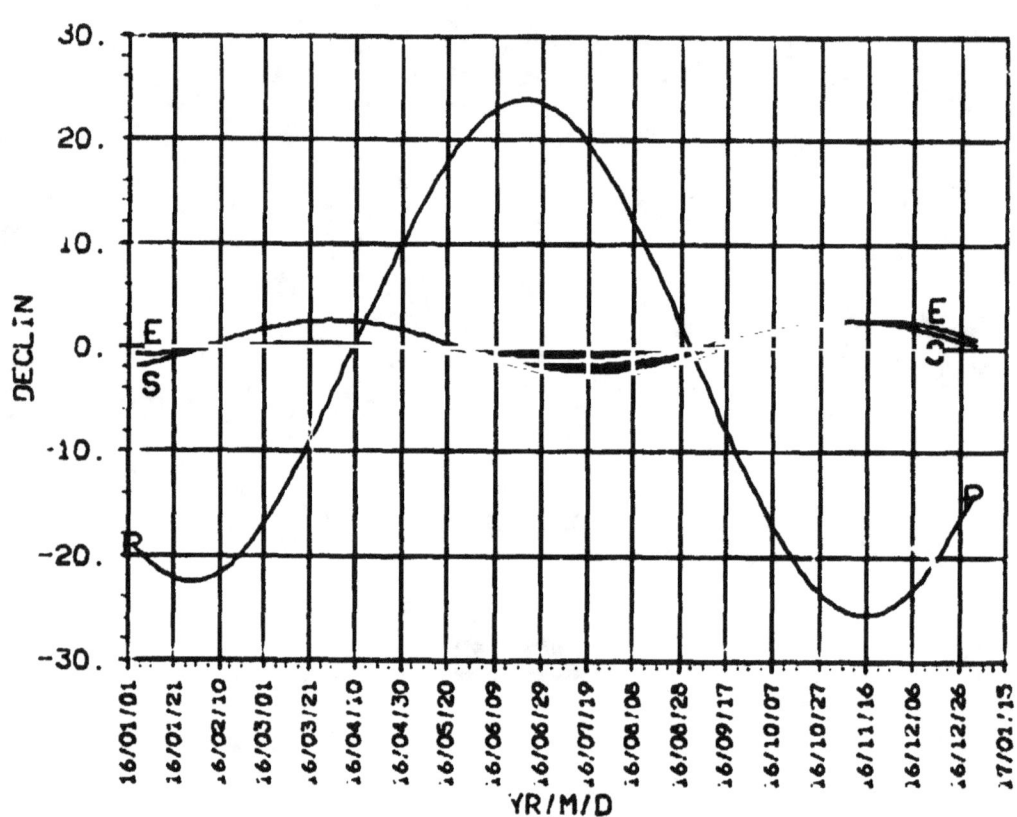

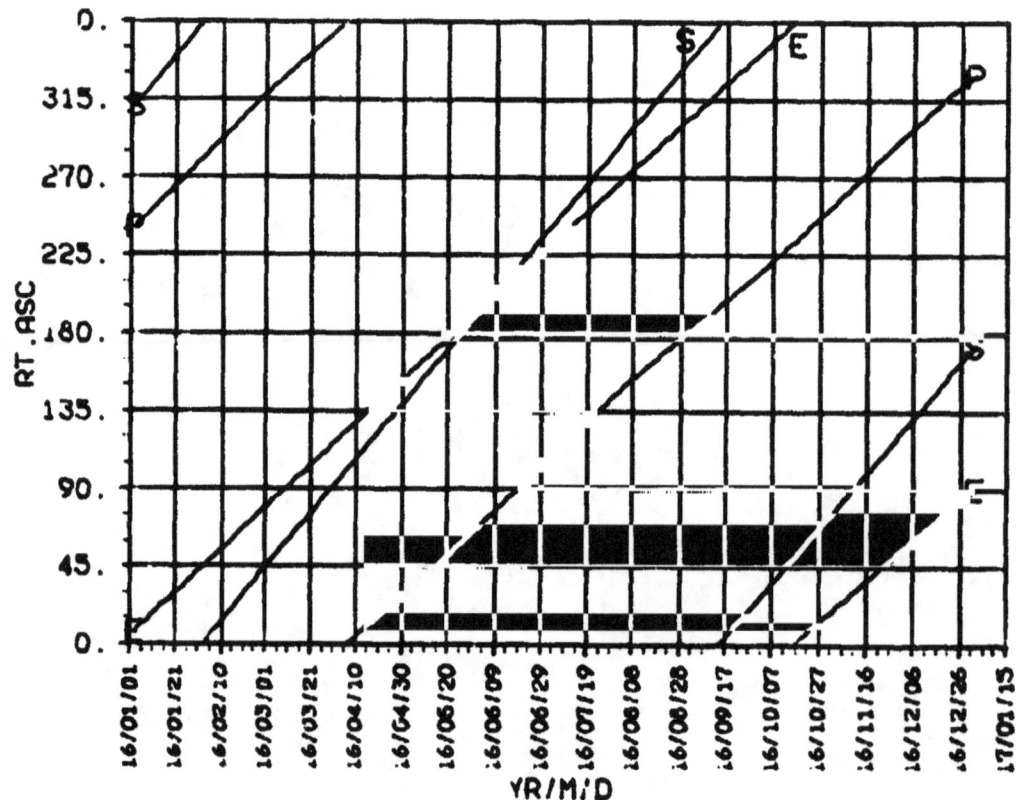

VENUS 2016

DISTANCE
EC.LON
2016

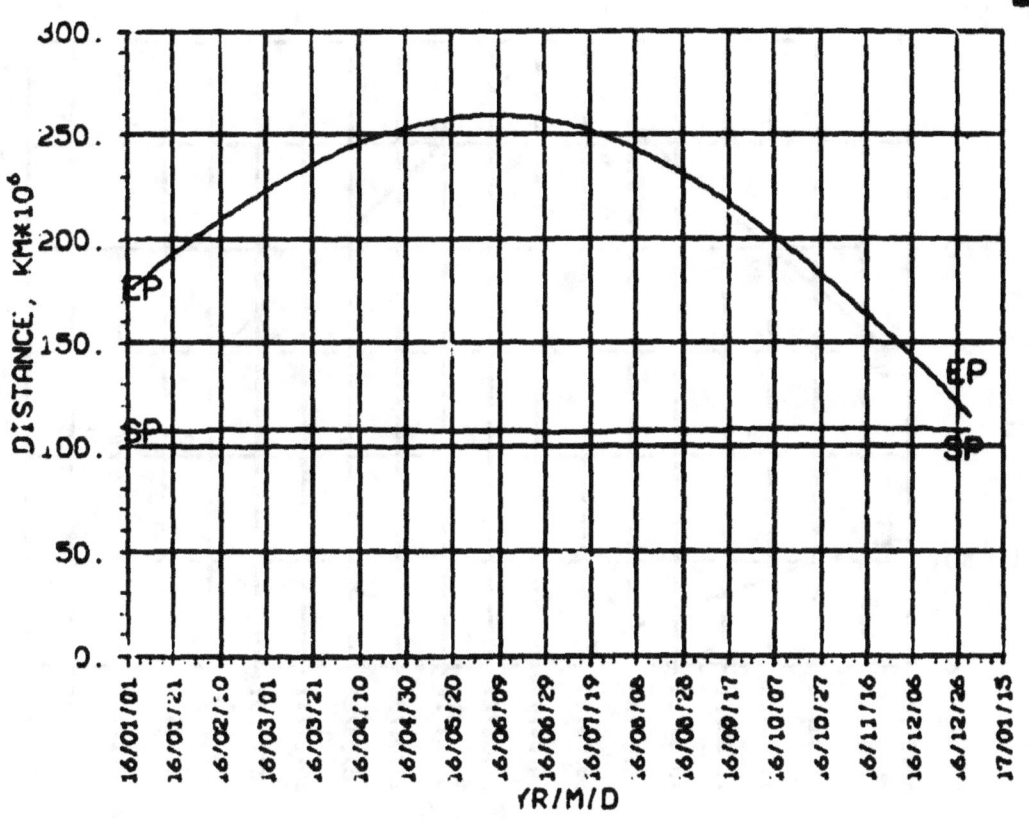

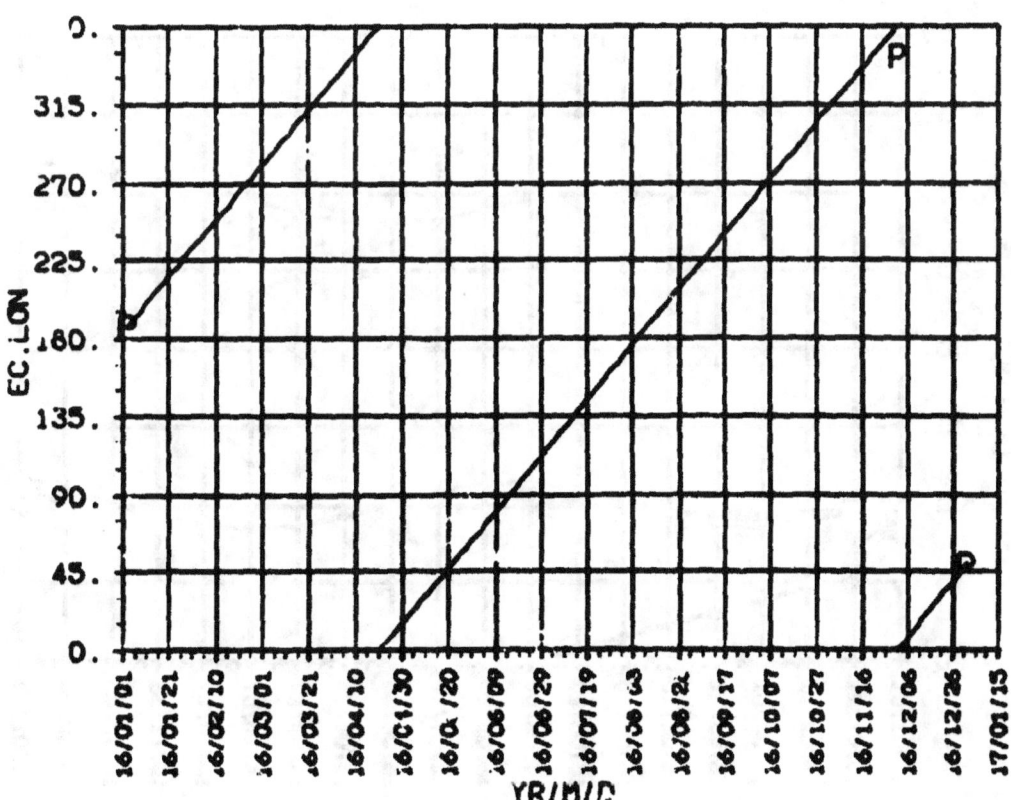

SEP, ESP CA, KA 2016 — VENUS 2016

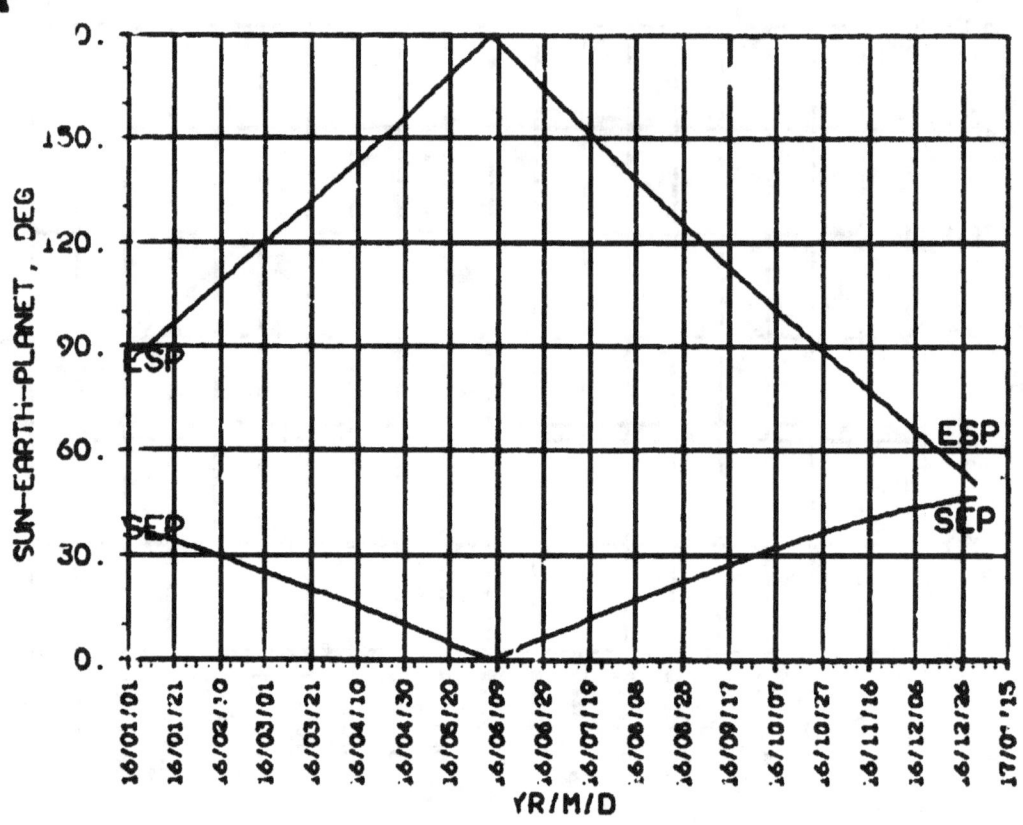
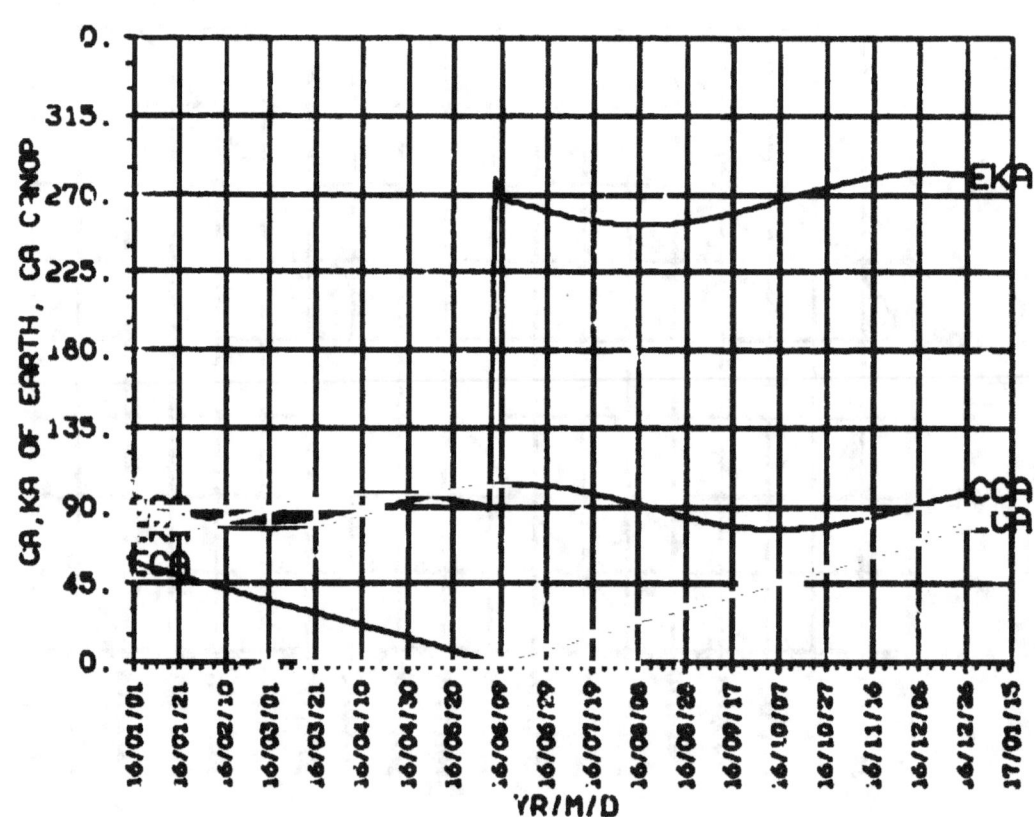

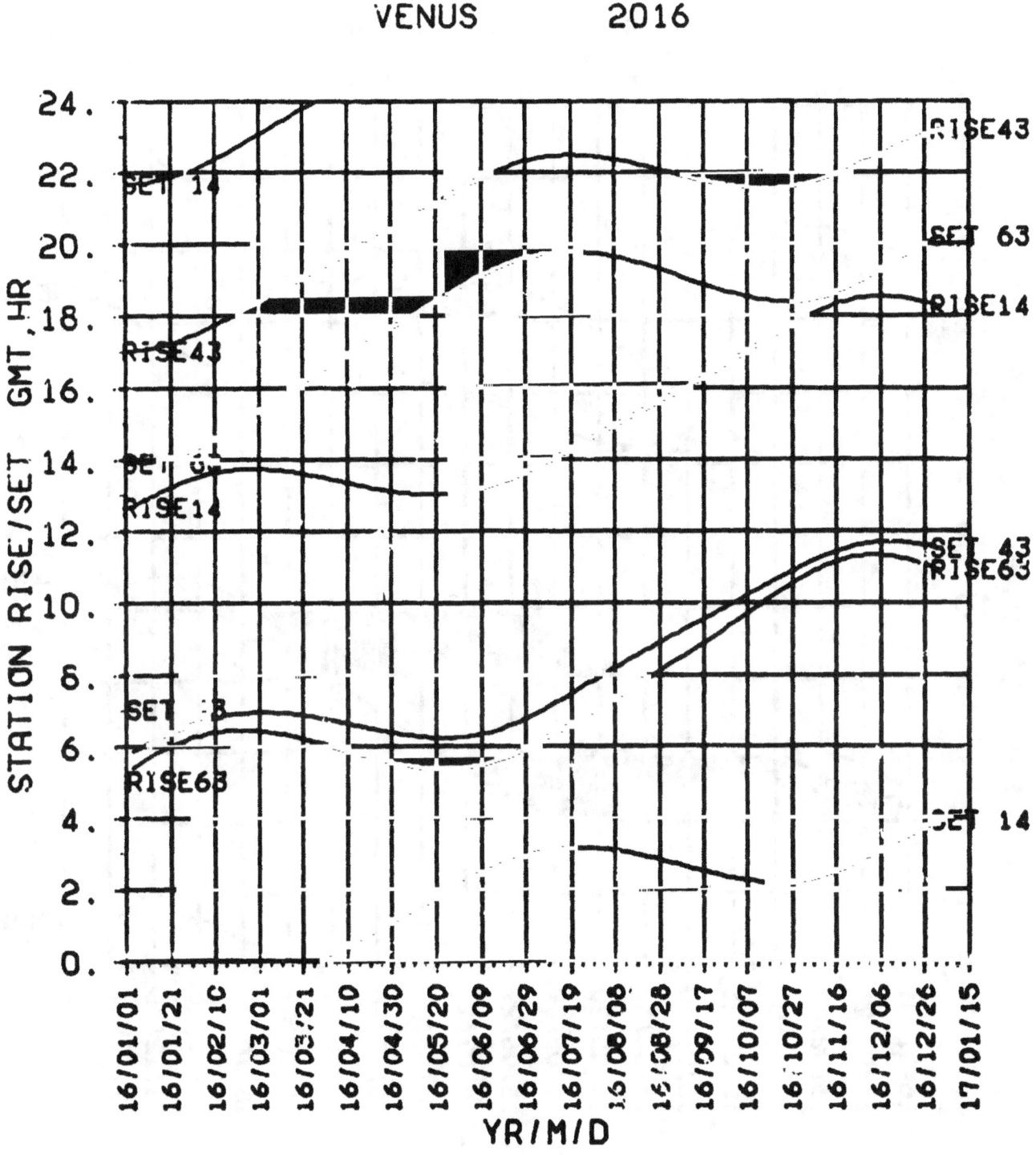

STA R/S NON-DSN 2016

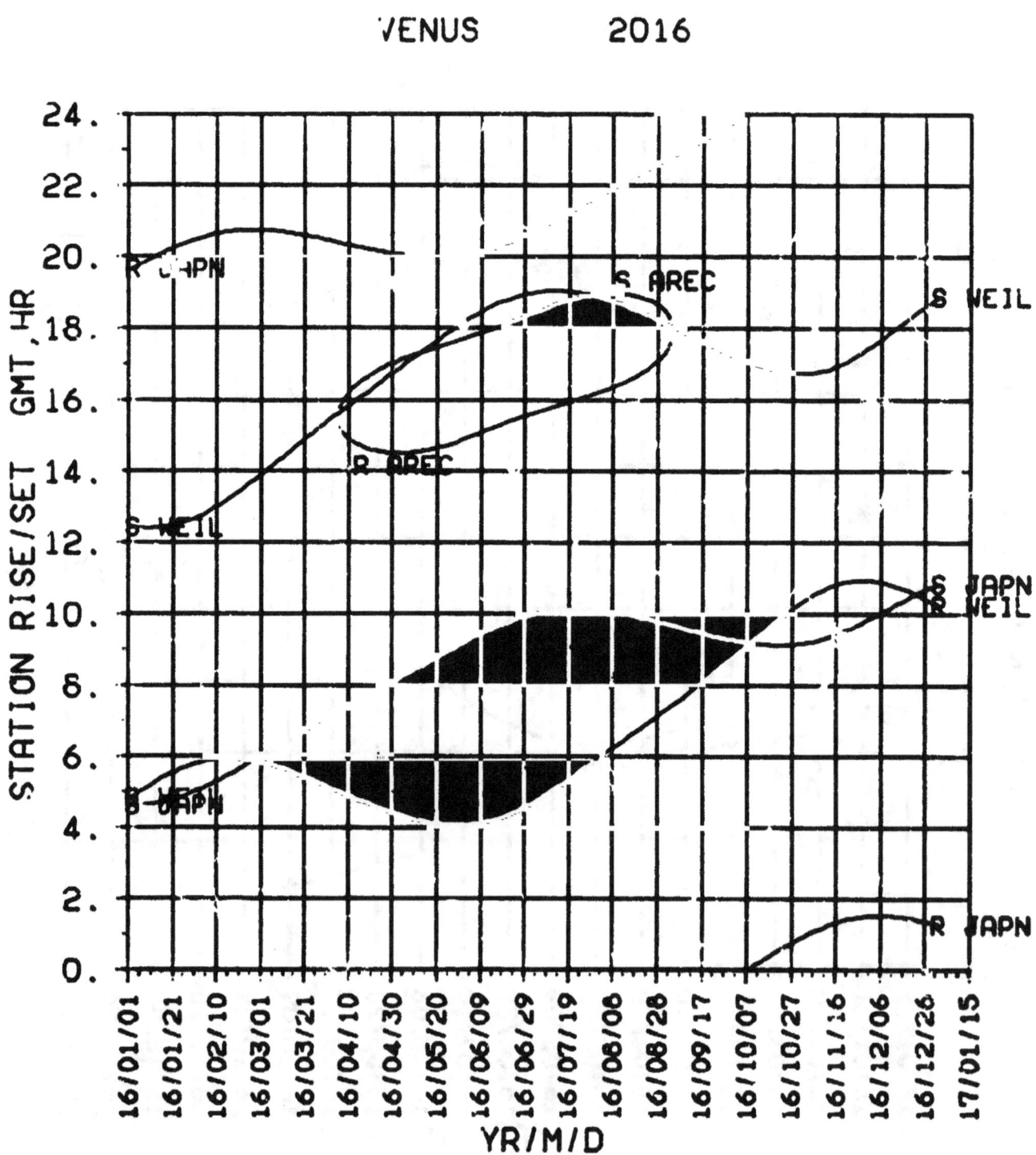

Venus

2017

DECLIN RT.ASC 2017

VENUS   2017

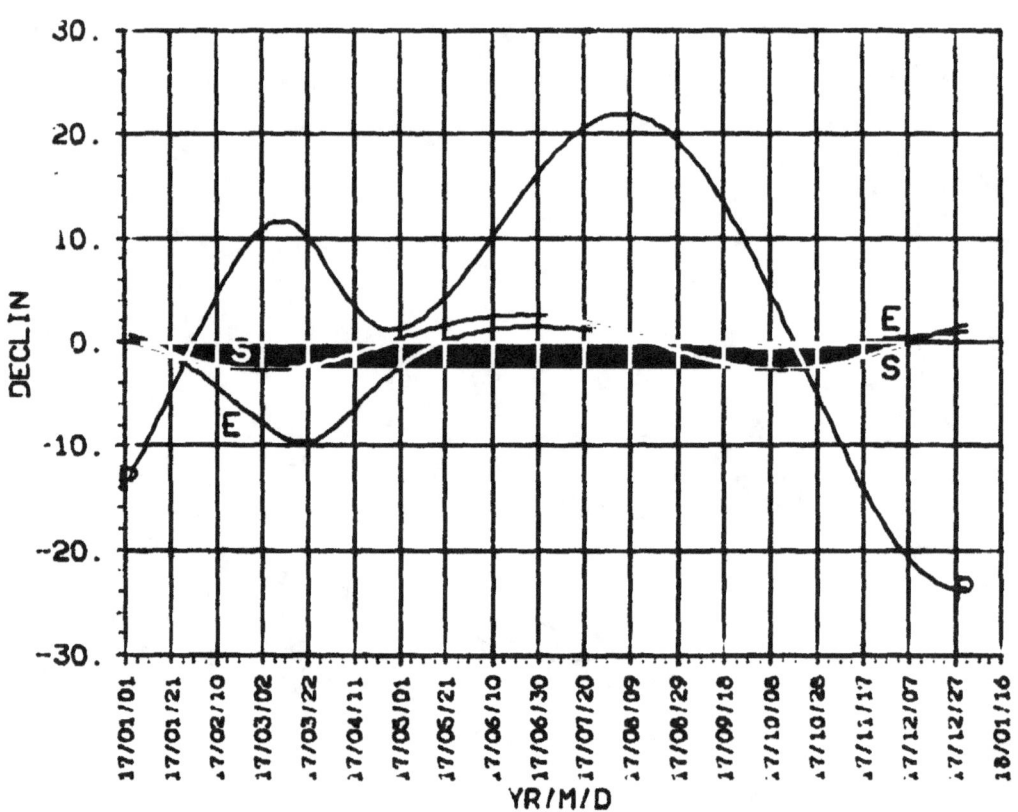

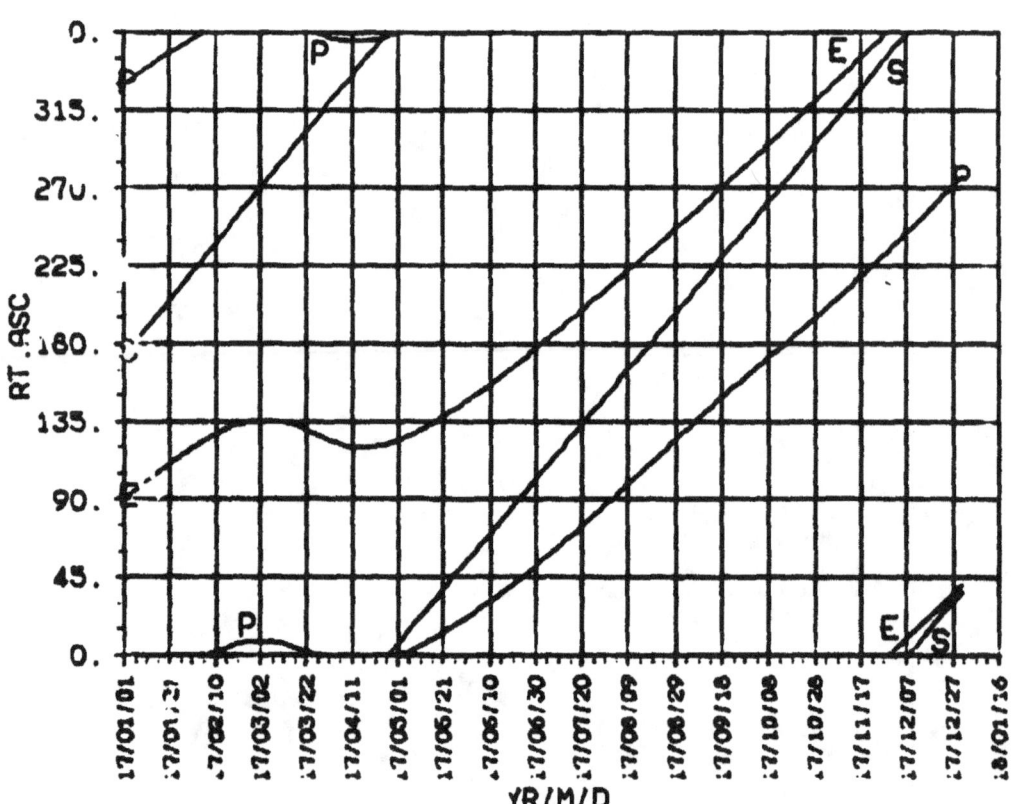

VENUS 2017

DISTANCE
EC.LON
2017

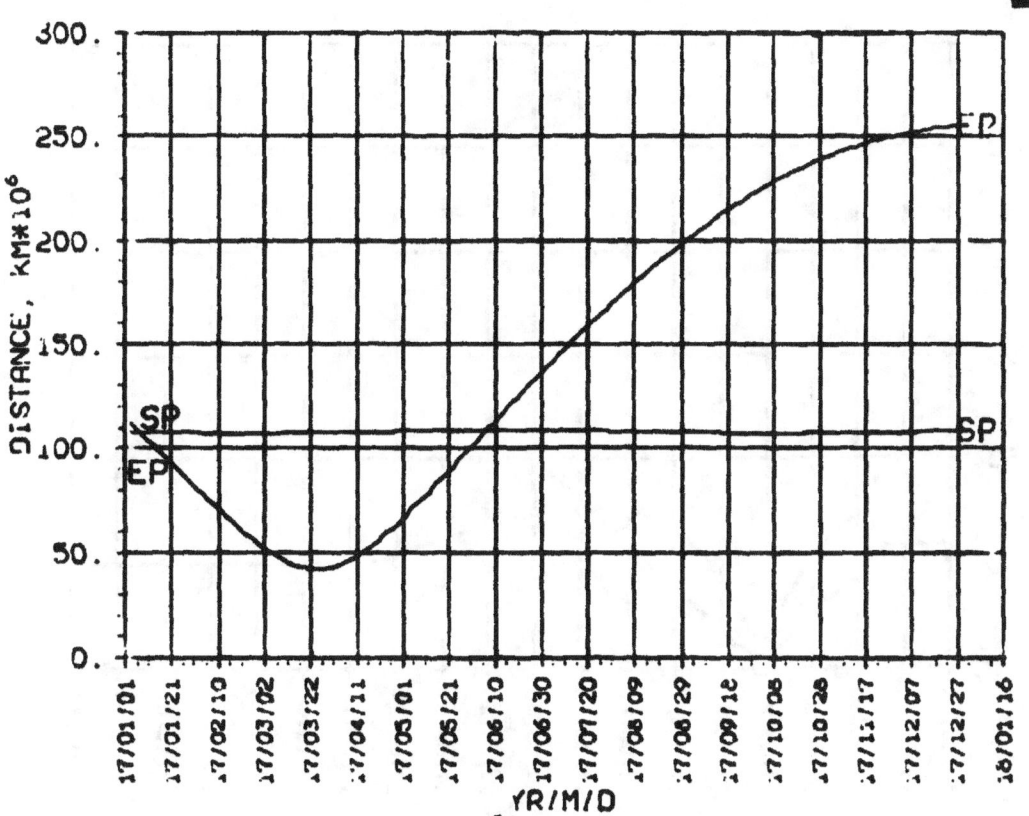

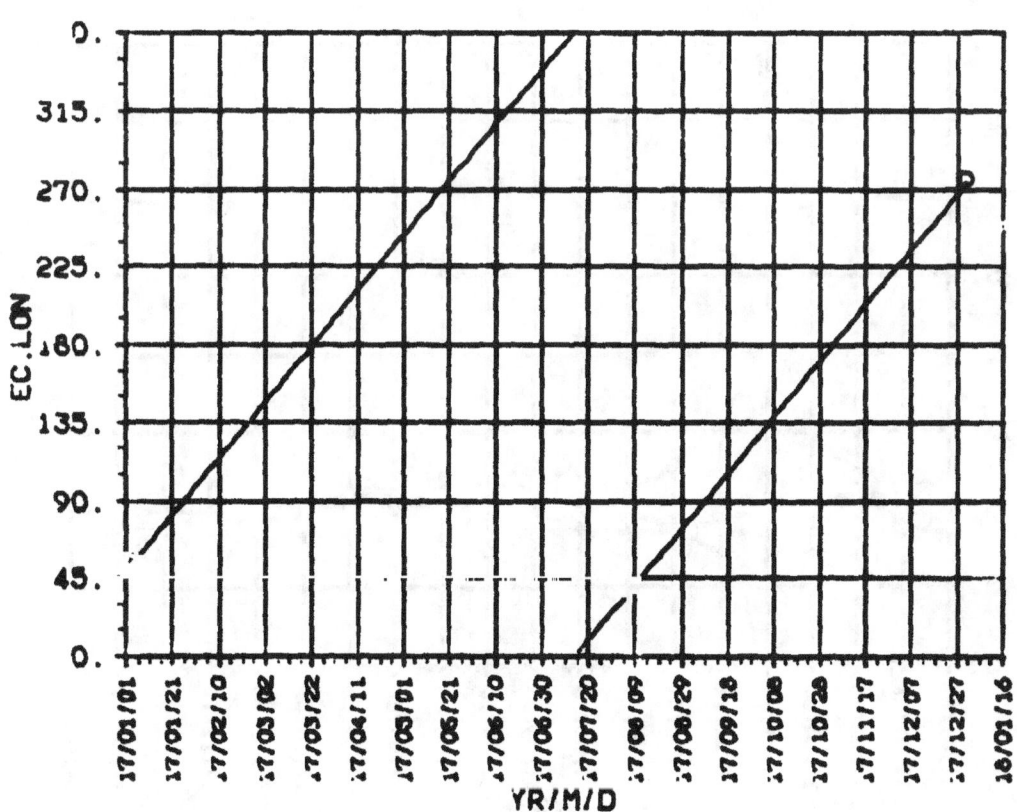

SEP, ESP CA, KA 2017

VENUS 2017

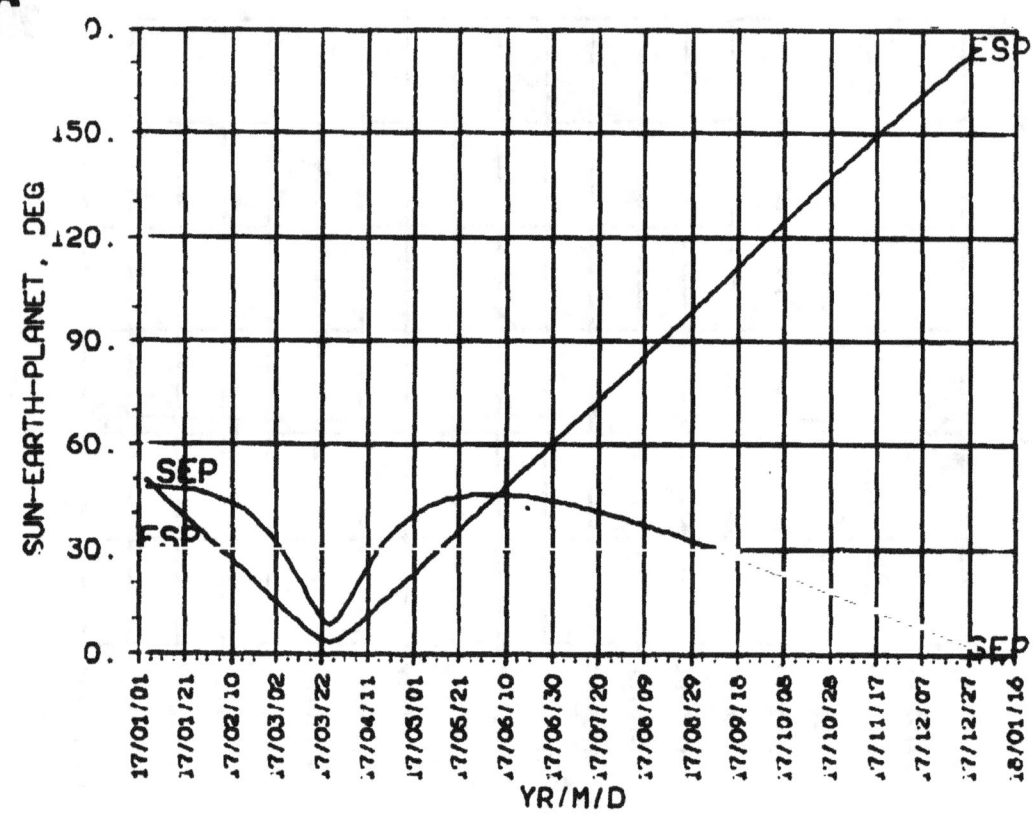

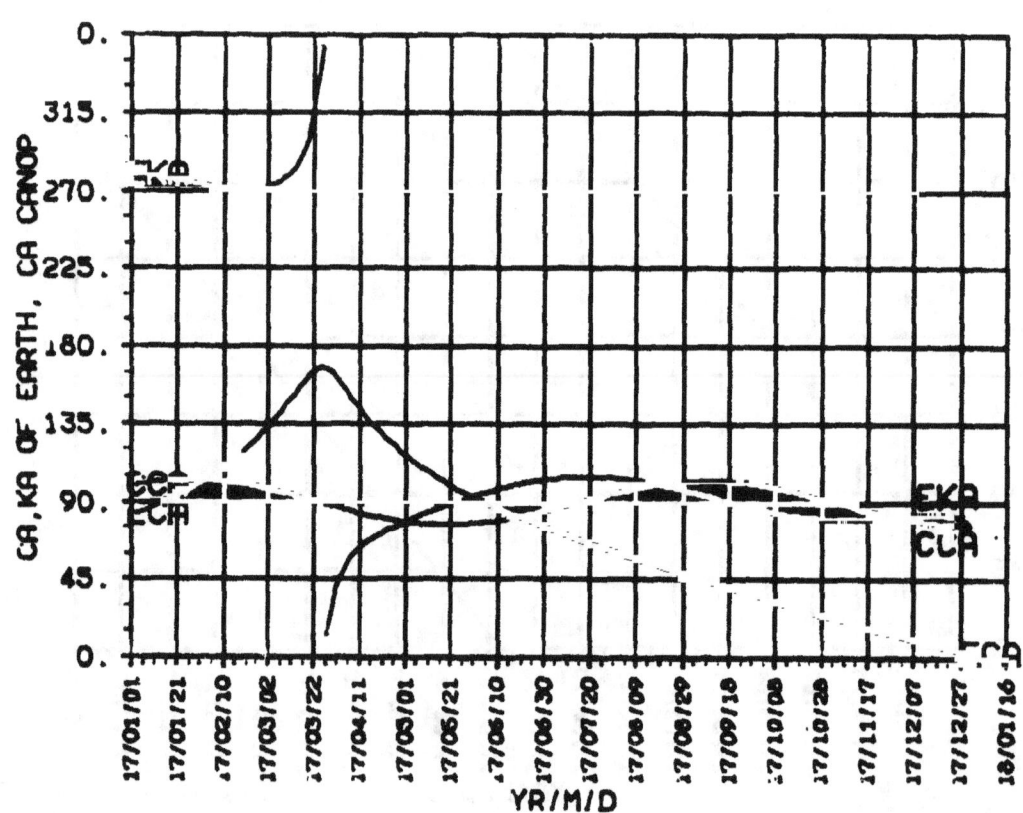

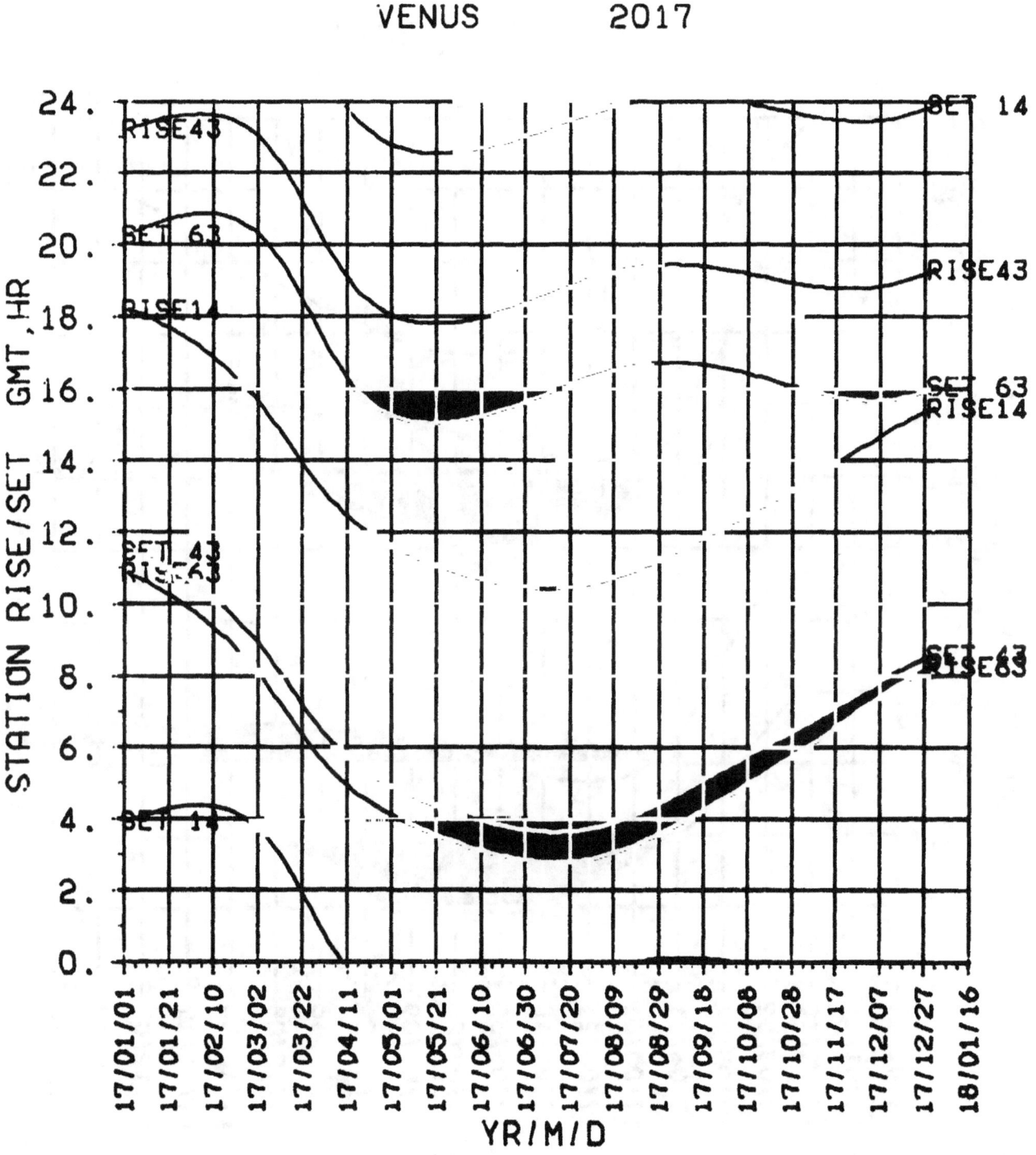

STA R/S
NON-DSN
2017

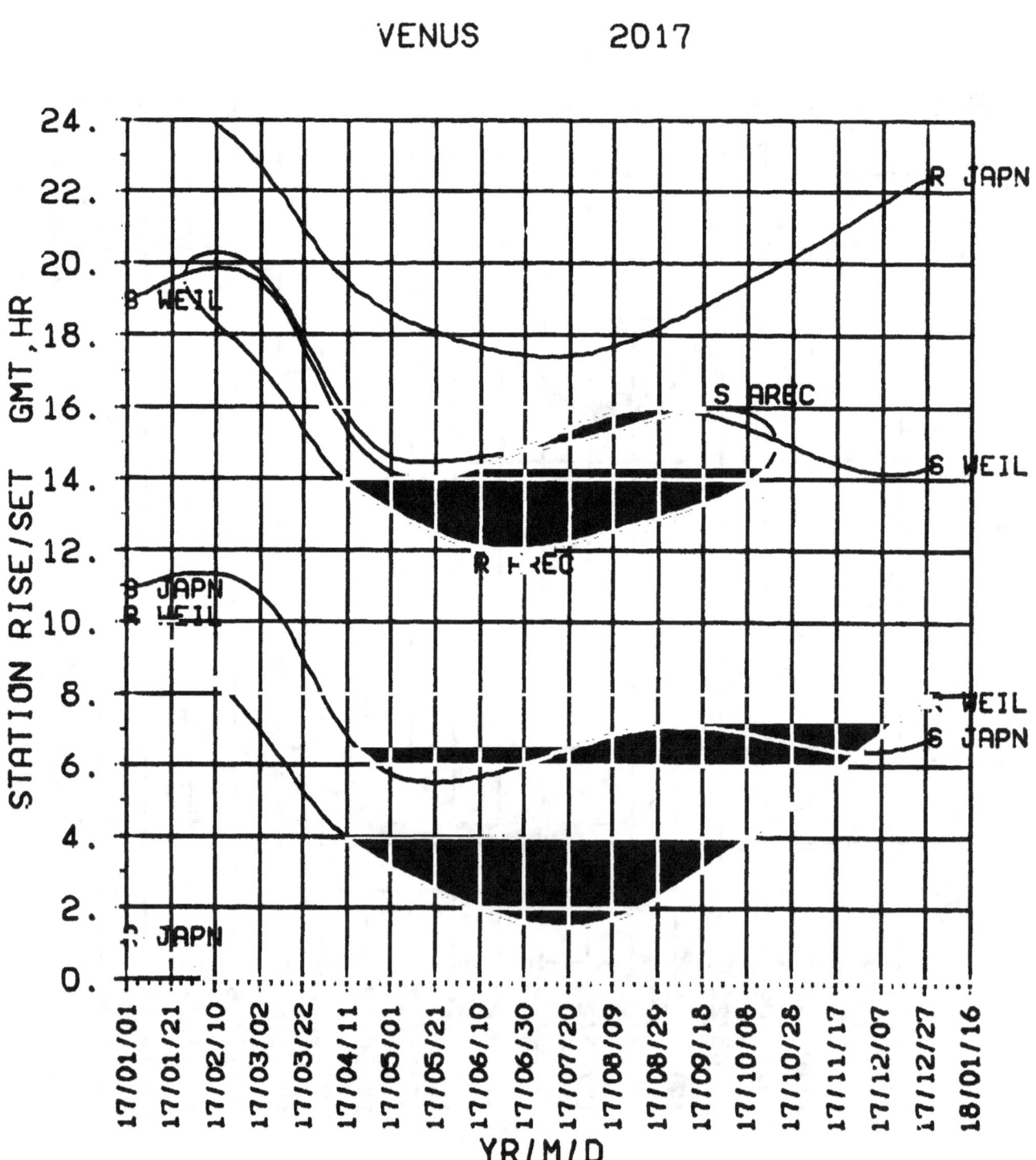

Venus

2018

DECLIN RT.ASC 2018

VENUS 2018

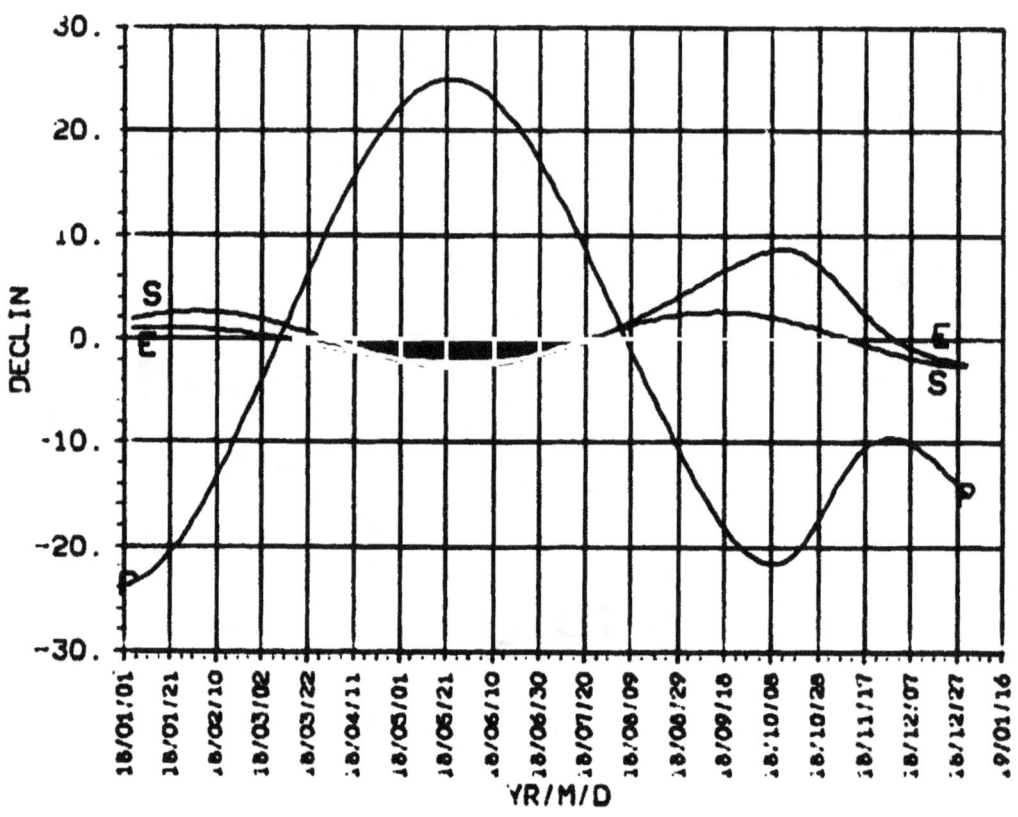

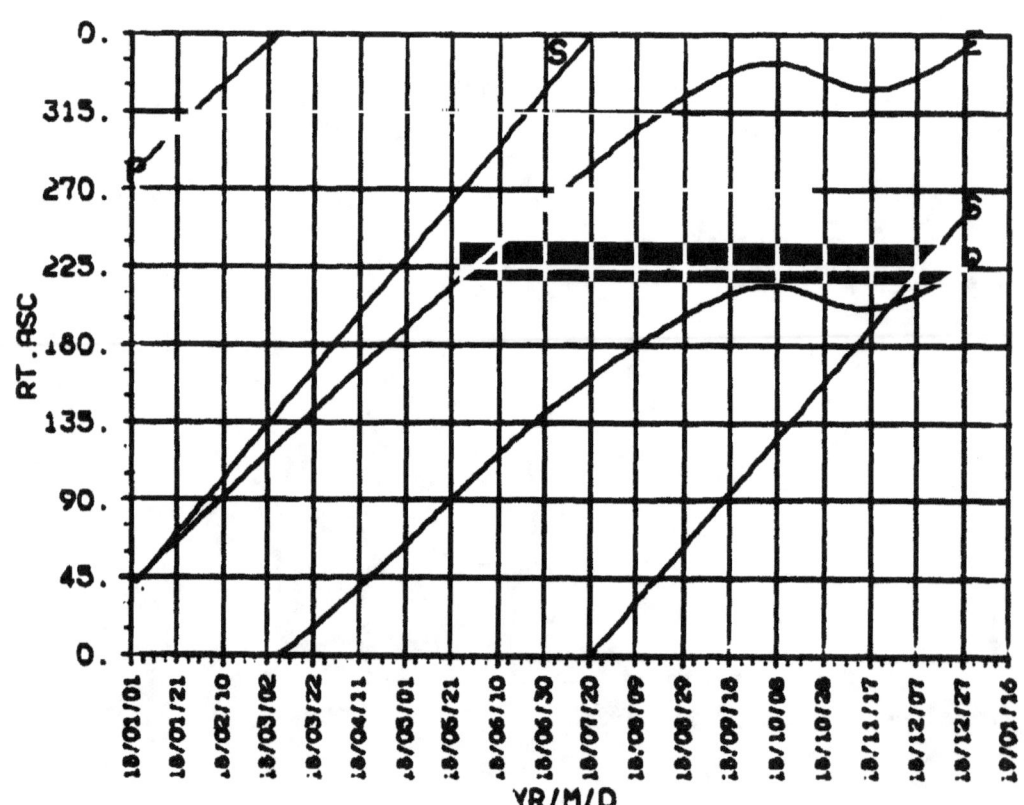

VENUS 2018

DISTANCE EC.LON 2018

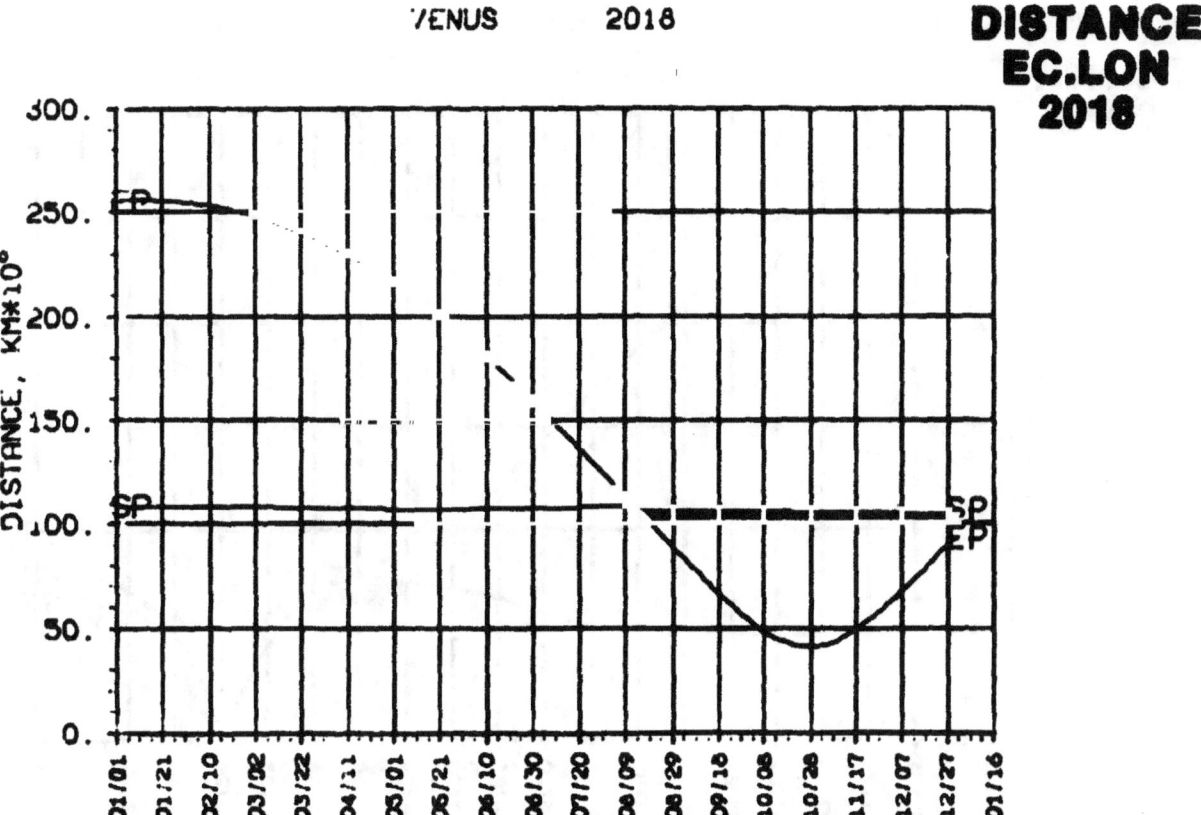

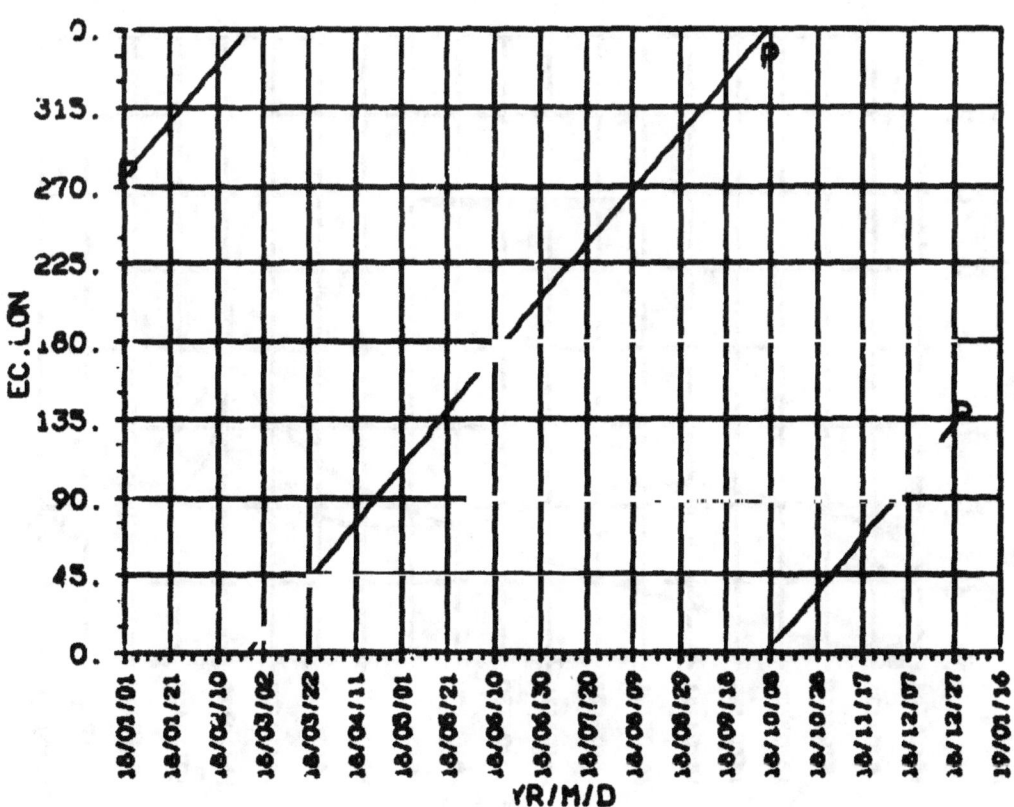

SEP, ESP CA, KA 2018 — VENUS 2018

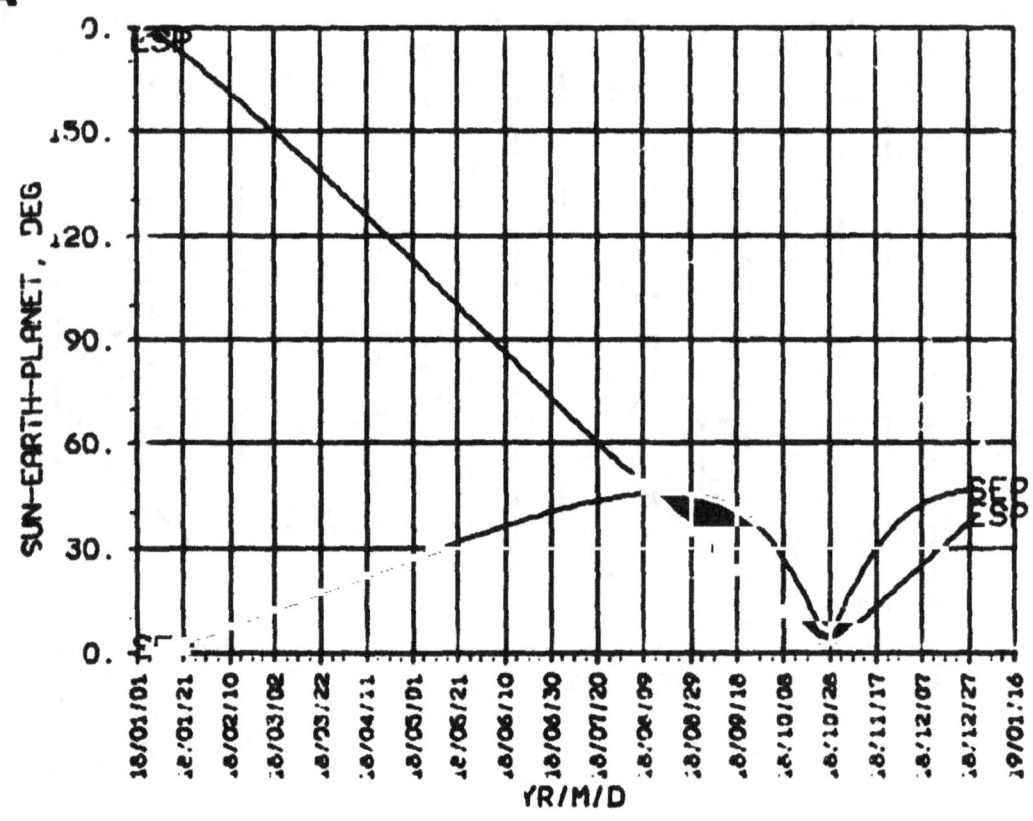

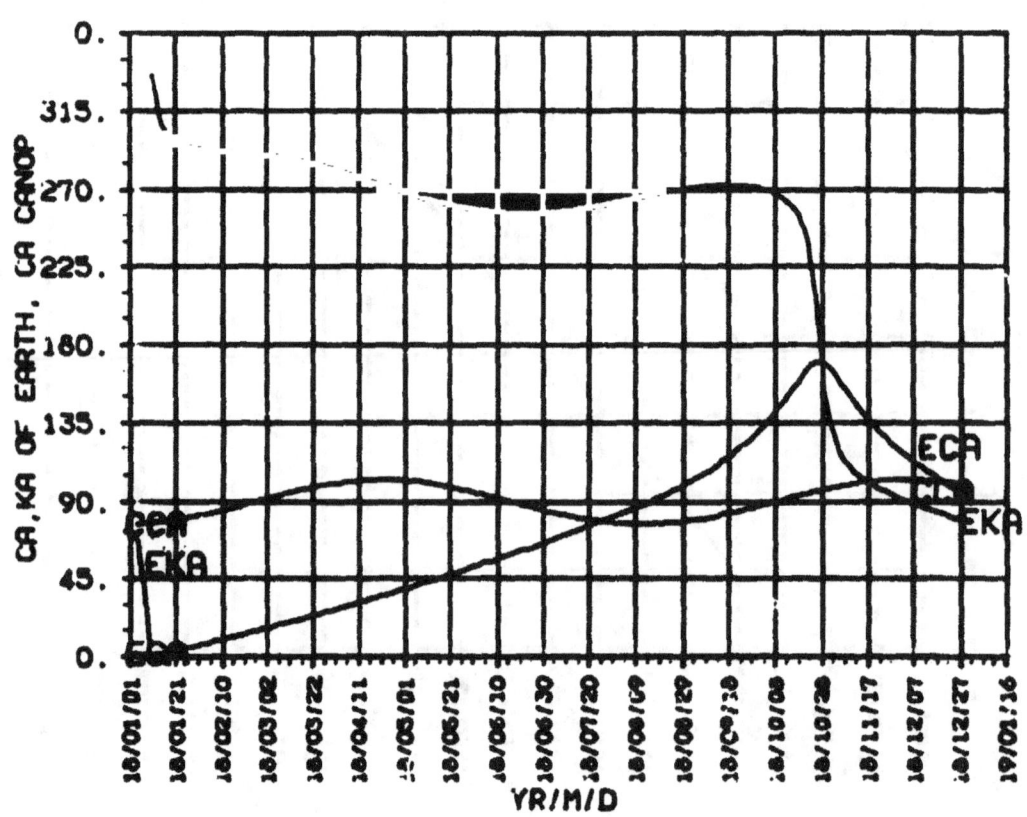

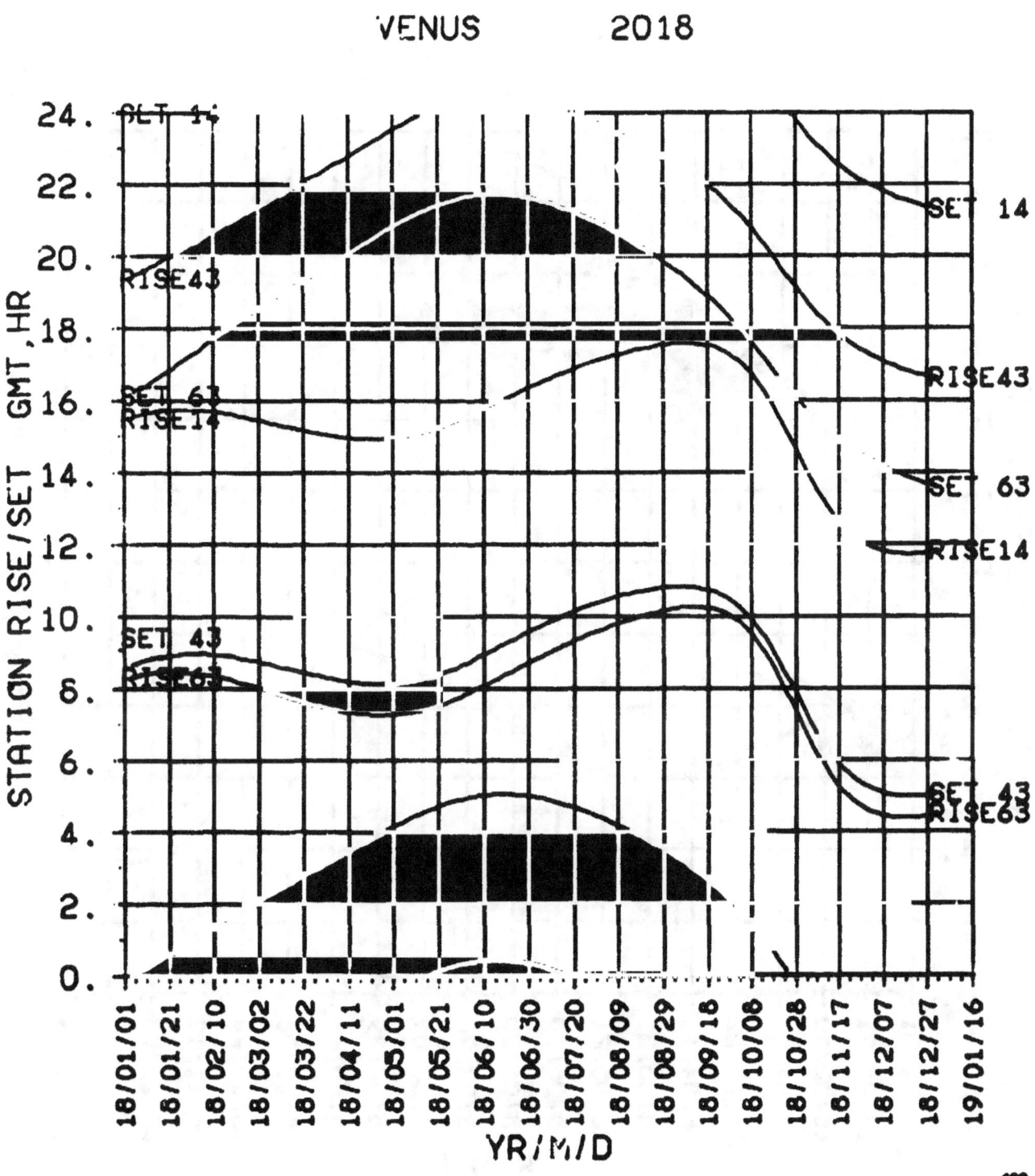

STA R/S NON-DSN 2018

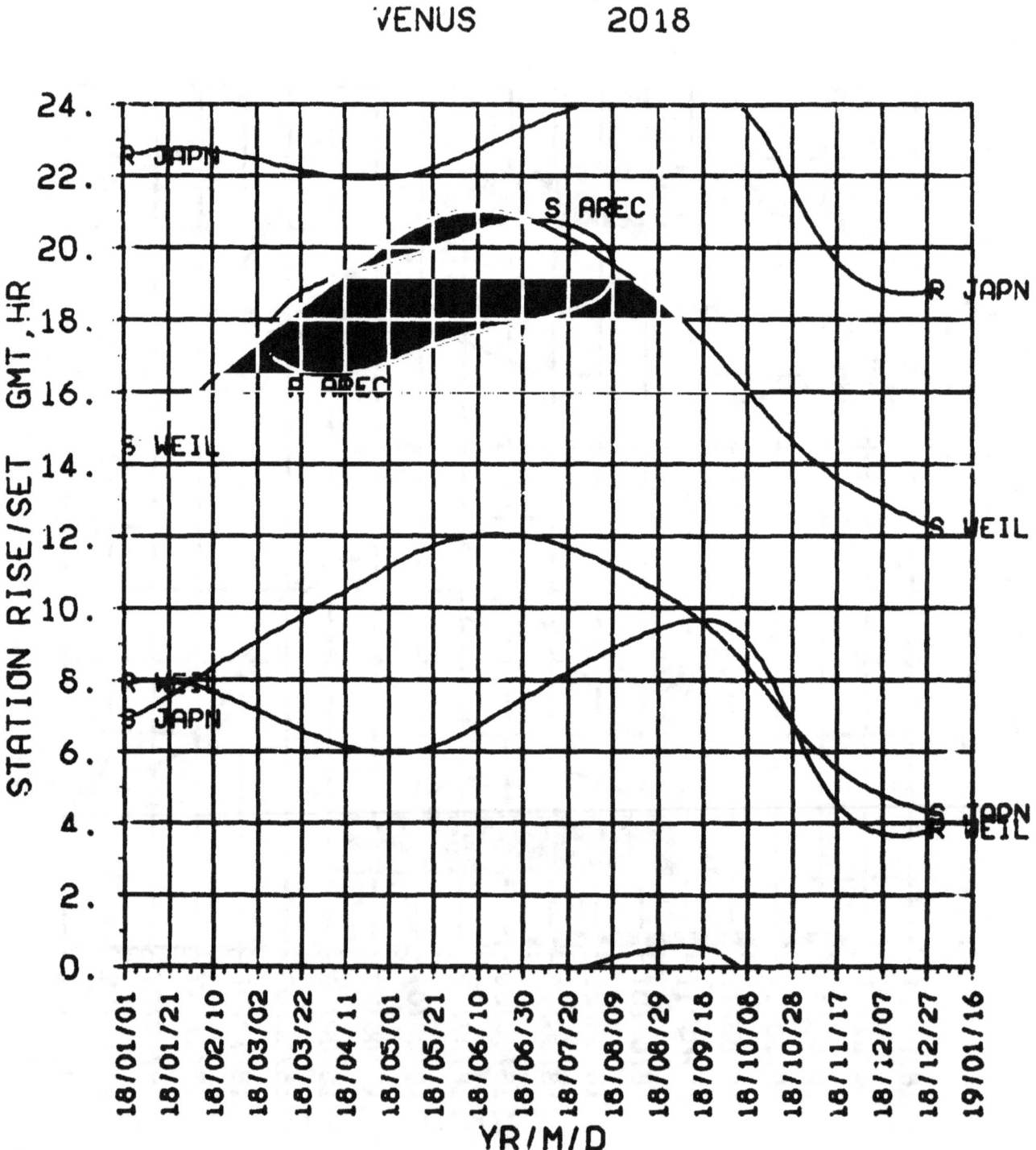

Venus

2019

DECLIN RT.ASC 2019

VENUS 2019

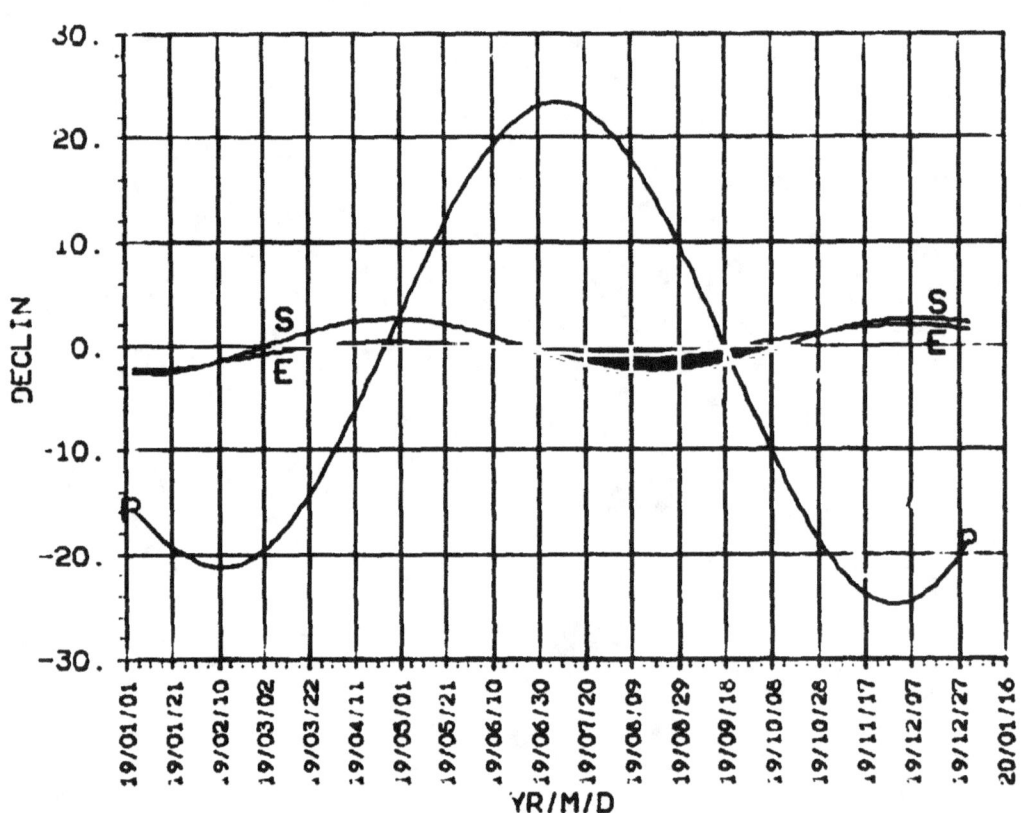

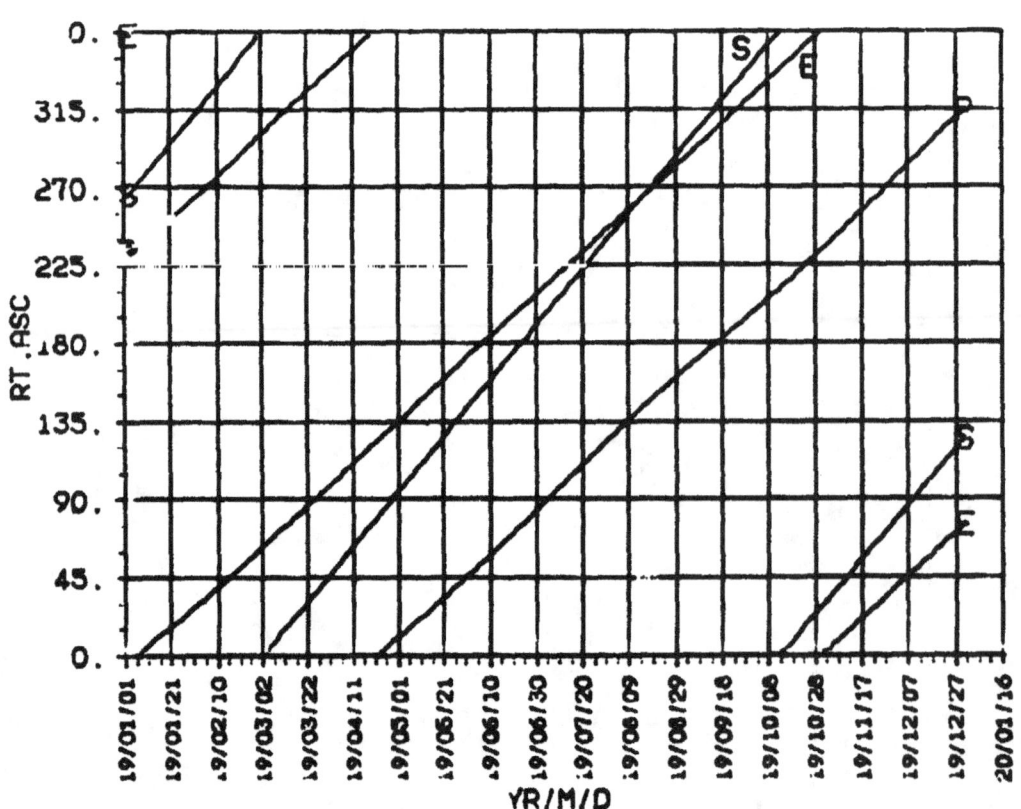

VENUS 2019

DISTANCE
EC.LON
2019

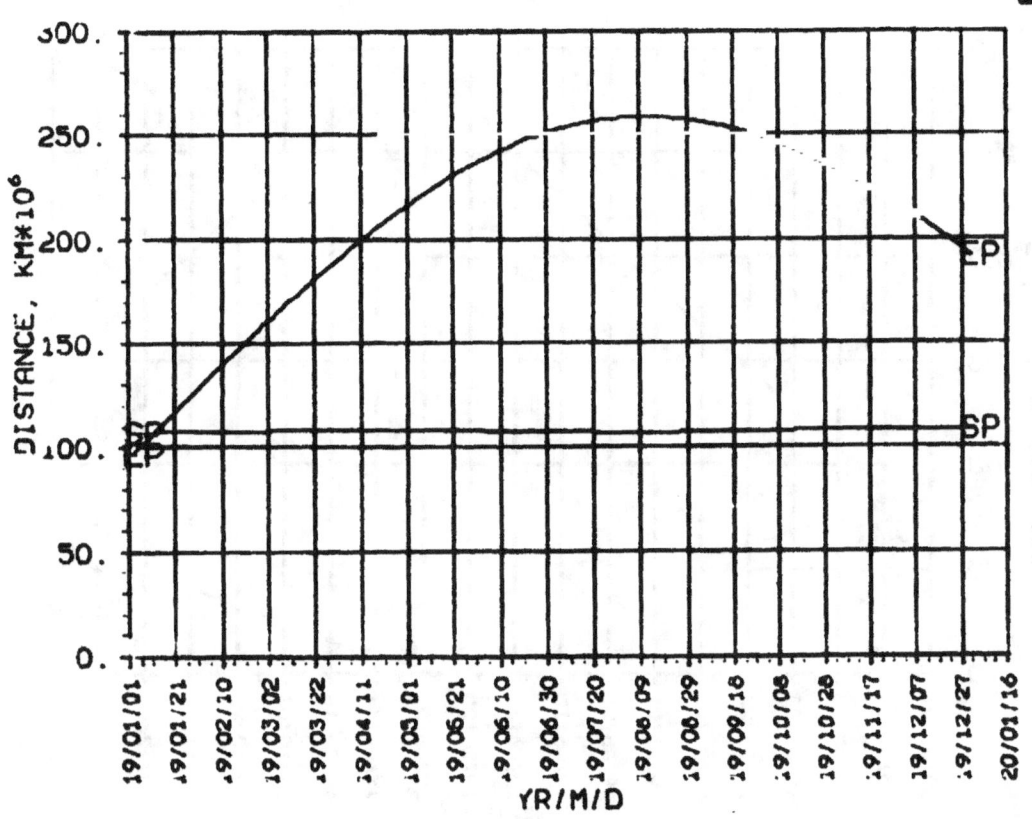

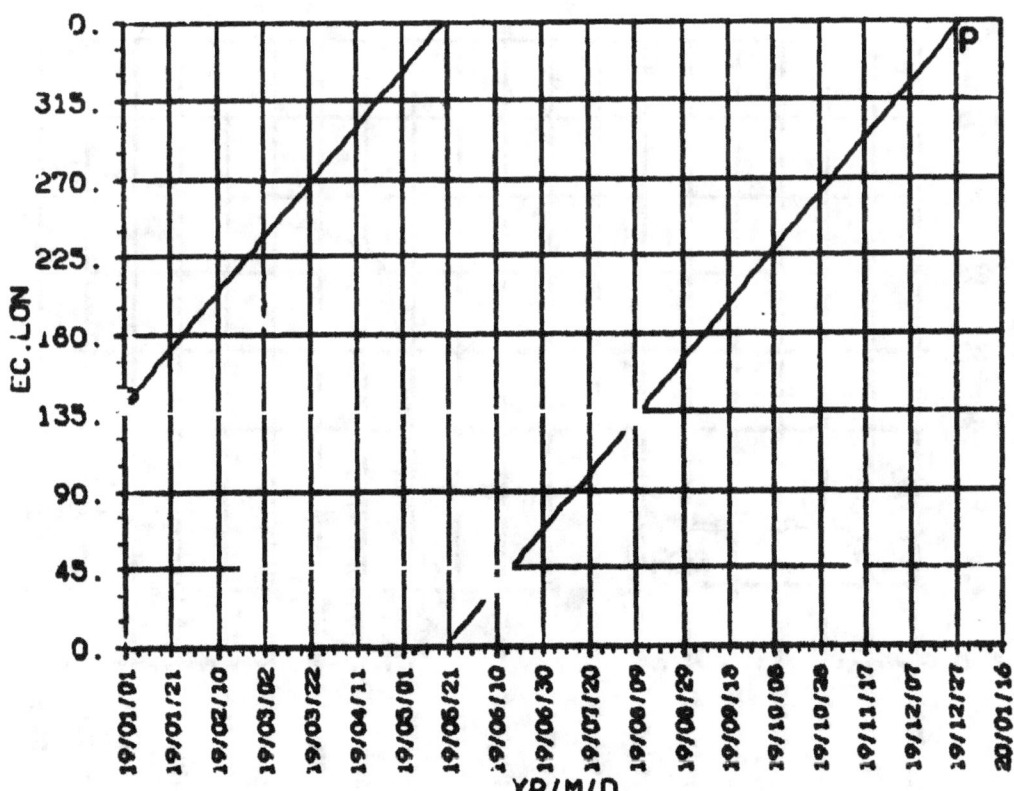

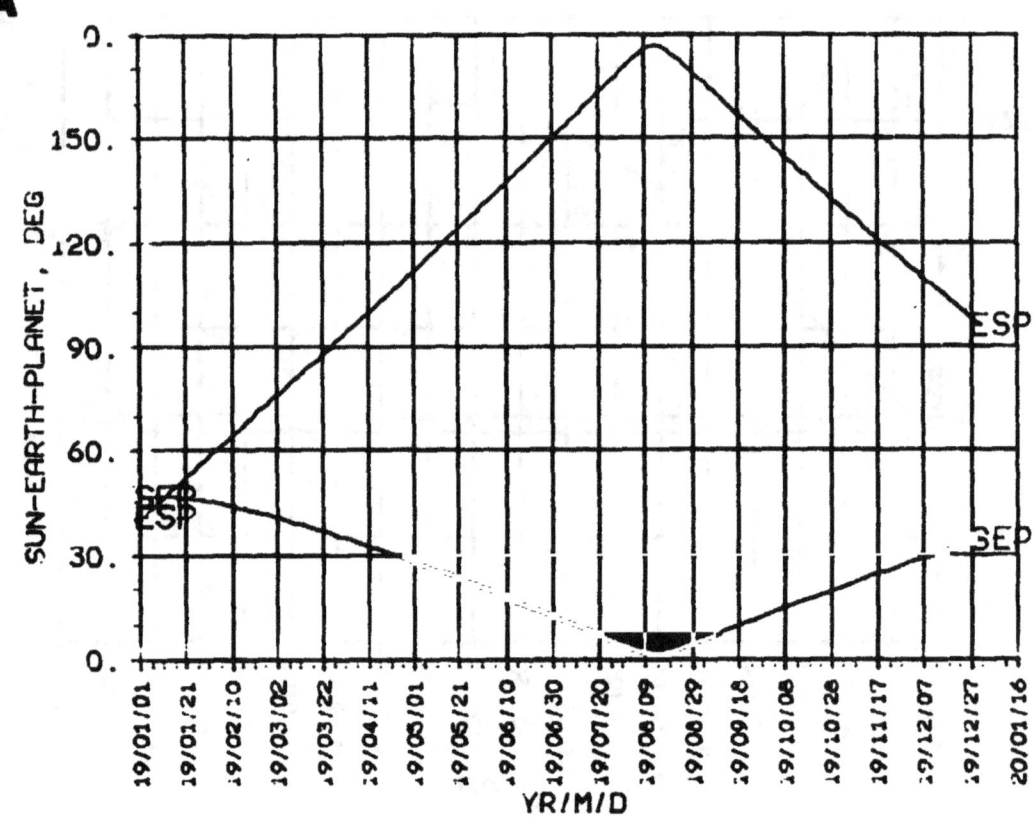
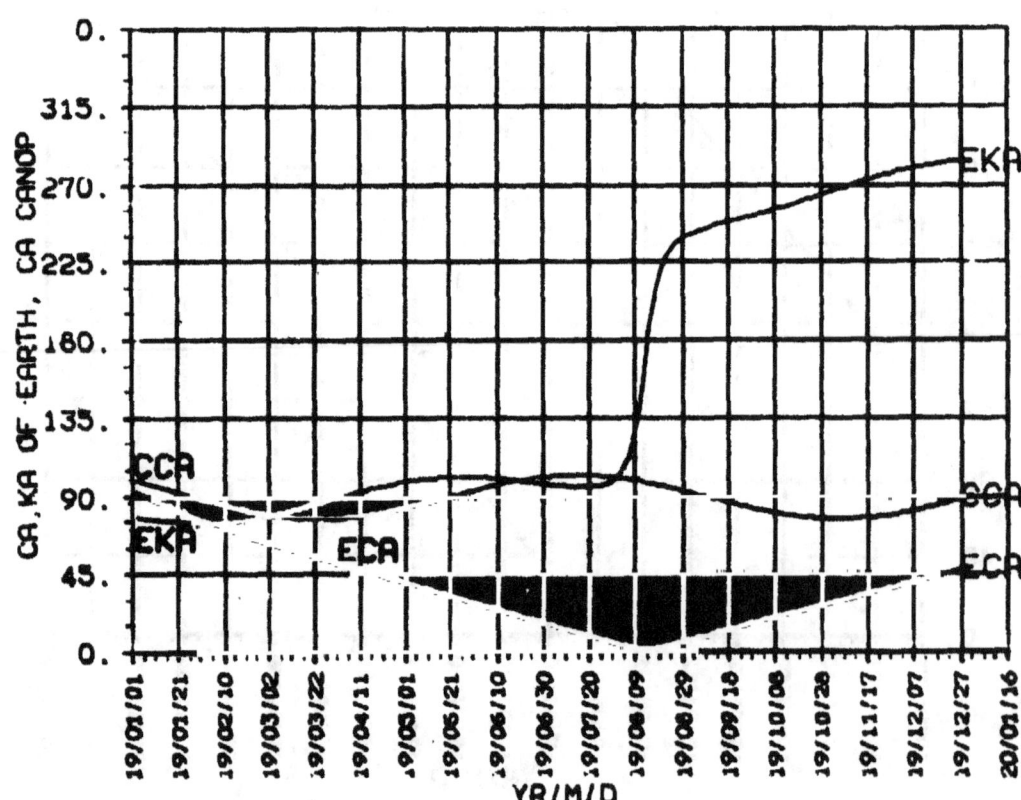

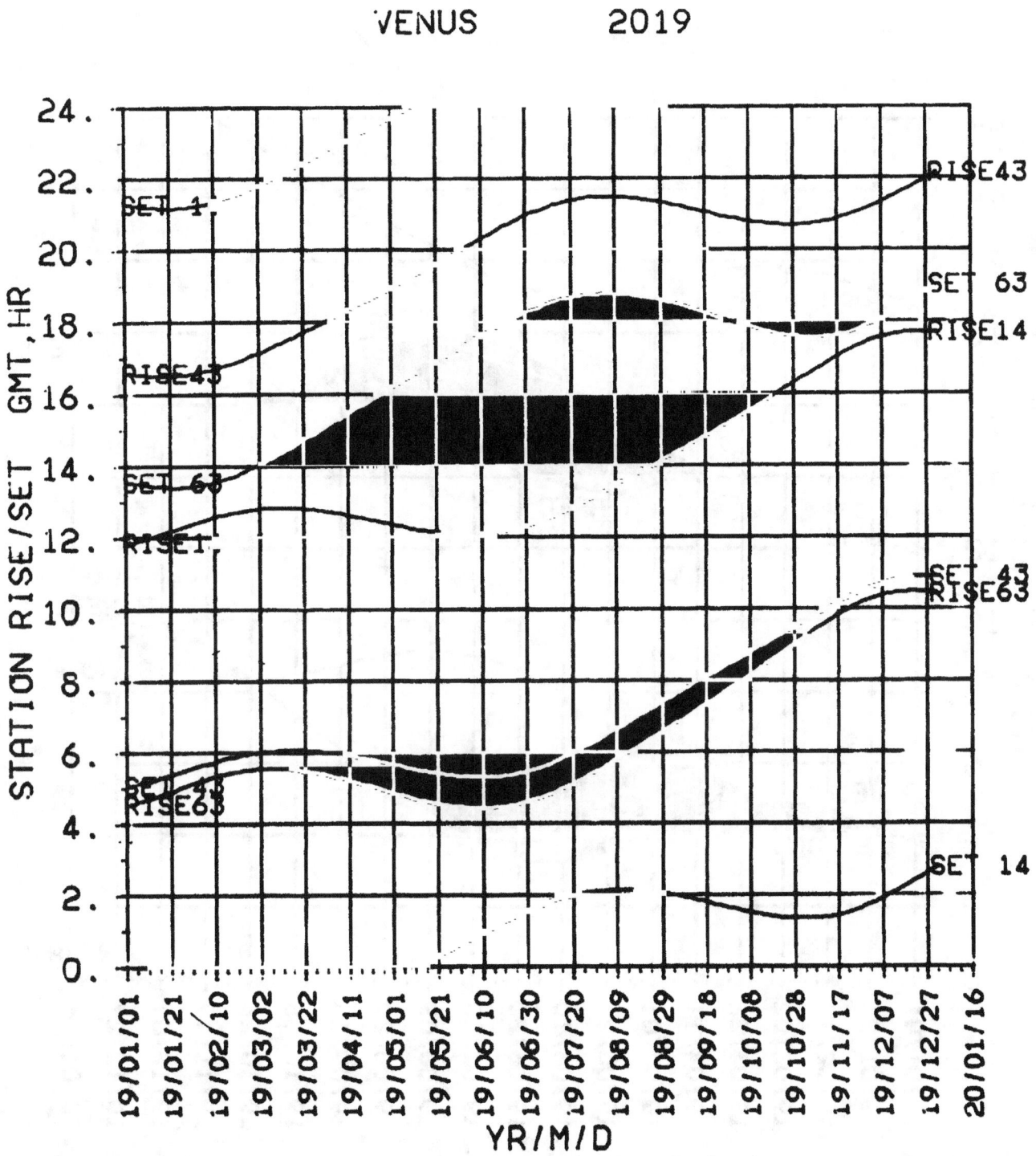

STA R/S NON-DSN 2019

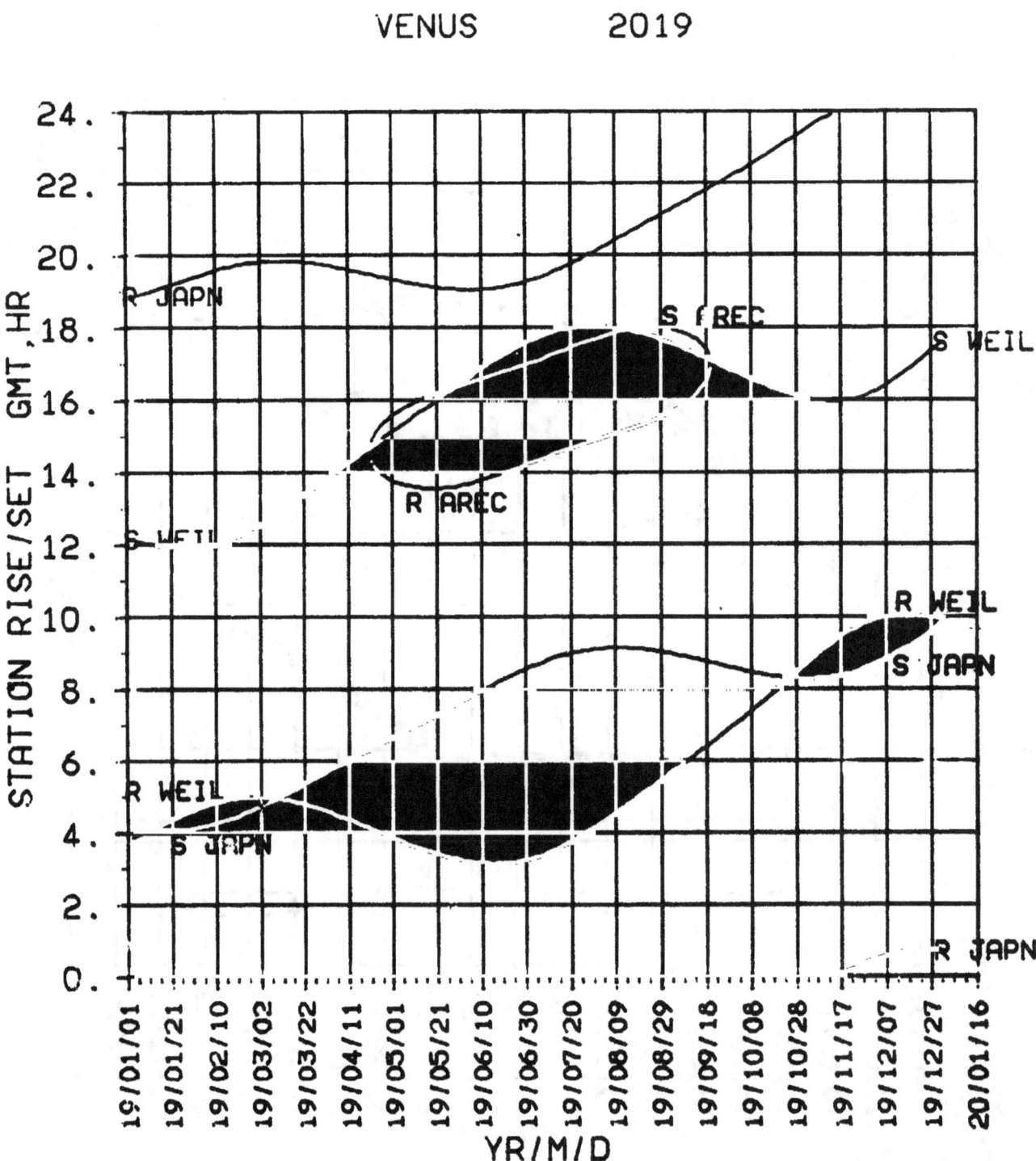

Venus

2020

DECLIN RT.ASC 2020

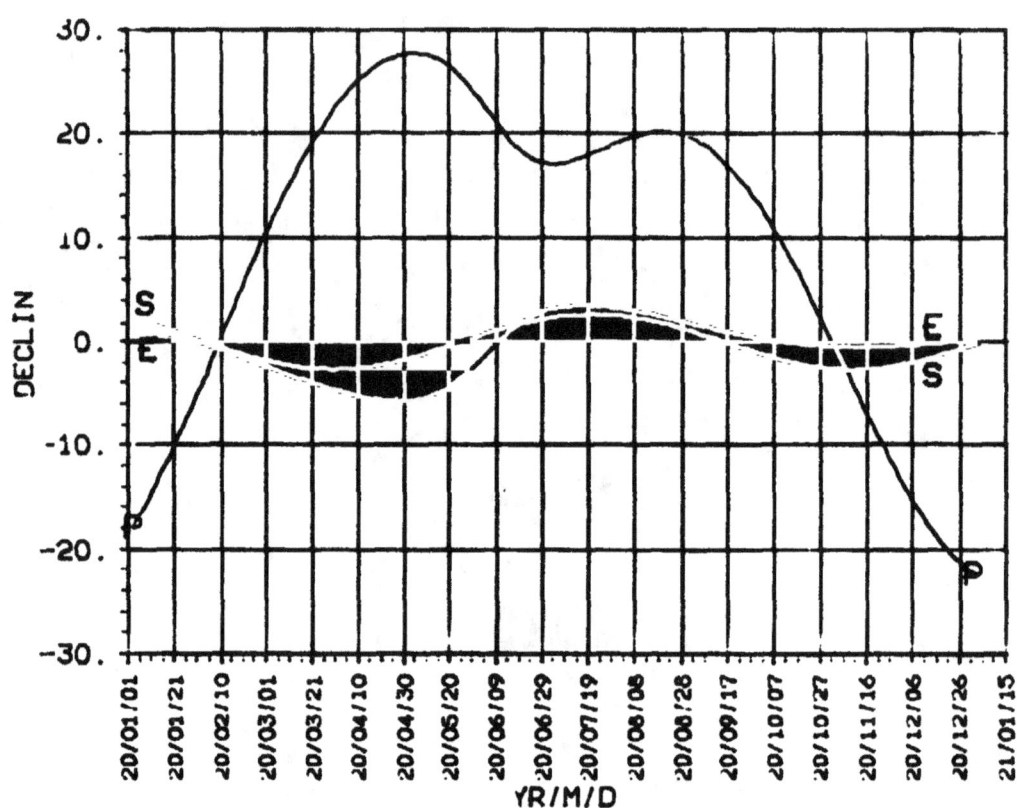

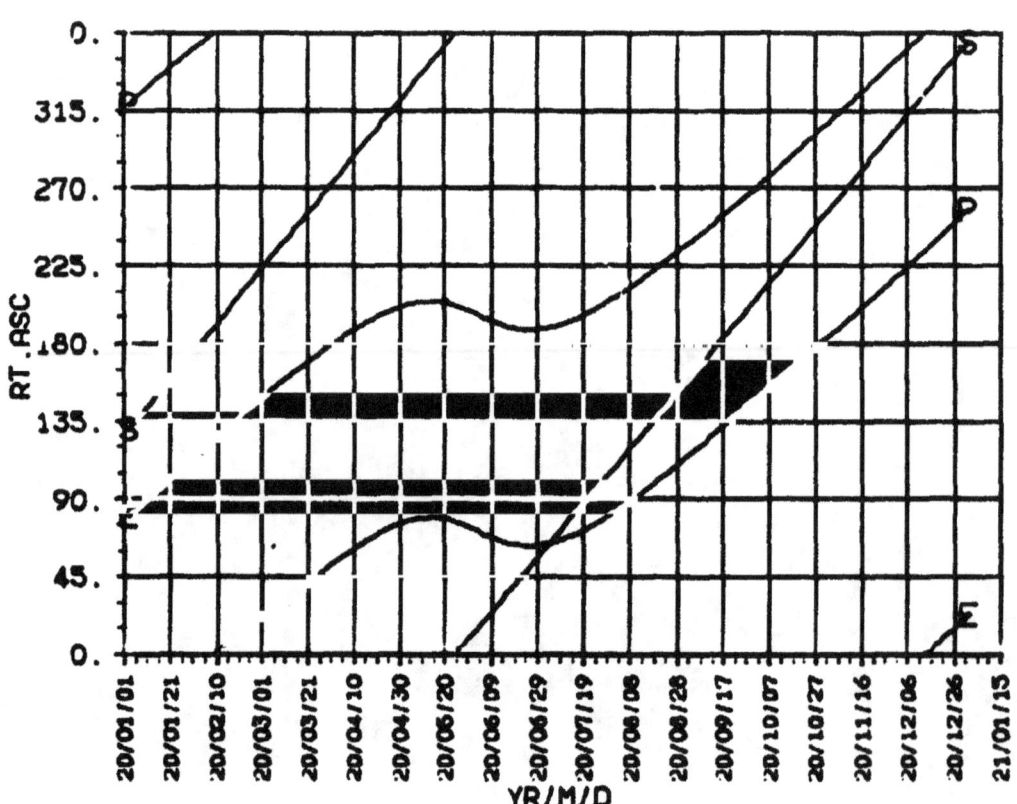

VENUS 2020

DISTANCE EC.LON 2020

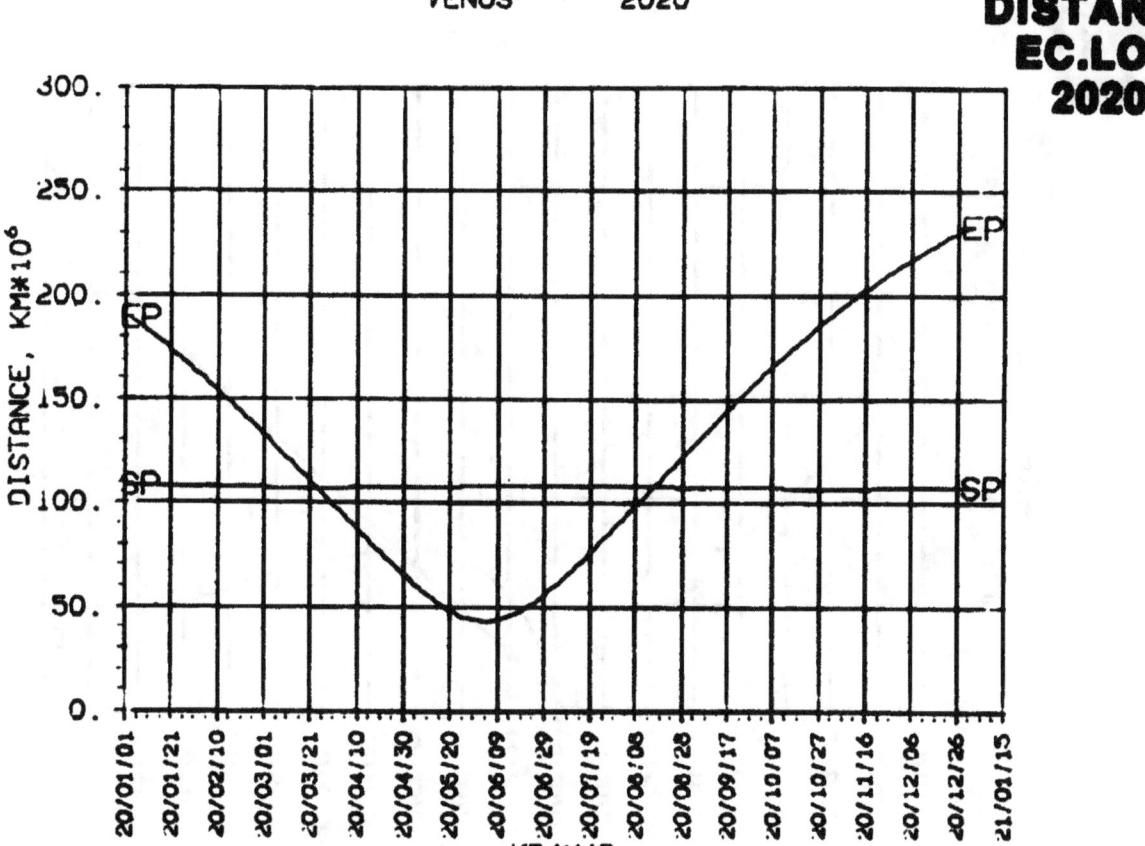

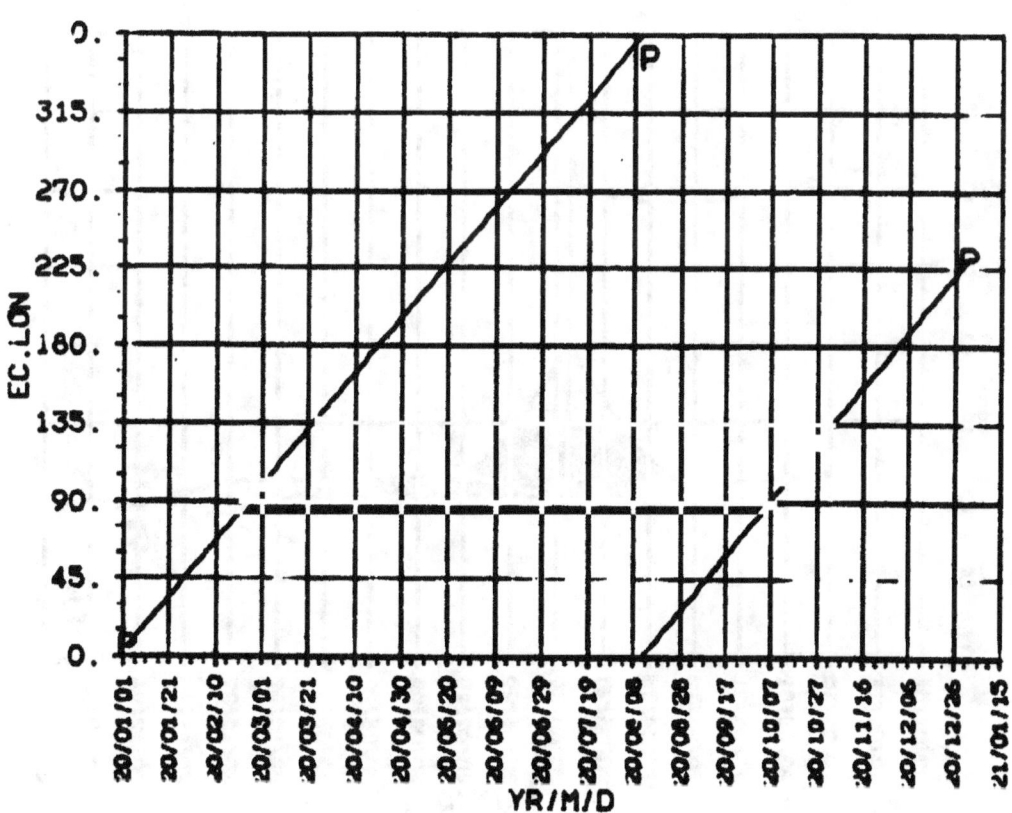

SEP, ESP CA, KA 2020 — VENUS 2020

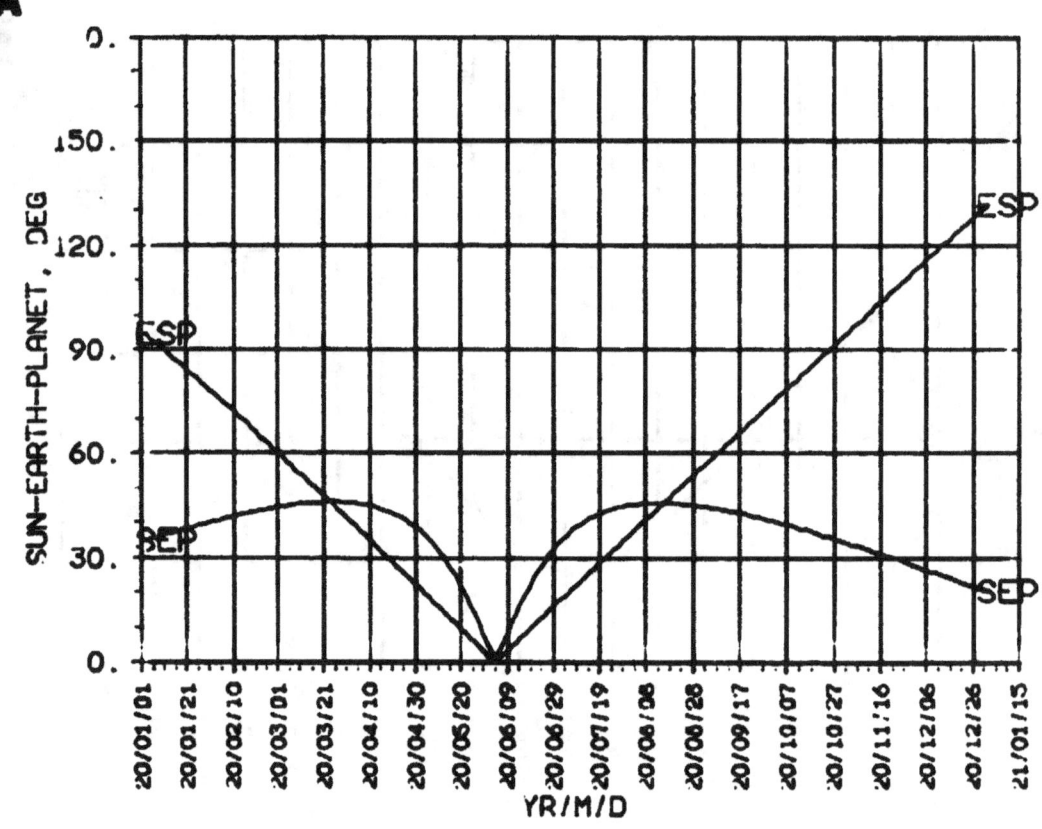

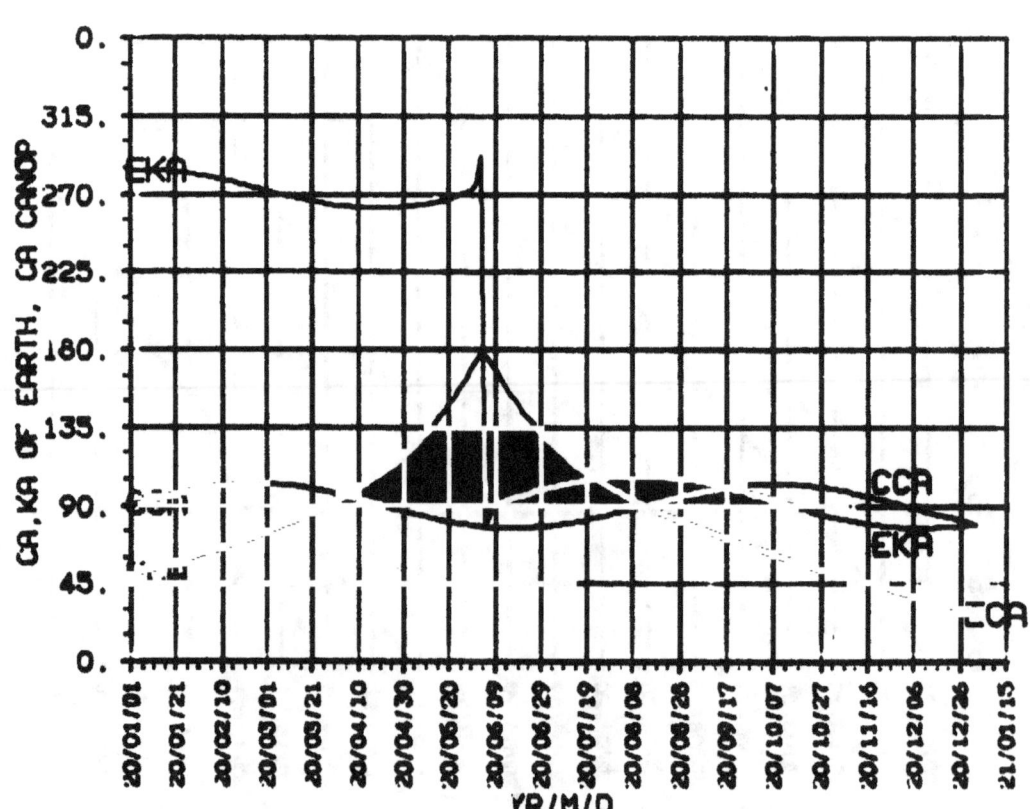

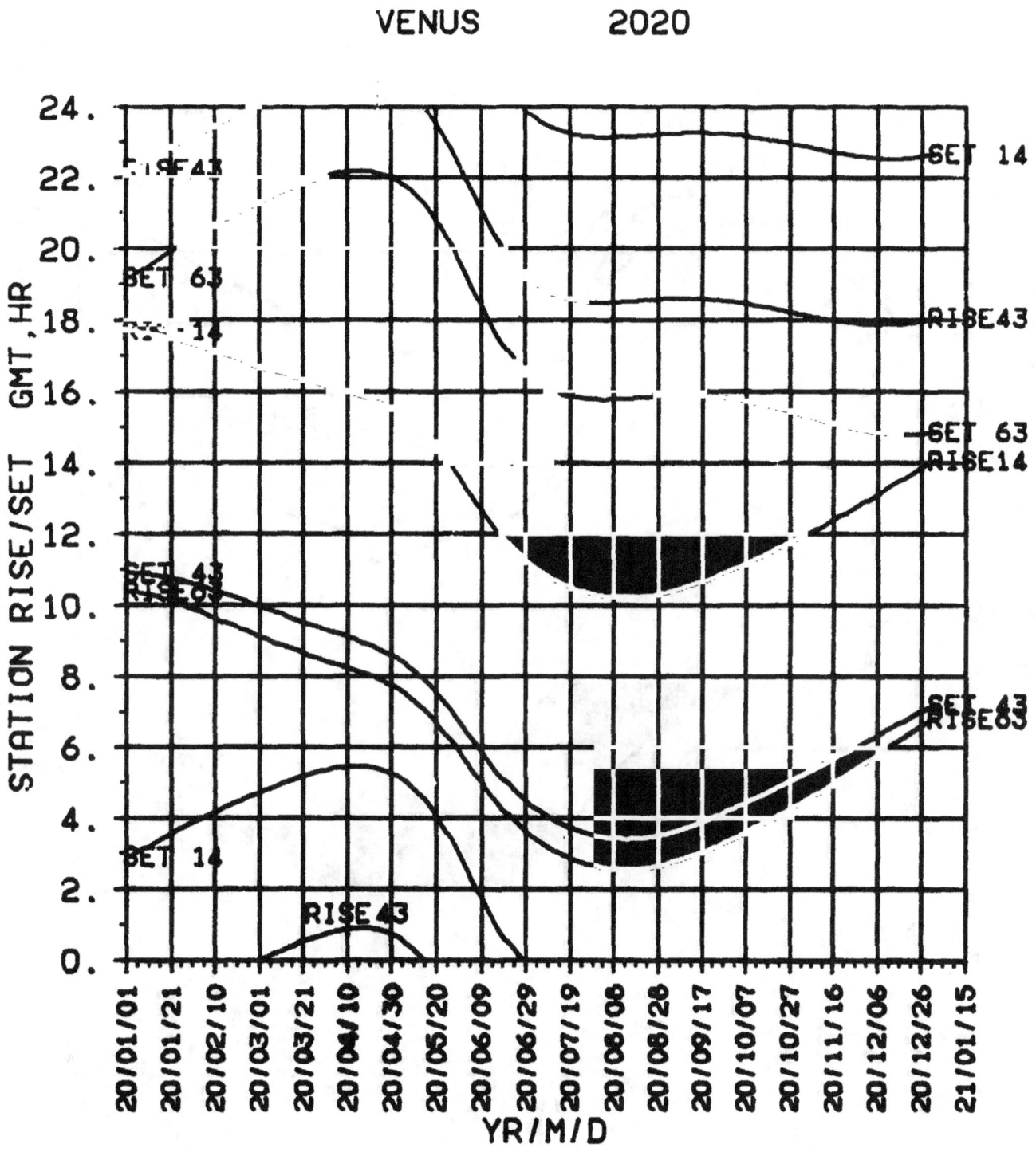

**STA R/S
NON-DSN
2020**

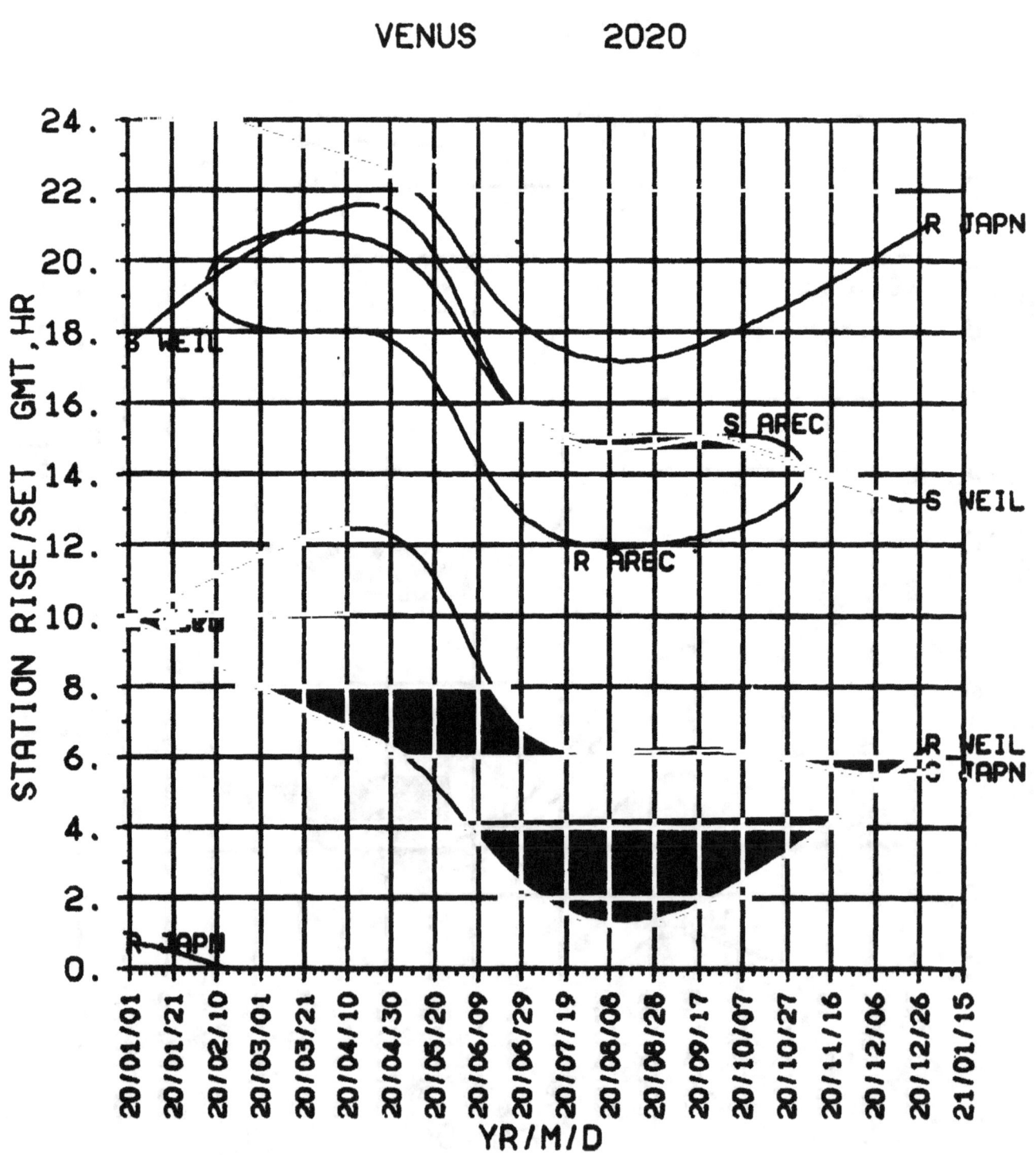

Positional Data

Earth
1985—2020

**EC.LON
EARTH
ALL YEARS**

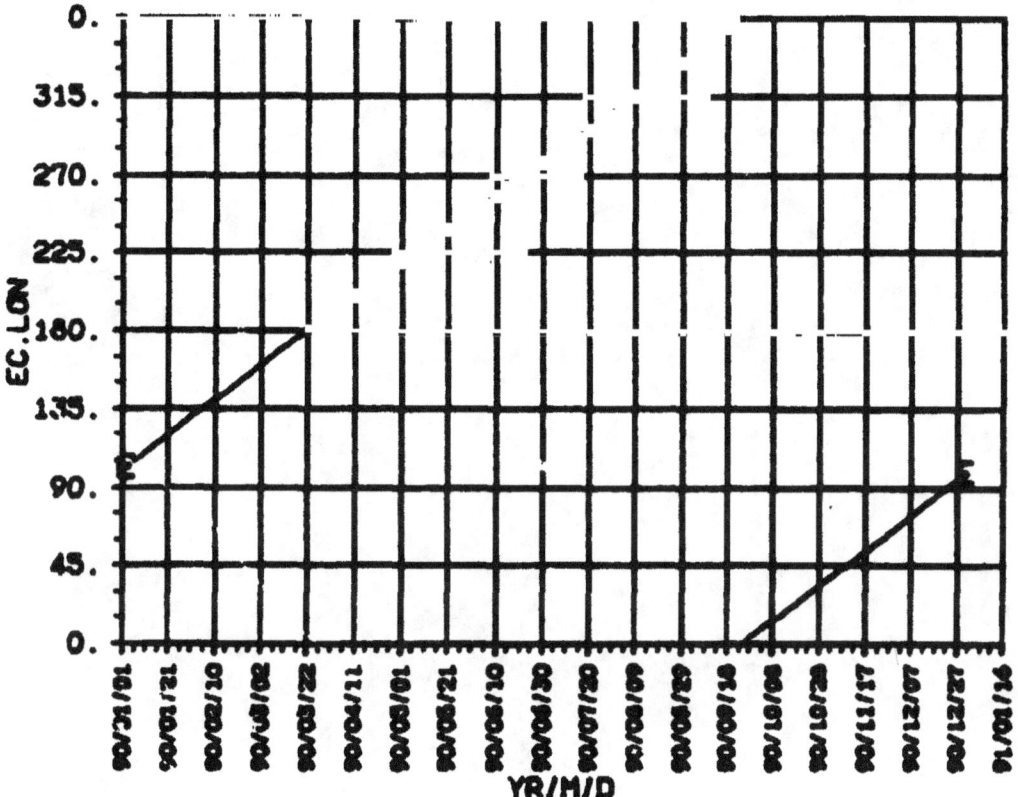

www.ingramcontent.com/pod-product-compliance
Lightning Source LLC
Chambersburg PA
CBHW081723170526
45167CB00009B/3674